Modelling the Dynamics of Biological Systems

Springer Series in Synergetics

Editor: Hermann Haken

Synergetics, an interdisciplinary field of research, is concerned with the cooperation of individual parts of a system that produces macroscopic spatial, temporal or functional structures. It deals with deterministic as well as stochastic processes.

36 **Temporal Disorder in Human Oscillatory Systems** Editors: L. Rensing, U. an der Heiden, M. C. Mackey

37 **The Physics of Structure Formation** Theory and Simulation Editors: W. Guttinger, G. Dangelmayr

38 **Computational Systems – Natural and Artificial** Editor: H. Haken

39 **From Chemical to Biological Organization** Editors: M. Markus, S. C. Müller, G. Nicolis

40 **Information and Self-Organization** A Macroscopic Approach to Complex Systems By H. Haken

41 **Propagation in Systems Far from Equilibrium** Editors: J. E. Wesfreid, H. R. Brand, P. Manneville, G. Albinet, N. Boccara

42 **Neural and Synergetic Computers** Editor: H. Haken

43 **Cooperative Dynamics in Complex Physical Systems** Editor: H. Takayama

44 **Optimal Structures in Heterogeneous Reaction Systems** Editor: P. J. Plath

45 **Synergetics of Cognition** Editors: H. Haken, M. Stadler

46 **Theories of Immune Networks** Editors: H. Atlan, I. R. Cohen

47 **Relative Information** Theories and Applications By G. Jumarie

48 **Dissipative Structures in Transport Processes and Combustion** Editor: D. Meinköhn

49 **Neuronal Cooperativity** Editor: J. Krüger

50 **Synergetic Computers and Cognition** A Top-Down Approach to Neural Nets By H. Haken

51 **Foundations of Synergetics I** Distributed Active Systems By A. S. Mikhailov

52 **Foundations of Synergetics II** Complex Patterns By A. S. Mikhailov, A.Yu. Loskutov

53 **Synergetic Economics** By W.-B. Zhang

54 **Quantum Signatures of Chaos** By F. Haake

55 **Rhythms in Physiological Systems** Editors: H. Haken, H. P. Koepchen

56 **Quantum Noise** By C. W. Gardiner

57 **Nonlinear Nonequilibrium Thermodynamics I** Linear and Nonlinear Fluctuation-Dissipation Theorems By R. Stratonovich

58 **Self-organization and Clinical Psychology** Empirical Approaches to Synergetics in Psychology Editors: W. Tschacher, G. Schiepek, E.J. Brunner

59 **Nonlinear Nonequilibrium Thermodynamics II** Advanced Theory By R. Stratonovich

60 **Limits of Predictability** Editor: Yu. A. Kravtsov

61 **On Self-Organization** An Interdisciplinary Search for a Unifying Principle Editors: R. K. Mishra, D. Maaß, E. Zwierlein

62 **Interdisciplinary Approaches to Nonlinear Complex Systems** Editors: H. Haken, A. Mikhailov

63 **Inside Versus Outside** Endo- and Exo-Concepts of Observation and Knowledge in Physics, Philosophy and Cognitive Science Editors: H. Atmanspacher, G. J. Dalenoort

64 **Ambiguity in Mind and Nature Multistable Cognitive Phenomena** Editors: P. Kruse, M. Stadler

65 **Modelling the Dynamics of Biological Systems** Editors: E. Mosekilde, O. G. Mouritsen

Volumes 1–35 are listed at the end of the book

Erik Mosekilde Ole G. Mouritsen (Eds.)

Modelling the Dynamics of Biological Systems

Nonlinear Phenomena and Pattern Formation

With 125 Figures

Springer

Professor Dr. Erik Mosekilde

Physics Department, Technical University of Denmark, Building 309
DK-2800 Lyngby, Denmark

Professor Dr. Ole G. Mouritsen

Department of Physical Chemistry, Technical University of Denmark, Building 206
DK-2800 Lyngby, Denmark

Series Editor:

Professor Dr. Dr. h. c. Hermann Haken

Institut für Theoretische Physik und Synergetik der Universität Stuttgart,
D-70550 Stuttgart, Germany and
Center for Complex Systems, Florida Atlantic University,
Boca Raton, FL 33431, USA

ISBN 3-540-58480-3 Springer-Verlag Berlin Heidelberg New York

Library of Congress Cataloging-in-Publication Data

Modeling the dynamics of living systems: nonlinear phenomena and pattern formation / Erik Mosekilde, Ole G. Mouritsen, eds. p. cm. – (Springer series in synergetics; v. 65) Includes bibliographical references and index. ISBN 0-387-58480-3 (New York: alk. paper). – ISBN 3-540-58480-3 (Berlin: alk. paper) 1. Biological systems-Mathematical models. 2. Biological control systems-Mathematical models. 3. System theory. I. Mosekilde, Erik. II. Mouritsen, Ole G. III. Series. QH323.5.M628 1994 574' .01'–dc20 94-37002 CIP

Printed in Germany

Typesetting: Camera ready copy from the editors using a Springer TEX macro package
SPIN 10129068 55/3140 - 5 4 3 2 1 0 - Printed on acid-free paper

Preface

The development of a proper description of the living world today stands as one of the most significant challenges to physics.

A variety of new experimental techniques in molecular biology, microbiology, physiology and other fields of biological research constantly expand our knowledge and enable us to make increasingly more detailed functional and structural descriptions. Over the past decades, the amount and complexity of available information have multiplied dramatically, while at the same time our basic understanding of the nature of regulation, behavior, morphogenesis and evolution in the living world has made only modest progress.

A key obstacle is clearly the proper handling of the available data. This requires a stronger emphasis on mathematical modeling through which the consistency of the adopted explanations can be checked, and general principles may be extracted.

As an even more serious problem, however, it appears that the proper physical concepts for the development of a theoretically oriented biology have not hitherto been available. Classical mechanics and equilibrium thermodynamics, for instance, are inappropriate and useless in some of the most essential biological contexts. Fortunately, there is now convincing evidence that the concepts and methods of the newly developed fields of nonlinear dynamics and complex systems theory, combined with irreversible thermodynamics and far-from-equilibrium statistical mechanics will enable us to move ahead with many of these problems. Synergetics represent an integration of essential aspects of these fields, and from the very beginning a better understanding of living systems has been one of the foremost goals of this field.

In the present volume we have collected contributions from leading experts in a broad spectrum of disciplines ranging from physics and chemistry over physiology and biology to large scale ecology. The purpose has been to present some of the most recent results in the physical and mathematical description of living systems. Some contributions are rather mathematical in style, while others are more descriptive, reflecting to a large extent the level that the modeling effort has reached in the various areas.

We want to express our gratitude to Michael Wiinberg Olesen for the significant help he has given us in the editing of the manuscripts. We also want to thank each of the contributors for the emphasis and care they have devoted to the project.

Copenhagen
June 1994

E. Mosekilde
O.G. Mouritsen

Contents

Part V Complex Ecologies and Evolution

List of Contributors

Olaf Sparre Andersen
Department of Physiology
and Biophysics,
Cornell University Medical College,
1300 New York, NY 10021-4896, USA

Per Bak
Brookhaven National Laboratory
Department of Physics
P.O. Box 5000
Upton, NY 11973-5000, USA

Anne Beuter
Neurokinetics Laboratory
Department of Kinanthropology
University of Quebec at Montreal
CP 8888 Succ. Centre Ville
Montreal, Qc H3C 3P8, Canada

Pierre Borckmans
Service de Chimie-Physique
CP 231, Université Libre
de Bruxelles
1050 Bruxelles, Belgium

K.H. Chon
Department of Biomedical
Engineering
University of Southern California
Los Angeles, CA, USA

Wm Cowan
Departments of Psychology
and Computer Science
University of Waterloo
Waterloo, Ontario, Canada

Ph. Daviet
Department of Physics
McGill University
3600 University Street
Montreal, Quebec, Canada H3A 2T8

Guy Dewel
Service de Chimie-Physique
CP 231, Université Libre
de Bruxelles
1050 Bruxelles, Belgium

K.R. Elder
Department of Physics
McGill University
3600 University Street
Montreal, Quebec, Canada H3A 2T8

Henrik Flyvbjerg
The Isaac Newton Institute
for Mathematical Sciences
20 Clarkson Road
Cambridge CB4 0EH, UK

Anne de Geoffrey
Neurokinetics Laboratory
Department of Kinanthropology
University of Quebec at Montreal
CP 8888 Succ. Centre Ville
Montreal, Qc H3C 3P8, Canada

Jeffrey Girsham
Department of Physiology
and Biophysics,
Cornell University Medical College,
1300 New York, NY 10021-4896, USA

Martin Grant
Department of Physics
McGill University
3600 University Street
Montreal, Quebec, Canada H3A 2T8

Hanspeter Herzel
Institut für Theoretische Physik
Technische Universität Berlin
Hardenbergstr. 36
10623 Berlin, Germany

N.-H. Holstein-Rathlou
Department of Medical Physiology
University of Copenhagen, Denmark
& Department of Physiology
Brown University, Providence, USA

John Hjort Ipsen
Department of Physical Chemistry
Technical University of Denmark
DK-2800 Lyngby, Denmark

M.H. Jensen
The Niels Bohr Institute
Blegdamsvej 17
DK-2100 Copenhagen Ø
Denmark

Ole Jensen
Physics Department
Technical University of Denmark
DK-2800 Lyngby, Denmark

Kent Jørgensen
Department of Physical Chemistry
Technical University of Denmark
DK-2800 Lyngby, Denmark

Jens A. Lundbæk
Department of Physiology
and Biophysics,
Cornell University Medical College,
1300 New York, NY 10021-4896, USA

D.J. Marsh
Department of Physiology
Brown University, Providence, RI,
USA

Miloš Marek
Department of Chemical Engineering
Prague Institute
of Chemical Technology
166 28 Prague, Techniká
Czech Republic

V.Z. Marmarelis
Department of Biomedical
Engineering
University of Southern California
Los Angeles, CA, USA

Jacqueline M. McGlade
Ecosystems Analysis
and Management Group,
Biological Sciences,
University of Warwick,
Coventry CV4 7AL,
UK

Erik Mosekilde
Physics Department
Technical University of Denmark
DK-2800 Lyngby, Denmark

Lis Mosekilde
Department of Cell Biology
Institute of Anatomy
University of Århus
DK-8000 Århus, Denmark

Ole G. Mouritsen
Department of Physical Chemistry
Technical University of Denmark
DK-2800 Lyngby, Denmark

S.C. Müller
Max-Planck-Institut
für Molekulare Physiologie
Rheinlanddamm 201
D-44139 Dortmund, Germany

Lennart Nilsson
Karolinska Institute Center
for Structural Biochemistry
NOVUM Research Park
S-141 57 Huddinge
Sweden

V.O. Pannbacker
Physics Department
Technical University of Denmark
DK-2800 Lyngby, Denmark

Igor Schreiber
Department of Chemical Engineering
Prague Institute
of Chemical Technology
166 28 Prague, Techniká
Czech Republic

Henrik Seidel
Institut für Theoretische Physik
Technische Universität Berlin
Hardenbergstr. 36
10623 Berlin, Germany

Kim Sneppen
The Niels Bohr Institute
Blegdamsvej 17
DK-2100 Copenhagen Ø
Denmark

G.Soga
Department of Physics
McGill University
3600 University Street
Montreal, Quebec, Canada H3A 2T8

M.M. Sperotto
Department of Physical Chemistry
Technical University of Denmark
DK-2800 Lyngby, Denmark

J.S.Thomsen
Department of Physics
Technical University of Denmark
DK-2800 Lyngby, Denmark

J.R.Thomson
Department of Physics
McGill University
3600 University Street
Montreal, Quebec, Canada H3A 2T8

A. Warda
Max-Planck-Institut
für Molekulare Physiologie
Rheinlanddamm 201
D-44139 Dortmund, Germany

A. De Wit
Service de Chimie-Physique
CP 231, Université Libre
de Bruxelles
1050 Bruxelles, Belgium

Z. Zhang
Department of Physics
McGill University
3600 University Street
Montreal, Quebec, Canada H3A 2T8

Martin J. Zuckermann
Department of Physics
McGill University
3600 University Street
Montreal, Quebec, Canada H3A 2T8

V.S. Zykov
Max-Planck-Institut
für Molekulare Physiologie
Rheinlanddamm 201
D-44139 Dortmund, Germany

Introduction

Erik Mosekilde and Ole G. Mouritsen

Dynamical phenomena in living systems occur on an extremely broad range of length and time scales.

The fastest molecular modes, such as small conformational changes in a macromolecule, that may eventually influence a process in a living system typically occur in the range of $10^{-12} - 10^{-9}$ sec whereas some evolutionary processes take place on geological time scales as long as $10^{12} - 10^{16}$ sec. Within this enormous span of time, life processes of all sorts outplay themselves; a protein may fold into an active state in less than a second, processes and flow regulation in the nephrons of the kidney take place with time delays of several seconds, and the bones in a mammalian subject undergo remodelling over periods of years.

Similarly, life processes occur and biological systems organize themselves over length scales ranging from those characterizing individual biological molecules, $\sim 10^{-9}$ m, to the size of a planetary ecosystem, $\sim 10^{7}$ m. Within this range, living organisms and phenomena involved in life processes are organized in space on all scales; a eucaryotic cell extends spatially over a range of $\sim 10^{-5}$ m, the size of the various organs in a living organism may be of the order of $\sim 10^{-2}$ m, and spatio-temporal organization in the ecology of an isolated island may occur on scales of $\sim 10^{4}$ m.

The main characteristics of living systems are temporal change together with the creation of spatial order out of disorder. Living systems are generally not in thermodynamic equilibrium but usually far from equilibrium. The biological state is a state of spatio-temporal organization which is a subtle consequence of the intricate interplay between the many (molecular) constituents of the system and their interactions with the environment through the fluxes of energy and matter. Hence living systems are in principle driven dissipative systems and their dynamics may be seen as a consequence of the complex non-linear behavior arising from the interplay between the system and its environment. However, the many-body nature of the individual system itself, be it a protein-polynucleotide complex, a cell membrane, or an organ like the kidney, imparts the system with complex behavior which often cannot be understood or described in terms of the individual constituents of the system alone. As an example, highly correlated and ordered dynamic structures evolve in membrane molecular assemblies due to the cooperativity in the intermolecular interactions, and similarly ecological systems may spontaneously evolve into a self-organized critical state due to the very nature of their internal dynamics.

The complexity of any living system, even the simplest one, and the fact that living systems evolve into dynamic states controlled by the intricate interactions between the system constituents and environmental driving forces, calls for a theoretical description in terms of mathematical *models* which are based on a reduction of the number of system variables and interactions to a minimum consisting of those that are most relevant for the dynamical phenomenon under consideration. This is obviously a tricky approach since it may overlook important (hidden) variables. The validity of such a reductionist's approach is ultimately judged by its ability to reproduce experimental results as well as by its usefullness in guiding more well-focussed experimental endeavours.

Obviously the model of any given dynamic phenomenon in a living system has to be chosen so as to correspond to a level of description that on the one hand involves variables and interactions that have a sufficient degree of physical realism but on the other hand are sufficiently coarse grained to allow properties to be be derived from the model that can be related to experimentally observable quantities of interest. This implies for example, that a description of the interaction between a molecular receptor and a ligand sometimes can be based on the level of, say, molecular groups like amino-acid residues rather than on individual atoms, and that a description of the dynamics of an ecosystem can be formulated by variables characterizing species and their mutations rather than single individuals. In other situations it is essential to start out from microscopic or "atomistic" description.

Once a model of a biological system has been formulated, the next step is to determine those properties of the model that are non-trivial consequences of the nature of the variables and their mutual couplings. One standard strategy is to derive macroscopic properties from a model formulated in terms of some kind of microscopic variables and forces. Another strategy is to calculate the space and time dependence of those macroscopic densities and fields that enter the theory. In any case, non-linearities and the many-body nature of the problem usually call for a numerical treatment, for example computer-simulation techniques by which the behavior of the model in space and time is simulated either by a numerical solution to sets of coupled dynamical non-linear equations or by a direct statistical mechanical simulation of the interacting assembly of particles.

In modelling dynamical phenomena in living systems there are two major different types of model approaches. One is aimed at describing, in as great detail as possible, the properties of a specific system or phenomenon. The models developed for this purpose are usually highly specific with only a small sphere of applicability. The other approach, which is aimed at describing the generic properties of a system or a phenomenon, attempts to construct the simplest possible model for the problem under consideration. The latter approach holds a promise for a more universal application of the model as well as a rationalization of apparently diverse phenomena in terms of unified principles.

The present volume covers a wide range of dynamical phenomena in living systems taking place on a great range of time and length scales. The first part of the volume is concerned with general properties of spatio-temporal organization in chemical-reaction systems which underlie most biochemical processes from the microscopic to the macroscopic level. Then in the second part dynamic pattern-formation phenomena in mesoscopic systems like cell membranes and in whole tissues like the visual cortex are described. The third part deals with dynamics of macromolecules, such as proteins, polynucleotides, and ion channels as well as the interactions of these molecules with other molecules and substrates. The fourth part is concerned with large-scale physiological systems and the dynamical control of their functionality. Finally, ecological systems and dynamical models of evolution are the subjects of the fifth and final part.

For several of the systems and phenomena under consideration in the present volume, a description is presented in terms of a theoretical model formulated by means of system variables and interactions that are chosen at an appropriate level of coarse graining. The models, which are both of the more system-specific type as well as the more generic type, are constructed using different types of experimental observations together with general conceptual considerations concerning the phenomenon under consideration. The properties of the models are derived by various theoretical techniques, involving computer-simulation methods, and the results are compared with experimental data in order both to provide an interpretation of the experimental observations as well as to improve and correct the model description in an continuous attempt to deepen our understanding of dynamic phenomena in living systems.

Part I

Pattern Formation in Chemical Systems

Spiral Waves in Bounded Excitable Media

S.C. Müller, A. Warda and V.S. Zykov

Abstract

The dynamics of spiral waves interacting with an impermeable boundary of an excitable medium is investigated. Experimental data on spiral waves rotating in a small piece of an excitable medium are presented. A wide variety of spatio-temporal scenarios of the spiral wave placed into a small disk is observed in computer simulations using a reaction-diffusion system. A simplified kinematical model describing wave fronts that move in the vicinity of an impermeable boundary is elaborated and used to analyze both stationary and nonstationary dynamic regimes.

1 Introduction

Many living systems (e.g. nerve and muscle tissue or colonies of microorganisms) exhibit the properties of so-called excitable media [1, 2]. The common features of excitable media are the autocatalytic local kinetics of individual cells and their interaction by diffusion. Each individual cell can form a pulse of excitation in response to an external signal. Due to the local coupling between neighboring elements excitation waves can propagate through the medium.

In the one-dimensional case such waves propagate without any decrement of their amplitude and velocity until they reach the boundary of the medium. If the excitable medium has the shape of a ring channel, the excitation wave runs along this ring for an infinitely long time. In a two-dimensional medium one can create a wave of excitation circulating around some obstacle [3]. In this case, the wave front has the form of a spiral that rotates around the boundary of the obstacle. Still the dynamics of such a rotation is rather simple and can be described, in fact, in terms of a one-dimensional wave phenomenon.

Spiral waves are also easily created in two-dimensional homogeneous media without artificial obstacles. These rotating wave patterns are a fascinating example of self-organization processes in nonlinear distributed systems. Their unusual properties attract the attention of many investigators. Spiral waves of excitation are observed in heart muscle [4, 5], in the retina of the eye [6], in colonies of Dictyostelium discoideum [7], in Xenopus laevis oocytes [8] and in many other systems. There is a remarkable similarity of such waves to the concentration waves found in thin layers of the Belousov-Zhabotinsky (BZ) reaction [9].

The detailed analysis of the spiral wave structure [10] showed that it rotates around a small central region (spiral wave core). If the size of the excitable medium exceeds by far the core diameter, the boundary conditions do not influence the dynamics of the spiral wave. Usually one tries to use a sufficiently large piece of excitable medium in order to avoid the undesirable influence of the boundary which complicates the consideration. But all of the real excitable media are bounded and sometimes it is impossible to remove the boundary effects. On the other hand, one can use the boundary effects as an efficient means to influence spiral waves to study their properties. For these reasons the simulation of bounded excitable media is now treated with increasing interest [11–13].

In order to investigate the basic properties of spiral waves, as a rule, an excitable medium is described by a two-component system of nonlinear equations of the "reaction-diffusion" type:

$$\begin{aligned} \frac{\partial u}{\partial t} &= D_u \nabla^2 u + \frac{1}{\epsilon} F(u,v), \\ \frac{\partial v}{\partial t} &= D_v \nabla^2 v + G(u,v). \end{aligned} \tag{1}$$

The components u and v can represent the concentrations of chemical species or the transmembrane potential and the conductivity of a cell membrane. The "excitable property" of the system (1) is determined by a nonlinear function F. Function G can be monotonic or even linear. Usually the value of the parameter ϵ is very small, $\epsilon \ll 1$. It determines the different time scales for the faster component u and the slower component v. The direct integration of (1) with appropriate initial and boundary conditions yields the time-space evolution of the waves for the concrete model of an excitable medium (i.e., for given functions F and G).

To generalize the data obtained for different kinds of models one can use the kinematical description of a propagating wave [14, 15]. The main idea of such an approach is to describe the evolution of the waves as variation in time of the position of an excited domain in which the variable u exceeds some level u_l. The kinematical approach allows rather easily to study the basic features of wave structures even in nonstationary or nonhomogeneous media [15].

The first part of this paper is devoted to the experimental data obtained for spiral waves rotating in a small disk with a thin layer of the BZ reaction. In the second part we describe the computational techniques and the results of computer simulations using a reaction-diffusion model of an excitable medium. The kinematical description of the spiral wave rotating near an impermeable boundary is given in the third part.

2 Experimental Observation of Spiral Waves in a Bounded Medium

The BZ reaction has been widely used to investigate spatial patterns in an excitable medium [9, 10, 16, 17]. This chemical reaction is accompanied by variation of the optical properties (e.g. the color) of the solution and one can visually observe two-dimensional concentration waves propagating in a thin layer of the solution placed in a petri dish.

Usually the radius of the core of the spiral wave is very small with respect to the size of the petri dish and in common chemical experiments the boundary conditions do not influence the tip motion. A special technique was used in our measurements to study the role of the boundary.

We used the ferroin–catalyzed BZ solution with the following concentrations of the reagents: 0.33 M $NaBrO_3$, 0.24 M malonic acid, 0.06 M NaBr, 0.41 M H_2SO_4, and 0.003 M ferroin. A volume of reactive solution was spread out in a large petri dish (diameter 7 cm) resulting in a layer depth of approximately 0.5 mm. The possibility of hydrodynamic disturbances was inhibited by gelling the reagent with agarose gel (2%). The 2D distribution of the intensity of the transmitted light (wavelength, 490 nm) was registered by a video recorder (Umatic, Sony). The front of the concentration wave appears as bright band due to low absorption of ferriin (the oxidized form of the catalyst). The spiral tip was determined as the highly curved open end of this narrow band. The traces of the spiral tip were then analyzed visually on a video screen.

A small disk (diameter 2 – 3 mm) of the gel was cut by a special tool (a plexiglass cylinder with a thin wall). For the cut we selected a piece of the medium surrounding the tip of a spiral wave created in the dish. Then the circulation of the excitation wave occurred in this small gel disk only (see Fig. 1). Significant oxygen influence on the reaction (which inhibits the reaction) was prevented by covering the gel by a thin layer of transparent, chemically inert silicon oil.

The subsequent evolution of the spiral is subjected to a strong influence of the boundary due to the small size of the disk and depends on the initial location of the wave inside the disk. As a rule, the boundary attracts the spiral wave core. Sometimes it leads to a fast motion towards the boundary followed by the death of the spiral (see Fig. 2a) . But rather often the center of

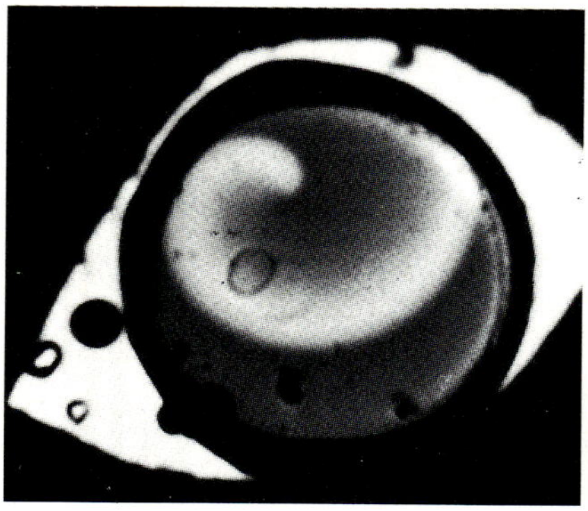

Fig. 1. Spiral wave rotating in a small gel disk of a Belousov-Zhabotinsky reagent. Diameter 2 mm.

a spiral wave started to move along the impermeable boundary. In this case the spiral tip describes a rather complicated trajectory (see Fig. 2b) that looks like an epicycloid. One can distinguish the rotation of the tip around a circular core and the drift of the core along the boundary. During the drift the characteristic distance from the center of the core to the boundary is about 0.3 mm. The drift occurs with a velocity of about 0.15 mm/min.

Note that the direction of the drift coincides with the direction of the rotation of the spiral wave. For instance, in Fig. 2b both the drift and the rotation of the spiral wave proceed in counterclockwise direction.

3 Reaction-diffusion Model and Simulation Techniques

The essential features of the BZ reaction have been extracted and condensed into differential equations, that are known as the Oregonator model [18, 19]. This classic scheme includes the autocatalytic production of $HBrO_2$ with Ce(IV) used as the catalyst of the reaction. We used the two-component version of the Oregonator having the structure (1):

$$\begin{aligned} \frac{\partial u}{\partial t} &= \nabla^2 u + \frac{1}{\epsilon}\left[u - u^2 - f\,v\,\frac{u-q}{u+q}\right], \\ \frac{\partial u}{\partial t} &= \sigma \nabla^2 v + k(u - v), \end{aligned} \tag{2}$$

where the variables u and v describe the evolution of $HBrO_2$ and the catalyst, respectively.

The coefficient D_u in (1) reduces to 1 in (2) due to the special choice of space units, and the coefficient $\sigma = D_v/D_u$. The parameters ϵ, f and q are common for the Oregonator and their values are indicated below. For the classic Oregonator model the value of the coefficient $k = 1$ is derived. We assume that this value depends on the state of the medium:

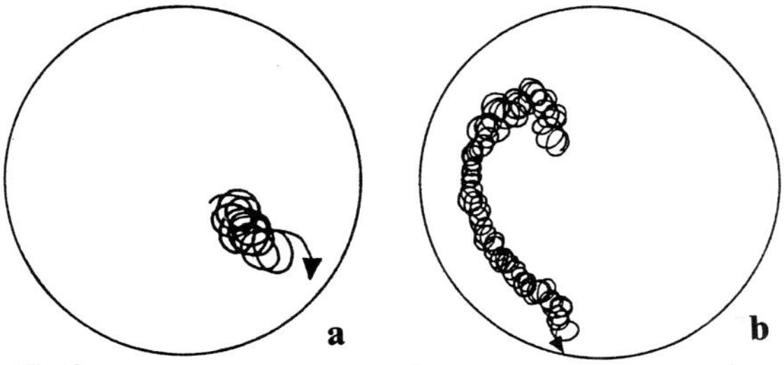

Fig. 2. Trajectories of the tip of the wave rotating in a small disk of a Belousov-Zhabotinsky reagent. Diameter 3 mm. (a) After a few rotations (4 min) the spiral wave dies by collision with the boundary. (b) The spiral wave rotates and drifts along the boundary (trace about 20 min).

$$k = \begin{cases} 1, & u - v \geq 0, \\ k_\epsilon, & u - v < 0. \end{cases} \tag{3}$$

The reason for this modification is the following: The classic Oregonator model reproduces qualitatively the dispersion relation measured for wave trains in the BZ solution [20], but results in some quantitative discrepancies. The additional parameter k_ϵ introduced into system (1) efficiently controls the refractory properties of the model medium [14] and can be used to remove these discrepancies. We used in our computations the value $k_\epsilon > 1.0$ to investigate the role of the refractoriness and to obtain data that are more consistent with the experimental measurements.

The system (2) was integrated for a ring domain with radius R and non-flux boundary conditions:

$$\left.\frac{\partial u}{\partial r}\right|_R = 0; \left.\frac{\partial v}{\partial r}\right|_R = 0. \tag{4}$$

Due to the central symmetry of the problem it is natural to use a polar coordinate system to integrate (2). For this aim the two-dimensional Laplacian is presented as:

$$\nabla^2 w = \frac{\partial^2 w}{\partial r^2} + \frac{1}{r}\frac{\partial w}{\partial r} + \frac{1}{r^2}\frac{\partial^2 w}{\partial \phi^2}, \tag{5}$$

where r and ϕ are the polar radius and the polar angle, respectively. The difficulty in using the polar coordinate system is well known: the simplest discrete scheme which uses a constant angular step $\Delta\phi$ results in a nonuniform distribution of the nodes of the computational grid, because the distance d between two adjacent nodes increases with $r : d = \Delta\phi \times r$. Hence, to obtain high space resolution for large r it is necessary to choose very small $\Delta\phi$. But then the density of the nodes near the center becomes too high and one needs to choose very small time steps Δt. The superfluous numbers of the nodes and the small time steps strongly increase the computational time.

To avoid this difficulty we applied the nonuniform computational grid shown in Fig. 3. All the nodes are located on concentric circles with radii $r_n = n \times \Delta r$, where $n = 1, 2, \ldots, N$ is the number of the circle and Δr is a constant. The value $\Delta\phi$ is a constant for a given circle, but decreases with r, since the number m_n of the nodes located on the circle with number n is

$$m_n = 6\,n. \tag{6}$$

Hence $\Delta\phi = \pi/(3n)$ is inversely proportional to the radius of the circle, and the distance d between the adjacent nodes $d = \Delta r \times \pi/3$ does not depend on the number of the circle.

The value of the Laplacian in the central point is given by the formula:

$$\nabla^2 w\Big|_{r=0} = \frac{2}{3\Delta r^2}\sum_{i=1}^{6}(w_i - w_0)\,, \tag{7}$$

where w_0 is the value of variable w in the central point and the w_i determine the values of w for the six nodes located on the first circle surrounding the center.

In accordance with ref. [21] we used in our computations an implicit computational scheme with the space steps $\Delta r = 0.25$ and the time steps $\Delta t = 0.005$ and the restriction of variable $u : u \geq q$. This allows to decrease considerably the computational time without significant changes of the results.

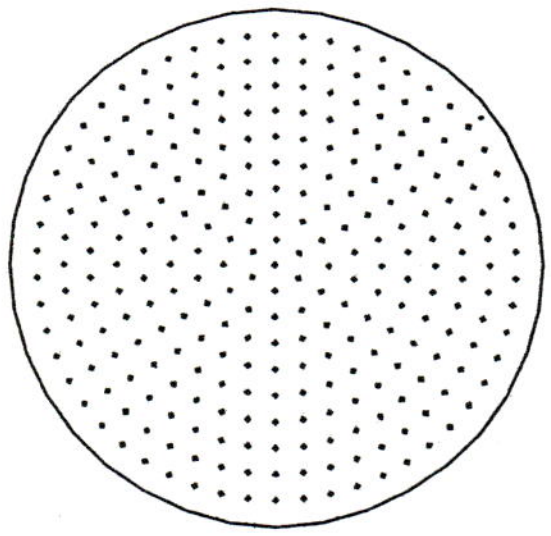

Fig. 3. Location of nodes of the discrete scheme used for numerical integration of the reaction-diffusion equations within a ring domain.

4 Dynamics of Spiral Waves Rotating Within a Small Disk

The system (2) has a spatially uniform steady state at the point (u_{ss}, v_{ss}) determined by the equations

$$F(u_{ss}, v_{ss}) = 0; \; G(u_{ss}, v_{ss}) = 0. \tag{8}$$

This steady state is stable with respect to small perturbations if $f > 1 + 2^{1/2}$ [20]. For $f < 1 + 2^{1/2}$ a limit cycle is a stable regime and in the case of uniform initial conditions one can observe bulk oscillations in the spatially extended system (2).

For creating a spiral wave one has to set a specific nonuniform distribution of the variables as the initial conditions for (2). In our computations these were:

$$u(r,\phi) = \begin{cases} u_A, & r_0 < r < R \text{ and } 0 < \phi < \phi_0 \\ u_{ss}, & \text{otherwise}, \end{cases}$$

$$v(r,\phi) = \begin{cases} v_0, & 0 < r < r_0 \\ v_{in}(\phi), & r_0 < r < R \text{ and } 0 < \phi < \phi_0 \\ v_{out}(\phi), & r_0 < r < R \text{ and } \phi_0 < \phi < 2\pi \end{cases} \tag{9}$$

with

$$v_{in}(\phi) = v_b - (v_b - v_f)\phi/\phi_0,$$
$$v_{out}(\phi) = v_f + (v_b - v_f)(\phi - \phi_0)/(2\pi - \phi_0).$$

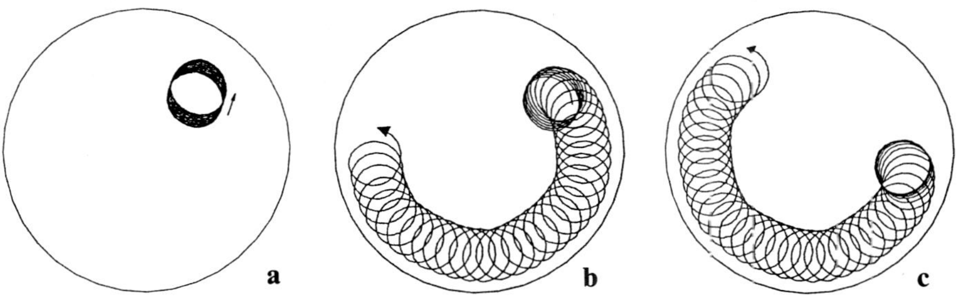

Fig. 4. The trajectory of the spiral tip computed for the Oregonator model (2) in a disk of radius $R = 20$. The given coefficients correspond to a weakly excitable medium: $f = 3.6$, ϵ=0.05, σ=0.6. (a) and (b) k_ϵ=1.0, (c) k_ϵ=1.5.

In accordance with these conditions the variable u exceeds a level u_l within a narrow sector of the ring, where $r_0 < r < R$ and $0 < \phi < \phi_0$. The distribution of the second variable v along the circle with radius $r > r_0$ looks like the one observed for a one-dimensional impulse propagating along

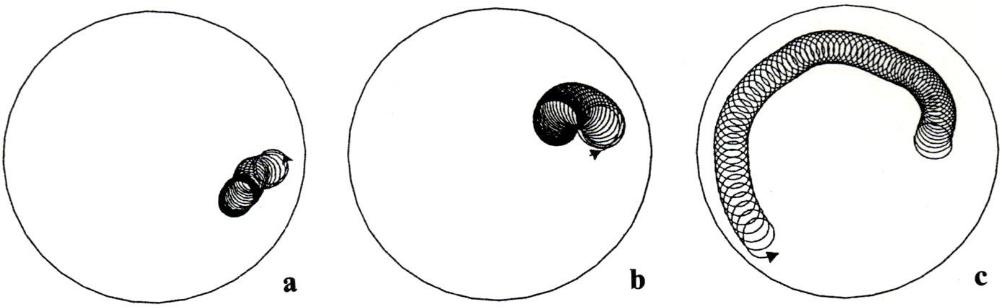

Fig. 5. The trajectory of the spiral tip computed for the Oregonator model (2) in a disk of radius $R = 20$. The given coefficients correspond to a highly excitable medium: $f = 2.0$, ϵ=0.1. (a) k_ϵ=1.0, σ=0.6; (b) k_ϵ=1.0, σ=1.0; (c) k_ϵ=1.5, σ=1.0.

a circular path. Two constants v_f and v_b specify the values of v for the front and the back of this impulse, respectively. The central part of the ring, where $r < r_0$, is placed in a refractory state determined by a constant $v_0 > v_{ss}$. The variation of r_0 allows to control the initial location of the spiral wave core with respect to the boundary of the ring.

For given initial conditions the boundary of the excited sector consists of two parts which are the front (here $du/dt > 0$) and the back (here $du/dt < 0$) of the wave. The tip of the wave is determined as the point at which $du/dt = 0$. It is known that in the Oregonator model of an unbounded medium the trajectory of the wave tip can be a circle or a more complicated meandering curve [22]. The shape of the trajectory depends on the choice of the parameters in (2). For instance, the circular trajectory is observed for two quite different sets of parameters. The first case corresponds to a weakly excitable medium near the propagation boundary. The second one is observed for a highly excitable medium with relatively small values of f. In the following we study the specific features of both of these regimes in a small ring.

At the beginning we chose the values $f = 3.6$ an $\epsilon = 0.05$ in the vicinity of the propagation boundary that was determined for $\sigma = 0.6$ in ref. [22]. These values correspond to a rigid rotation of the spiral wave in an unbounded medium. The trajectory of the spiral wave in a small ring is also practically a circle as long as the center of the spiral core is placed rather far from the boundary. An example of such a trajectory is shown in Fig. 4a. The trajectory looks like a circle but its center shifts extremely slowly towards the boundary. The evolution of the spiral wave placed near the boundary is more complicated. It drifts along the boundary (see Fig. 4b). The drift velocity and the distance from the boundary to the core center approach some constant values in the course of time. Variation of the refractoriness of the medium by increasing k_ϵ practically does not affect this regime (see Fig. 4c).

To investigate the case of a highly excitable medium we fixed the parameter values of system (2) to $f = 2.0$ and $\epsilon = 0.1$. Fig. 5a shows a tip trajectory computed for $k_\epsilon = 1.0$ and $\sigma = 0.6$. In this case the impermeable boundary pushes the spiral wave core away. A drift induced by the influence of the boundary vanishes in the course of time and at the end the trajectory approaches a circle.

An increase of the diffusion constant for the second variable v up to $\sigma = 1.0$ results in the a reversal of the direction of the initial drift, but the limit trajectory is still a circle (Fig. 5b). This regime is very sensitive to the variations of the refractoriness. By increasing k_ϵ up to $k_\epsilon = 1.5$ we obtained a drift of the spiral wave along the boundary (Fig. 5c).

It is important to stress that the drift direction in Fig. 5c is opposite to the drift direction in Fig. 4b but that it coincides with the one observed experimentally in the BZ system (see Fig. 2b).

5 Basic Kinematical Equations of the Wave Front Motion

The most essential factors that determine the evolution of spiral waves rotating near a nonflux boundary of an excitable medium can be extracted by using a simplified kinematical description of this process [14, 15].

We consider the front of a spiral wave as an oriented curve which has a broken end (tip). Each point of this curve moves in normal direction with a velocity depending on the arc length s measured from the tip. The tip can have some tangential velocity v_0 along the curve. The curvature of the curve $K = K(s,t)$ depends on arc length s and time t. It was shown in [15] that the function $K(s,t)$ obeys the following differential equation

$$\frac{\partial K}{\partial t} - \frac{\partial K}{\partial s}\left(v_0 + \int_0^s K\theta\, ds'\right) = \frac{\partial^2 \theta}{\partial s^2} + K^2\theta. \tag{10}$$

Equation (10) describes the purely geometrical properties of any moving curve. The analysis of the system (1) gives us an additional information related to the kinematics of waves in excitable medium. In particular, it was shown in [14] that in a weakly excitable medium the normal velocity θ of a wave depends on the local curvature K according to

$$\theta(K) = \theta_0 + DK \quad \text{for} \quad K > K_c, \tag{11}$$

where the coefficient $D \approx D_u$ for $\epsilon \ll 1$.

The tangential velocity v_0 depends on the local curvature K_0 near the tip [15]:

$$v_o = \gamma(K_c - K_0) \tag{12}$$

with the positive constant γ. The curvature K_0 near the broken end changes with time due to the displacement of the tip along the curve:

$$\frac{dK_0}{dt} = v_0 \left.\frac{\partial K}{\partial s}\right|_{s=0} . \tag{13}$$

Equations (10)-(13) determine the curvature $K(s,t)$ of the curve, and thus the evolution of its shape. The Cartesian coordinates of the wave front can be defined by integration of the following system

$$\begin{aligned} \frac{dX}{ds} &= \cos\alpha \\ \frac{dY}{ds} &= \sin\alpha \\ \frac{d\alpha}{ds} &= K, \end{aligned} \tag{14}$$

where α is the angle between the tangent vector to the curve and the X-axis. The initial value of the tangent angle $\alpha_0 = \alpha(0,t)$ should be determined from the corresponding system of kinematical equations as well as the values $X_0 = X(0,t)$ and $Y_0 = Y(0,t)$

$$\begin{aligned} \frac{dX_0}{dt} &= -\theta(0,t)\sin\alpha_0 + v_0\cos\alpha_0 \\ \frac{dY_0}{dt} &= \theta(0,t)\cos\alpha_0 + v_0\sin\alpha_0 \\ \frac{d\alpha_0}{dt} &= \left.\frac{\partial\theta}{\partial s}\right|_{s=0} + v_0 K_0. \end{aligned} \tag{15}$$

In accordance with our definition the wave front is a line on which $u(x,y,t) = u_l =$ constant. This line should be orthogonal to the impermeable boundary due to (4). Hence the tangent angle $\alpha(L,t)$, where L is the length of the front, should be equal to the angle ϕ_B which determines the direction of the normal to the boundary at the point B where the front touches the boundary. The Cartesian coordinates of the point B are determined as the intersection of the boundary with the front curve given by (14). Furthermore, using the last equation from (14) one can write

$$\alpha_0 + \int_0^L K(s,t)ds = \phi_B. \tag{16}$$

In fact, this equation determines the value of $K(L,t)$. Hence, it is the second boundary condition for equation (10) (the first is equation (13)). Note in addition that the variation of the whole length of the front, L, with time t is determined by the simple equation

$$\frac{dL}{dt} = -v_0 - \int_0^L K\theta ds. \tag{17}$$

The system of equations (10)-(16) completely determines the wave front evolution in a bounded medium and can be used to study nonstationary regimes as well as stationary ones.

6 Kinematics of Stationary Circulation

To consider the stationary rotation of a spiral wave around the center of a disk we can integrate equation (10). Indeed, for the stationary case two conditions are imposed

$$\frac{\partial K}{\partial t} = 0, \quad \text{and} \quad v_0 = 0. \tag{18}$$

Inserting (18) into (10), after integration we obtain

$$\frac{d\theta}{ds} + K \int_0^s K\theta ds' = \omega, \tag{19}$$

where ω is the angular velocity of stationary rotation. The solution of equation (19) should satisfy two conditions:

$$K(0) = K_c, \tag{20}$$

which follows from (12), and

$$\int_0^L K\theta ds = 0, \tag{21}$$

which corresponds to (17) for the stationary case.

Hence, equation (19) together with (11), (20) and (21) formulate a nonlinear eigenvalue problem for the unknown parameter. The numerical solution of this problem yields the dependence of the angular velocity ω_R on the radius R of the disk

$$\omega_R = \omega_0 \left\{ 1 + \left[\frac{\beta(p) D}{\theta_0 R} \right]^4 \right\}^{1/4}, \tag{22}$$

where ω_0 is the angular velocity in an infinite medium, and the function $\beta(p)$ of dimensionless parameter $p = |DK_c/\theta_0|$ and is presented as

$$\beta(p) = \frac{1.46}{p^{3/2}} + 3.28. \tag{23}$$

According to (22) the angular velocity ω_R decreases monotonously with R, but is practically constant for $R > \beta(p)D/\theta_0$. The radius of the core also decreases with the radius of the disk.

7 Nonstationary Circulation Near an Impermeable Boundary

The evolution of a spiral wave placed near an impermeable boundary was studied numerically by direct integration of the basic kinematical system (10)-(16). Let us assume that the boundary is a straight line, and the initial position of the front is also a straight line that is orthogonal to the boundary. The trajectory of the spiral wave tip is shown in Fig. 6a.

One can see that the boundary attracts the core of the spiral wave, but after a relaxation process some distance between the core and the boundary is conserved. Subsequently, the core drifts along the boundary. The reason for such an evolution is found by using the results obtained above. Indeed, the arc length L increases with the radius of the disk. Hence, the angular velocity and the curvature of the tip trajectory should decrease with the arc length. If the tip is far away from the boundary, the front length L is longer and the curvature of the trajectory is smaller. When the tip is close to the boundary the curvature of the trajectory increases (see Fig. 6b). These alternations of the curvature lead to a deformation of the core and create the picture of the cycloid-type motion along the boundary.

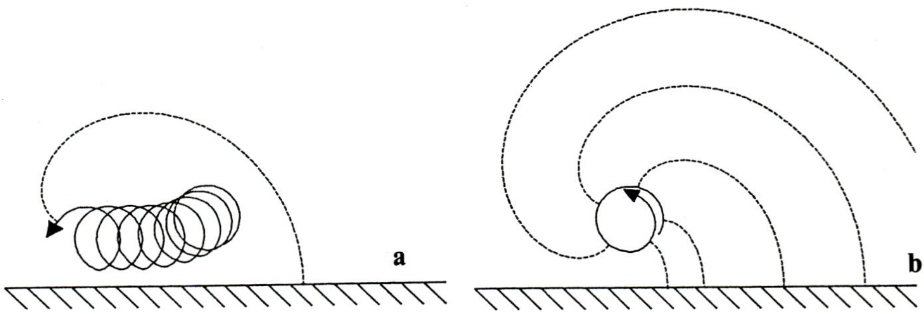

Fig. 6. Trajectory of the tip (a) and the evolution (b) of a spiral front placed near a rectilinear impermeable boundary. The trajectories of the tip are plotted as a solid curve and snapshots of the front shape are given by a dotted curves. The parameter values are: θ_0=1.0, K_c=–1.0, D=0.5, γ=2.0.

In Fig. 7 the results of similar computations are shown for the case of a circular domain of an excitable medium. If the core is placed near the center of this domain (see Fig. 7a) the spiral wave rotates practically with a constant angular velocity and the tip describes a circle. If the initial position of the core is shifted toward the boundary, it starts at first to drifts slowly in radial direction (see Fig. 7b). Then the velocity of the drift increases and finally the core moves along the boundary. Note, that the drift leads to some kind of quasiperiodic regime in the excitable medium. One frequency corresponds

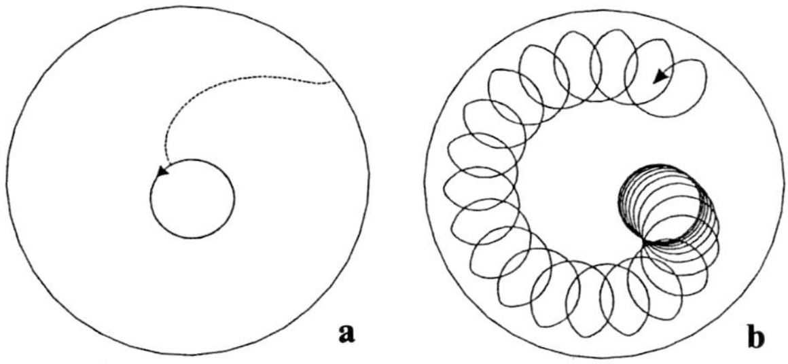

Fig. 7. Trajectory of the tip of a spiral wave placed near the center of a disk with a radius R = 6.0 (a) and in a vicinity of the impermeable boundary (b). The parameter values are: $\theta_0 = 1.0$, $K_c = -1.0$, D=0.5, γ=2.0.

to the rotation of the spiral wave around the core and the other one arises due to the circulation along the boundary.

8 Conclusion

The experimental data obtained for a small disk containing the BZ reaction show that spiral waves rotating in a homogeneous solution are strongly influenced by a boundary. The simplest regime of rigid rotation can be easily destroyed when the spiral tip is placed near the boundary of a small disk.

The two-component Oregonator model for the BZ reaction was used to reproduce the experimentally observed dynamics of spiral waves in the vicinity of a no-flux boundary. The drift of the spiral wave core was observed for a weakly excitable medium as well as for the case of high excitability. But the directions of the drift in these two types of excitable media are opposite to each other.

In the case of low excitability a small variation of the model parameters does not change the basic features of the drift phenomenon. The effect of the drift is explained in the framework of a simplified kinematical description which, in fact, takes into account only the dependence of the propagation velocity on the local curvature of the wave front. The main conclusion of the kinematical consideration is that the angular velocity of a spiral wave rotating in a stationary regime inside a small disk increases when the radius of the disk decreases. Under nonstationary circulation the distance to the boundary is modulated which results in a periodic variation of the curvature of the tip trajectory. This in turn leads to a drift towards the boundary and then along it.

The evolution of the spiral wave in a small domain of a highly excitable medium is more complicated. At first glance the shape of the trajectory in Fig. 5c is very similar to the one in Fig. 4b. But the properties of this regime are quite different. The evolution of the spiral wave in a highly excitable medium is very sensitive to small variations of the model parameters. A comparison between Fig. 5b and Fig. 5c proves that in this case the refractoriness of the medium is very essential, whereas it plays no role in weakly excitable media (see Fig. 4b and Fig. 4c). In the Oregonator model the refractory properties are determined by the dynamics of the second variable v. That is why the evolution of the spiral wave is very sensitive to the diffusion constant of this component (see Fig. 5a and b).

We emphasize here that the direction of the experimentally observed drift of the spiral wave placed near an impermeable boundary has been reproduced correctly only by the modified Oregonator model. Many of our attempts to obtain the same drift direction in the "classical" Oregonator were unsuccessful. Interestingly the same drift direction was observed for the Ginzburg-Landau model used for this problem in ref. [13]. On the other hand, the drift direction in a modified FitzHugh-Nagumo model [12] is the same as for weakly excitable media. Which of the properties of these models are responsible for the opposite results on the drift direction is still not clear.

Thus, the evolution of the spiral wave near the nonflux boundary can be used to distinguish and to classify different models of excitable media. Up to now this interesting problem is rather far from a rigorous solution. It is obvious that in the case of a weakly excitable medium all the models exhibit similar properties. The presented kinematical model is a powerful tool to study this case. In this approach the dynamic behaviour of spiral waves is described in terms of a restricted number of kinematical parameters. These are the propagation velocity θ_0 of a flat front, the critical curvature K_c, and the coefficients D and γ. For instance, the values of the drift velocity along the boundary and the characteristic distance between the boundary and the spiral core can be expressed in terms of these kinematical values. For applying the results obtained in a quite general form to the concrete model of an excitable medium one needs to provide estimates of these parameters. This can be done analytically for some special models or numerically for any given model. Furthermore, these parameters can be measured directly or estimated indirectly from the data obtained in real experiments.

Why does this approach not work when applied to the case of highly excitable media? Firstly, the refractory properties of the medium are now very essential and have to be taken into account. In addition, the refractory tail behind the wave of excitation should be different at different points of the medium, and the refractoriness should depend on the distance from the boundary. In fact, we deal with a functionally nonhomogeneous medium. The two factors were considered separately for different applications of the kinematical approach [15]. This calls for a modified kinematical description that simultaneously takes into account the refractoriness and the nonhomogeneity of the excitable media.

Acknowledgments

The authors thank Dr. K.I. Agladze for useful discussions of the experimental part of the work. V.S.Z. acknowledges support from the WE-Heraeus-Stiftung, Hanau.

References

1 A.T. Winfree: When time breaks down, Princeton Univ. Press, Princeton 1987
2 J.D. Murray: Mathematical biology, Springer Verlag, New York 1978
3 N. Wiener, A. Rosenblueth: The mathematical formulation of conduction of impulses in a network of connected excitable elements, specifically in cardiac muscle. Arch. Inst. Cardiol. Mexico **16**, 205–265 (1946)
4 M.A. Allessie, F.I.M. Bonke, F.J.G. Schopman: Circus movement in rabbit atrial muscle as a mechanism of tachycardia. Circul. Res., **33** 54–62 (1973)
5 J.M. Davidenko, A.V. Pertsov, R. Salomonsz, W. Baxter, J. Jalife: Stationary and drifting spiral waves of excitation in isolated cardiac muscle. Nature **355**, 349–351 (1992)
6 N.A. Gorelova, J. Bures: Spiral waves of spreading depression in the isolated chicken retina. J. Neurobiology **14**, 353–363 (1983)
7 G. Gerisch: Stadienspezifische Aggregationsmuster bei Dictyostelium discoideum. Wilhelm Roux Archiv Entwicklungsmech. Organismen **156**, 127–144 (1965)
8 J. Lechleiter, S. Girard, E. Peralta, D. Clapham: Spiral calcium wave propagation and annihilation in Xenopus laevis oocytes. Science **252**, 123–126 (1991)
9 A.T. Winfree: Spiral waves of chemical activity. Science **175**, 634–636 (1972)
10 S.C. Müller, Th. Plesser, B. Hess: The structure of the core of the spiral wave in the Belousov-Zhabotinsky reagent. Science **230**, 661–663 (1985)
11 E. A. Yermakova, A. M. Pertsov: Interaction of rotating spiral waves with a boundary. Biophysics (USSR) **31**, 855–859 (1986)
12 V. A. Davydov, V.S. Zykov: Spiral waves in a small disk and on a small sphere, Zh. Eksp. Teor. Phys. **76**, 414–424 (1993)
13 J.A. Sepulchre, A. Babloyantz: Motions of spiral waves in oscillatory media and in presence of obstacles. Phys. Rev. E **48**, 187–195 (1993)
14 V. S. Zykov: Simulation of wave processes in excitable media. Manchester Uni. Press, Manchester 1988
15 A. S. Mikhailov, V. A. Davydov, V. S. Zykov: Complex dynamics of spiral waves and motion of curves. Physica D **70**, 1–39 (1994)
16 K. I. Agladze, V. I. Krinsky: Multi-armed vortices in an active chemical medium. Nature **296**, 425–426 (1982)
17 G.S. Skinner, H.L. Swinney: Periodic to quasiperiodic transition of chemical spiral rotation. Physica D **48**, 1–16 (1991)
18 R. J. Field, E. Körös, and R. M. Noyes J. Am. Chem. Soc. **94**, 8649–8664 (1972)
19 J. J. Tyson: A quantitative account of oscillations, bistability, and travelling waves in the Belousov-Zhabotinsky reaction. In: Oscillations and traveling waves in chemical systems. Eds. Field, R.J. Burger, M. (Wiley, New York, 1985) pp. 93–144
20 J. P. Keener, J. J. Tyson: Spiral waves in the Belousov-Zhabotinskii reaction. Physica D **21** 307–324 (1986)

21 W. Jahnke, W.E. Skaggs, A.T. Winfree: Chemical vortex dynamics in the Belousov-Zhabotinsky reaction and in the two-variable Oregonator model. J. Phys. Chem. **93**, 740–749 (1989)

22 W. Jahnke, A.T. Winfree: A survey of spiral-wave behaviors in the Oregonator model. Int. J. Bifurcation and Chaos **1**, 445–466 (1991)

Dynamics of Oscillatory Chemical Systems

Igor Schreiber and Miloš Marek

Abstract

Self-sustained oscillations and related phenomena (excitability, bursting, etc.) in chemical systems arise from special feedback interactions constrained by stoichiometric relations. This makes it possible to predict oscillatory, excitatory or multistable behavior of spatially homogeneous chemical systems directly from their underlying kinetics. We review the basic patterns of behaviour of three experimentally well-studied reactions and discuss consistence with the respective mathematical models. Using a reduction of the system to a phase mapping, we also discuss excitatory dynamics that emerge when such systems are coupled via mass transfer and/or are periodically stimulated. Analogies with biological oscillatory and excitatory systems are pointed out.

1 Introduction

Oscillations in living organisms are observed on time scales from milliseconds (neurons), over seconds (cardiac cells), minutes (oscillatory enzymes), and hours (pulsatile hormone secretion) to even longer periods, following a hierarchical organization. For example, insulin secreting beta-cells found in the pancreas oscillate [1] and, when coupled with other cells, burst [2]. Recently, a human pituitary cell line was genetically engineered to have the properties of an insulin-secreting beta cell [3]. The intentional modification of intermediary metabolism using recombinant DNA techniques is now called metabolic engineering [4, 5], and the modification of cell properties using the same techniques is called cell engineering [6]. Both metabolic and cell engineering require not only genetic information and knowledge about the physiology of the host organism, but also detailed information about the biochemistry of reaction pathways, stoichiometry and kinetic constraints. Detailed kinetic studies have been performed within the last two decades. Savageau [7], and Hayashi and Sakamoto [8] have developed a formal power law representation of kinetics in biochemical networks. Kacser and Burns [9] and Heinrich et al. [10] have developed a metabolic control theory. However, the utility of these approaches in metabolic and cell engineering is still hindered by incomplete knowledge of metabolic models and lack of kinetic data.

At the same time, there is an explosive increase of interest in nonlinear brain dynamics as studied both through neurobiology and formal neural networks theory [11, 12]. Actual dynamics of synaptic interactions depend on detailed kinetics of specific receptors. Recently, methods of computations of synaptic conductances based on a kinetic model of receptor binding were proposed [9], and in spite of insufficient information about the kinetics of corresponding biochemical pathways, neuron models with dynamically regulated conductances are analysed via computer modelling [13, 14]. We believe that the confrontation of model predictions with experimental observations of nonlinear dynamics in specific well controlled chemical systems may facilitate further development of models of far more complex biological systems.

In this review we use three examples of chemical systems, the well-known Belousov-Zhabotinsky (BZ) reaction, the minimal bromate (MB) oscillator and the peroxidase-oxidase (PO) enzymatic reaction system to show the variety of dynamics experienced in complex chemical reactions. We illustrate possibilities of *a priori* estimations of conditions for oscillations, excitability and multiple steady states in systems satisfying laws of mass action kinetics. We also discuss ways of constructing phase mappings which make it possible to predict the response to single or periodic stimulations of oscillatory and excitatory systems arranged in linear arrays. References are drawn both from the chemical literature and from the more biologically oriented literature.

Transmission of information in biological systems occurs via excitable media which may also become oscillatory (e.g., due to a stimulation). Excitable chemically reacting liquid ionic systems and excitable biological systems, such as chemical or electric synapses, specific receptors, neurons, neural networks, various excitable tissues (e.g. heart) etc., share a number of common properties. On a macroscopic level of description excitability is associated with rapid changes of temporal and spatiotemporal patterns of concentrations and fluxes of reacting and/or transported (via diffusion, ionic migration or convection) chemical species in response to a stimulation by an external perturbation of concentration, temperature, light etc. Particular mechanisms of excitation in biological systems are mutually coupled in a complex and often still not identified way [15–18]. Hence, relatively simpler chemical systems operating under well controlled conditions, such as a continuous flow-through reaction cell (CSTR) or a network of coupled CSTRs, can serve as a tool for testing different descriptions of excitable systems.

A widely used technique for studying the phase-resetting of biological clocks by an external stimulus is based on the measurements of phase shifts of oscillations caused by perturbing the system at various phases of the clock; see, for example, a recent study of circadian rhythms in prokaryotes [19]. The variation of the phase shift with the phase of the perturbation is represented by a one-dimensional mapping – *the phase response curve* (PRC) [20–22].

Such a mapping was also constructed from measurements of the BZ reaction run in a CSTR [23]. Repeated iterations of a map simply related to the PRC provide an excellent way of predicting the dynamics, periodic as well as aperiodic, observed in the CSTR when perturbed by pulsatile additions of a reactant, see [23] and [24, chapt. 5]. The dynamic response to single as well as repeated pulse stimulations of the excitable BZ reaction was also a subject of our previous work [25, 26]. A one-dimensional map analogous to the PRC, *the phase excitation curve* (PEC), was constructed from the responses to repeated stimuli and used to describe the dynamics of the periodically perturbed excitable BZ reaction in the same fashion as the PRC was applied in the case of the oscillatory BZ system [27, 28]. The experimental technique for the construction of the PEC was tested on several kinetic schemes for the BZ and other chemical systems and used for the construction of a resonance (or excitation) diagram – the plot of a firing number against the period and/or the amplitude of the perturbation. The firing number measures the average number of the events of excitation within one forcing period [29].

The above methods are equally applicable to biochemical and biological systems. There is a striking analogy between complex biological and simple chemical systems, manifested by a similar structure of the excitation diagrams of the BZ reaction to those reported for periodically stimulated giant squid axons [30]. In experimental studies of neurophysiological systems a two-stimulus paradigm is often used, implying that a first stimulus starts the oscillations and a second stimulus is applied at specific phases of the ongoing oscillations, [see 31, and references therein]. This is a procedure equivalent to that used for the PEC construction. Excitatory and oscillatory units in biological systems (receptors, neurons) [32] are frequently coupled, locally or globally, chemically or electrically [32, 33]. Information transmission and coding in such networks is tied with the properties of outcoming sequences of interspike intervals (firing patterns). Specific cells can be activated by a particular stimulus and the resulting firing pattern [15–18] may serve a physiological function, e.g. to encode visual images of objects or to provide a mechanism of short term memory or learning [34–36].

Coupled chemical oscillators and excitators have been the subject of several recent experimental and computer-assisted studies, see [37–40] for recent references and reviews. Formally derived models based on discrete space systems like cellular automata or coupled map lattices have been primarily used for studies of cooperative dynamics in coupled systems [41, 42]. The closely related problem of travelling waves and other reaction-diffusion phenomena in spatially extended excitable media is also of considerable interest [43, 44].

2 Mechanism of Oscillatory Behaviour and Classification

Chemical reaction systems generate oscillations for the very same reasons as other dynamical systems. Mathematically, the most important way for oscillations to arise is through a Hopf bifurcation from a steady state. The laws of reaction kinetics allow more detailed examination of the conditions necessary for the occurrence of oscillations in a homogeneous, open chemical system, such as the continuous stirred tank reactor (CSTR). At the same time, the oscillatory instability can be cast into chemical rather than mathematical terms. This is made possible by viewing the chemical systems as networks and applying the methods of reaction networks theory [45–48].

Assume that there are s species taking part in m chemical reactions (forward and reverse reactions are treated as separate) so that n species, $n \leq s$, are entering at least one of the reactions

$$\mathrm{R}_j : \sum_{i=1}^{n} \nu_{ij}^{L} \mathrm{X}_i \rightarrow \sum_{i=1}^{s} \nu_{ij}^{R} \mathrm{X}_i, \; j = 1, \cdots, m. \tag{1}$$

The first n species are reactants and intermediates and the remaining $s - n$ are products. Let $\nu = \{\nu_{ij}^R - \nu_{ij}^L\}$ be the $(n \times m)$ stoichiometric matrix, $x = (x_1, \cdots, x_n)$ the vector of the chemical species concentrations, and $r(x) = (r_1(x), \cdots, r_m(x))$ the vector of reaction rates. The chemical equations together with the vector function $r(x)$ define the mechanisms of the reaction, also called the stoichiometric network. Only the concentrations of the first n species are dynamical variables for which independent equations of motion can be set up. We assume that the reactions take place in a thermostated flow-through CSTR. Since the net rate of chemical production of each species is a linear combination of the relevant reaction rates, the time evolution of x is governed by the following mass balance equations,

$$\frac{dx}{dt} = f(x) = \nu r(x) + k_0(x_0 - x) = \mathcal{S} v(x) \tag{2}$$

where x_0 is the vector of feed concentrations and k_0 is the flow rate (reciprocal residence time). It is convenient to treat the $2n$ flow terms as additional $2n$ pseudoreactions. The augmented stoichiometric matrix $\mathcal{S}$ and the augmented rate function $v(x)$ accomodate these flow terms.

If the rank of ν is less than n there is a nonempty null space of ν^T of dimension $d_n = n - rank(\nu)$ and there are d_n independent linear combinations of x_i's such that, after an initial transient,

$$\gamma x = const. = \gamma x_0 \tag{3}$$

where each of the rows of the $(d_n \times n)$ matrix γ specifies a constraint condition that may be used to reduce the n Eqs. (2) by d_n. This condition differs from a true stoichiometric conservation constraint by the fact that the constant depends on the feed concentrations rather than on the initial composition in the reactor.

A stationary state x_s satisfies the equation $\mathcal{S}v(x) = 0$. Thus $v_s = v(x_s)$ is contained in the null space of $\mathcal{S}$. Moreover, all components of v must be nonnegative numbers. This narrows the set of all possible stationary reaction rate vectors v_s to an open convex d_m-dimensional cone, $d_m = m - \text{rank}(\mathcal{S})$, in the space of all v's. Moreover, because of the flow terms, $\text{rank}(\mathcal{S}) = n$. The (normalized) edges of this steady state cone can be found by standard numerical methods of linear algebra. They represent reaction rate vectors that may be conveniently used for selecting a basis generating all v_s's. So the edges represent major reaction pathways. If one of these pathways contains an unstable steady state (typically a saddle), a Hopf bifurcation giving rise to oscillations is expected near that unstable pathway, provided that the topology of the network admits a stabilizing negative feedback [49].

The determination of stability of a particular steady state reaction vector v_s relies on the analysis of the Jacobian J of Eq. (2) at x_s,

$$J = \left.\frac{df}{dx}\right|_{x=x_s} = \mathcal{S}\left.\frac{dv}{dx}\right|_{x=x_s} = \mathcal{S}(\text{diag } v_s)\kappa^T(\text{diag } x_s)^{-1}, \tag{4}$$

where the kinetic matrix $\kappa = \{\kappa_{ij}\} = \{\partial \ln v_j(x_s)/\partial \ln x_i\}$. The number κ_{ij} is the effective order of the j-th reaction with respect to the i-th species; κ is in general dependent on x_s. If the reaction rates obey a power law, $v_j(x) = k_j \prod_{i=1}^n x_i^{a_{ij}}, j = 1, \cdots, m$, where a_{ij}'s are the reaction orders and k_j's are the rate coefficients, then $\kappa_{ij} = a_{ij}$ and κ is independent of x_s.

Notice that the steady state rate vector v_s may be expressed as a linear combination of the (normalized) edges of the steady state cone and therefore may be independent of x_s (adjustable rate coefficients are assumed). Due to the outflow term $k_0 x$, a flow-through reaction network (2) determines x_s for given v_s and k_0. Conversely, a choice of x_s greatly reduces the number of edges that may be arbitrarily combined. Thus the flow-through arrangement significantly simplifies the stability analysis of the network.

The stability of x_s is determined by the eigenvalues of J at x_s. However, even without knowing x_s we can use the following estimation: if power law kinetics applies, the matrix $V = -\mathcal{S}(\text{diag } v_s)\kappa^T$ is constant and the stability of v_s is indicated by order k principal subdeterminants β_k of V, $k = 1, \cdots, n$; there are $\binom{n}{k}$ different β_k's related to all permutations of k species. If at least one β_k is negative, then at least one eigenvalue of J is unstable (i.e., it has a positive real part) provided that the values of the steady state concentrations of the corresponding k species are sufficiently small.

This method identifies the species that are crucial for the instability. Typically, such species belong to one of two classes: a) species located on a cycle in the reaction network generating an autocatalytic loop (a positive feedback), b) and species that regulate some of the autocatalytic species by an exit reaction removing the cycle species. There is another class of essential species, not necessarily present in a small concentration. They take part in a reaction pathway, providing a negative feedback. Thus there are *autocatalytic*, *exit* and *feedback* species [49]. Finally, there is an additional type, the *recovery* species, that play a role similar to the autocatalytic species without being a part of the autocatalytic cycle [49]. The above species are essential for oscillations, all others are nonessential. A concerted action of the reactions involving the essential species generates a Hopf bifurcation and oscillations. Other species may be held at a fixed concentration without losing the oscillatory behaviour. The outlined classification implies a categorization of chemical oscillators [49, 50] which are identified by characteristic phase shifts of the oscillating essential species (near a Hopf bifurcation) [50], by the sensitivity of the steady state x_s with respect to changes in the inflow concentrations x_0, and by some additional criteria [49].

Thus given a power law reaction network by Eqs. (1, 2), the source of a steady state instability leading to oscillations can be systematically searched for without knowing the numerical values of rate coefficients and steady state concentrations. The outcome is an identification of the source of the network's instability and a range of reaction rate coefficients and steady state concentrations where a Hopf bifurcation may occur.

In the following we use three well-known chemical oscillators to show various instability mechanisms and extend the analysis by examining their bifurcation behaviour [24, 51]. This method allows a more detailed comparison by localizing parameter regions where basic dynamical modes occur.

2.1 Belousov-Zhabotinskii Reaction

The Belousov-Zhabotinskii (BZ) reaction, an oxidation of bromomalonic acid by bromate catalyzed by certain metal ions (e.g., Ce^{3+}/Ce^{4+} or ferroin/feriin) in an acidic solution, is a paradigmatic example of nonlinear effects in chemical dynamics [52, 53]. Dynamics displayed by the BZ reaction include bistability, periodic oscillations, excitability and more complex temporal patterns as quasiperiodicity and chaos [53]. Spatio-temporal patterns arise when transport phenomena like diffusion are involved. The BZ reaction has been throughly studied for more than three decades. A complete reaction mechanism is still not available, but a number of models reproduce many aspects of the dynamics reasonably well. Notably, Oregonator type models have been applied succesfully [54, 55].

The simplified reaction scheme for the bromate–cerium–bromomalonic acid system is:

$$H^+ + Br^- + HOBr \rightleftharpoons Br_2 + H_2O, \tag{R1}$$
$$H^+ + Br^- + HBrO_2 \rightarrow 2HOBr, \tag{R2}$$
$$H^+ + Br^- + HBrO_3 \rightarrow HBrO_2 + HOBr, \tag{R3}$$
$$2HBrO_2 \rightarrow HBrO_3 + HOBr, \tag{R4}$$
$$2H^+ + HBrO_3 + HBrO_2 \rightleftharpoons 2HBrO_2^+ + H_2O, \tag{R5}$$
$$Ce^{3+} + HBrO_2^+ \rightleftharpoons Ce^{4+} + HBrO_2, \tag{R6}$$
$$Ce^{4+} + CHBr(COOH)_2 \rightarrow Ce^{3+} + qBr^- + \text{products}. \tag{J}$$

The step (J) is a simplified description of the overall production of bromide ions from bromomalonic acid. The stoichiometric factor q is expected to range from 0.5 to 1 depending on the external conditions; it has been shown that q is close to 1 in the presence of oxygen [56]. As a consequence of the complex nature of the process (J), the reaction (R1) may be neglected [55] leaving HOBr as a product. The concentrations H = $[H^+]$, A = $[HBrO_3]$ and B = $[CHBr(COOH)_2]$ are assumed effectively constant since these species are in surplus. The stoichiometric matrix for the five dynamical variables $x = [HBrO_2]$, $y = [Br^-]$, $z = [Ce^{4+}]$, $w = [Ce^{3+}]$ and $u = [HBrO_2^+]$ reads:

$$\nu = \begin{bmatrix} -1 & 1 & -2 & -1 & 1 & 1 & -1 & 0 \\ -1 & -1 & 0 & 0 & 0 & 0 & 0 & q \\ 0 & 0 & 0 & 0 & 0 & 1 & -1 & -1 \\ 0 & 0 & 0 & 0 & 0 & -1 & 1 & 1 \\ 0 & 0 & 0 & 2 & -2 & -1 & 1 & 0 \end{bmatrix}.$$

Thus, using mass balance equations and applying power law kinetics, the reactions (R2–R6, J) generate the following dynamical system:

$$\begin{aligned} \frac{dx}{dt} &= -r_2 + r_3 - 2r_4 - r_5 + r_{-5} + r_6 - r_{-6} - k_0 x, \\ \frac{dy}{dt} &= -r_2 - r_3 + q r_J + k_0(y_0 - y), \\ \frac{dz}{dt} &= r_6 - r_{-6} - r_J + k_0(z_0 - z), \\ \frac{dw}{dt} &= -r_6 + r_{-6} + r_J - k_0 w, \\ \frac{du}{dt} &= 2(r_5 - r_{-5}) - r_6 + r_{-6} - k_0 u, \end{aligned} \tag{5}$$

where the reaction rates are

$$\begin{aligned} r_2 &= k_2 Hxy, & r_5 &= k_5 HAx, & r_{-6} &= k_{-6} xz, \\ r_3 &= k_3 HAy, & r_{-5} &= k_{-5} u^2, & r_J &= k_J Bz. \\ r_4 &= k_4 x^2, & r_6 &= k_6 uw \end{aligned}$$

Since the rank of ν is 4 (third and fourth rows sum up to zero), there is a constraint,

$$z + w = z_0, \tag{6}$$

and either z or w may be eliminated.

By applying the network analysis we find that the pathway involving reactions (R2), (R3), (R5), (R6) and (J) is the unstable element and that x, u and y should be sufficiently small to achieve the instability. An inspection of the network reveals an autocatalytic cycle formed by (R5) and (R6) with x and u as cycle species; (R2) is the exit reaction involving x and y. Hence x and u are the autocatalytic species and y is the exit species. The role of the feedback species is played by z since it provides a negative feedback by producing y through (J) and thereby suppressing the autocatalytic production of x. The reduced form of the metal catalyst, w, is nonessential for the oscillations. The conditions for the occurrence of a Hopf bifurcation are satisfied for a wide range of reaction coefficients which includes, as expected, also the commonly used values derived from experimental measurements [57, 58]: $k_2 = 3 \times 10^6 \mathrm{M}^{-2}\mathrm{s}^{-1}$, $k_3 = 2\mathrm{M}^{-2}\mathrm{s}^{-1}$, $k_4 = 3 \times 10^3 \mathrm{M}^{-1}\mathrm{s}^{-1}$, $k_5 = 42\mathrm{M}^{-2}\mathrm{s}^{-1}$, $k_{-5} = 4.2 \times 10^7 \mathrm{M}^{-1}\mathrm{s}^{-1}$, $k_6 = 8 \times 10^4 \mathrm{M}^{-1}\mathrm{s}^{-1}$, $k_{-6} = 8.9 \times 10^3 \mathrm{M}^{-1}\mathrm{s}^{-1}$, $k_J = 0.4\mathrm{M}^{-1}\mathrm{s}^{-1}$. Typical values of the other parameters corresponding to experiments carried out in our laboratory [25, 26, 27, 59] are $A = 0.3$ M, $B = 0.1$ M, H $= 0.82$ M, $z_0 = 0.006$ M, $y_0 = 0.01$ M.

We are left with two free parameters, the flow rate k_0 and the stoichiometric factor q. In the presence of oxygen, $q \approx 1$. Assuming $q = 1$, the Hopf bifurcation occurs at $k_0^* = 1.2 \times 10^{-4}$ s^{-1}. The bifurcation is subcritical implying the occurrence of hysteresis. Below k_0^* there are stable large amplitude oscillations which overlap in a very narrow interval above k_0^* with a stable steady state. Just above the upper hysteresis limit the oscillations vanish and excitability occurs, i.e., a small but finite perturbation of the steady state triggers a large excursion, tracing a path resembling the extinct limit cycle.

It is interesting to examine the phase shifts of the oscillations, since these are helpful in separating the nonessential species and identifying the role of essential species. Fig. 1 shows the waveform of the oscillations during one period at $k_0 = 1 \times 10^{-4}$ s^{-1}. The concentrations relative to their average values are plotted. Generally, the nonessential species are marked by having small relative amplitudes of oscillations. The essential species of the same class tend to be mutually in-phase, and the phase relations of essential species in different classes determine the category [50]. Clearly, the autocatalytic species x and u are mutually in-phase, lagged by y and by z which itself is advancing y; this arrangement is consistent with a category called 1B [49].

Next we are interested in finding the regions of oscillations, excitability and multiplicity by allowing both q and k_0 to vary. This can be achieved by using numerical continuation techniques [24, 51]. The oscillations arise in a Hopf bifurcation, multiplicity is marked by the saddle-node bifurcations and the excitability is found near the simultaneous occurrence of the Hopf and

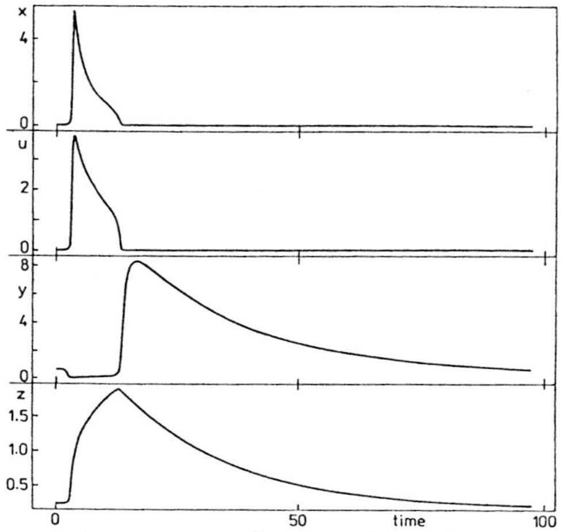

Fig. 1. Periodic oscillations of the dynamical variables x = [$HBrO_2$], y = [Br^-], u = [$HBrO_2^+$] and z = [Ce^{4+}] relative to their average values for the BZ system; $k_0 = 1.0 \times 10^{-4}$ s^{-1}, see text for other parameter values.

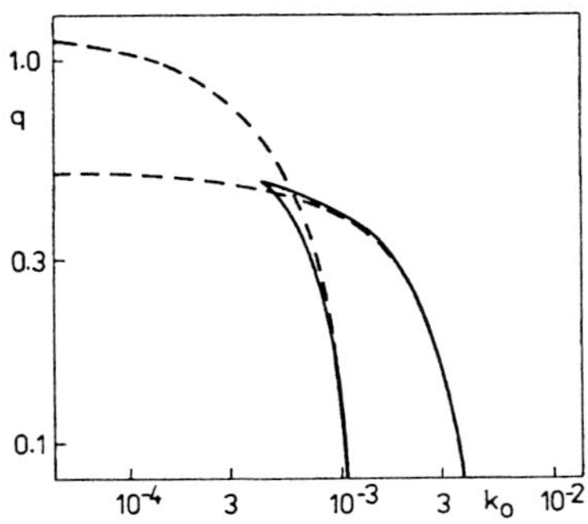

Fig. 2. Bifurcation diagram for the BZ system showing the curves of Hopf bifurcation (dashed lines) and the curves of saddle-node bifurcation (full lines) in the k_0-q parameter plane.

saddle-node bifurcations, provided that the former is subcritical. In a two-parameter plane, generic bifurcations occur along curves as shown in Fig. 2.

The bifurcation structure is characterized by a cusp point marking the wedge-like region of simultaneous existence of three steady states, and two Hopf curves that cross each other. Stable oscillations exist primarily between the two Hopf curves outside the region of three steady states, bistability of steady states is found well inside that region; the bifurcation structure is complex near the intersection of the Hopf curves. This is the well-known 'cross-shaped phase diagram' [60, 61]. Such a structure implies the existence of a subcritical Hopf bifurcation near the cusp and therefore excitability is expected, in the vicinity of the oscillatory region. The excitability and oscillations persist when k_0 tends to zero and the system operates in a (quasi)batch mode. This is typical of 1B category oscillators (B stands for batch).

Excitability is a precursor for pulse waves that occur if transport by diffusion is taken into account as, for instance, in a shallow layer of unstirred solution on a Petri dish. Since there is a symmetry in the bifurcation diagram, two kinds of excitability may exist, one related to an 'upper' steady state above the cusp and the other to a state below the cusp. The former is related to oxidation pulse waves occurring in the presence of oxygen (q is close to 1), the latter to reduction pulse waves (q is close to 1/2). In experiments, waves are observed mainly with ferroin as a catalyst; oxidation waves appear as blue fronts moving in red solution, whereas reduction waves are red fronts in blue solution [62–64].

2.2 Minimal Bromate Oscillator

The minimal bromate (MB) reaction system is the inorganic part of the BZ reaction; (bromo)malonic acid is not present in the reaction medium and the processes are described by the chemical equations (R1–R6) [65, 66]. The absence of a process equivalent to (J) implies that the bromide ions have to be continuously supplied from an external source. This is easy to arrange for in a CSTR. A second major difference is in the flow terms; there is an irreversible cycle between both forms of the catalyst due to the presence of reaction (J) in the BZ reaction. Therefore it does not matter in what form the catalyst is fed to the CSTR. The MB system, on the other hand, does not recycle the catalyst and the reduced form, rather than the oxidized one, has to be supplied. The flow-through arrangement is essential for maintaining the oscillatory regime. Due to the absence of bromomalonic acid the reaction (R1) is no longer negligible. As earlier, the concentrations H = $[H^+]$, A = $[HBrO_3]$ are assumed effectively constant. The stoichiometric matrix for the seven dynamical variables $x = [HBrO_2]$, $y = [Br^-]$, $z = [Ce^{4+}]$, $w = [Ce^{3+}]$, $u = [HBrO_2^+]$, $v = [HOBr]$ and $s = [Br_2]$ is

$$\nu = \begin{bmatrix} 0 & 0 & -1 & 1 & -2 & -1 & 1 & 1 & -1 \\ -1 & 1 & -1 & -1 & 0 & 0 & 0 & 0 & 0 \\ 0 & 0 & 0 & 0 & 0 & 0 & 0 & 1 & -1 \\ 0 & 0 & 0 & 0 & 0 & 0 & 0 & -1 & 1 \\ 0 & 0 & 0 & 0 & 0 & 2 & -2 & -1 & 1 \\ -1 & 1 & 2 & 1 & 1 & 0 & 0 & 0 & 0 \\ 1 & -1 & 0 & 0 & 0 & 0 & 0 & 0 & 0 \end{bmatrix},$$

and the dynamical equations are as follows:

$$\begin{aligned}
\frac{dx}{dt} &= -r_2 + r_3 - 2r_4 - r_5 + r_{-5} + r_6 - r_{-6} - k_0 x, \\
\frac{dy}{dt} &= -r_1 + r_{-1} - r_2 - r_3 + k_0(y_0 - y), \\
\frac{dz}{dt} &= r_6 - r_{-6} - k_0 z, \\
\frac{dw}{dt} &= -r_6 + r_{-6} - k_0(w_0 - w), \\
\frac{du}{dt} &= 2(r_5 - r_{-5}) - r_6 + r_{-6} - k_0 u, \\
\frac{dv}{dt} &= -r_1 + r_{-1} + 2r_2 + r_3 + r_4 - k_0 v, \\
\frac{ds}{dt} &= r_1 - r_{-1} - k_0 s.
\end{aligned} \tag{7}$$

The additional reaction rates are $r_1 = k_1 H y v$, $r_{-1} = k_{-1} s$. The rank of ν is 5; hence there are two constraints,

$$z + w = w_0, \tag{8}$$

$$2x + 6y + w + u + 4v + 10s = 6y_0 + w_0, \tag{9}$$

implying that two variables may be eliminated.

The network analysis indicates that there are two possible pathways causing an instability. The first one involves the forward parts of reactions (R1),(R5),(R6) and reaction (R2). In addition, an inflow of both y and w is required. The other unstable pathway does not include (R1). In both cases x, u and y must be sufficiently small for the instability to occur. In agreement with experiments, no instability is indicated if z is supplied instead of w. The classification of the species suggested by these results is as follows: x and u are the autocatalytic species as in the BZ reaction, y is again the exit species and w is a feedback species (instead of z which is no longer essential for oscillations). s is also nonessential whereas v may be essential or not, depending on the unstable pathway involved in the oscillations. Since experiments suggest that the reaction (R1) should not be neglected, v is likely to be essential and is then classified as the recovery species.

These suggestions can be verified by finding an actual Hopf bifurcation and examining the phase shifts. The values of the rate coefficients of (R1) consistent with experiments [66] are $k_1 = 8 \times 10^9 \mathrm{M}^{-2}\mathrm{s}^{-1}$ and $k_{-1} = 110\mathrm{s}^{-1}$. By taking the other relevant parameter values as in the BZ reaction (we have replaced z_0 for w_0 and slightly changed the value of y_0 from 0.01 to 0.009) a Hopf bifurcation is found at $k_0^* = 2.76 \times 10^{-3}\ \mathrm{s}^{-1}$. Comparing this value to that for the BZ system, a decrease in residence time by about one order of magnitude is found. On the other hand, the period of oscillations is increased by about the same order. The bifurcation is subcritical with large amplitude oscillations to the left of k_0^* and a narrow hysteresis interval to the right, above which excitability occurs. Fig. 3 shows the waveform of the periodic oscillations within one period at $k_0 = 2.7 \times 10^{-3}\ \mathrm{s}^{-1}$. The oscillations are again strongly nonlinear with x, u and v mutually in-phase, y in opposite phase with the first three species and z somewhat advancing y. This is in complete agreement with the analysis of the unstable pathways; the relative amplitude of v is large suggesting that v is essential and the pathway involving (R1) is the determining one. The MB system belongs to a category called 1C [49] (C stands for CSTR).

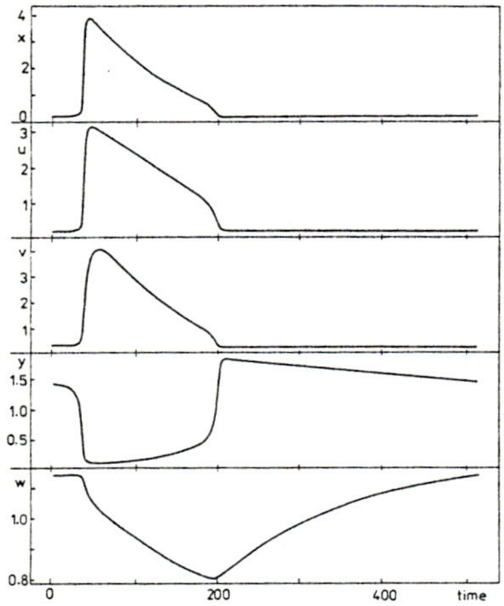

Fig. 3. Periodic oscillations of the dynamical variables $x = [HBrO_2]$, $u = [HBrO_2^+]$, $v = [HOBr]$, $y = [Br^-]$ and $w = [Ce^{3+}]$ relative to their average values for the MB system; $k_0 = 2.7 \times 10^{-3}\ \mathrm{s}^{-1}$, see text for other parameter values.

The next step is the construction of a bifurcation diagram comparable to that for the BZ reaction. The MB oscillator does not produce bromide ions internally and therefore y_0 is comparable to q rather than to y_0 in the BZ system. Thus the free parameters are y_0 and k_0. The Hopf and saddle-node bifurcation curves are shown in Fig. 4. The presence of the 'cross-shaped-phase diagram' signifies the presence of all the basic modes, oscillations, excitability and multiplicity. The major difference from Fig. 2 is that the oscillatory region does not extend to small values of k_0, stressing the CSTR conditions and ruling out any complex dynamics under (quasi) batch conditions. This kind of bifurcation behaviour is typical of 1C category oscillators.

Although the excitability is present in the MB reaction, the reaction-diffusion waves are hard to see experimentally since it is difficult to maintain a large continuous supply of Br^- ions in an unstirred medium.

2.3 Peroxidase-Oxidase System

Experimental observations of the dynamics of the peroxidase-oxidase (PO) system include bistability, periodic, quasiperiodic and chaotic oscillations [67–76].

The simplest of the realistic mechanisms for the PO system called model A [77, 78] is represented by the following set of reaction steps:

$$\mathrm{per}^{3+} + \mathrm{H_2O_2} \rightarrow \mathrm{coI}, \qquad \text{(PO1)}$$
$$\mathrm{coI} + \mathrm{NADH} \rightarrow \mathrm{coII} + \mathrm{NAD}\cdot, \qquad \text{(PO2)}$$
$$\mathrm{coII} + \mathrm{NADH} \rightarrow \mathrm{per}^{3+} + \mathrm{NAD}\cdot, \qquad \text{(PO3)}$$
$$\mathrm{coIII} + \mathrm{NAD}\cdot \rightarrow \mathrm{coI} + \mathrm{NAD}^{+}, \qquad \text{(PO4)}$$
$$\mathrm{per}^{3+} + \mathrm{O_2^-} \rightarrow \mathrm{coIII}, \qquad \text{(PO5)}$$
$$\mathrm{NAD}\cdot + \mathrm{O_2} \rightarrow \mathrm{NAD}^{+} + \mathrm{O_2^-}, \qquad \text{(PO6)}$$
$$\mathrm{NADH} + \mathrm{O_2^-} + \mathrm{H}^{+} \rightarrow \mathrm{H_2O_2} + \mathrm{NAD}\cdot, \qquad \text{(PO7)}$$
$$2\mathrm{NAD}\cdot \rightarrow \mathrm{NAD\text{–}NAD}, \qquad \text{(PO8)}$$

The native form of the enzyme (horseradish peroxidase), per^{3+}, is oxidized by superoxide ion radical O_2^- to form oxyperoxidase (compound III), which is successively reduced through enzymatic forms known as compound I and II, respectively, back to the native form. The first step in the reduction consumes the radical NAD·, while the second and third steps consume NADH and produce two NAD·. Hence, there is a net stochiometric production of NAD· in an autocatalytic double-cycle: NAD· → coI → NAD· and NAD· → coI → coII → NAD·. The cycle is driven by NADH via (PO2) and (PO3), and by O_2, through reactions (PO6), (PO5) and (PO4). As a byproduct, H_2O_2 is formed by (PO7). Hydrogen peroxide is a competitive substrate to superoxide ion for the native enzyme by which it is oxidized to compound I through reaction (PO1). H^+ is buffered, NAD^+ and NAD–NAD are products. The system is externally supplied with oxygen via interface mass transfer from the air. We assume CSTR arrangement with NADH and per^{3+} supplied via the inlet stream.

To make the system simpler we merge steps (PO2) and (PO3) and eliminate coII since this species is low in concentration and may be assumed at a quasi-steady state. The equations of motion for the eight dynamical variables x = [NAD·], $y = [O_2]$, z = [coIII], u = [coI], $v = [per^{3+}]$, $w = [O_2^-]$, p = $[H_2O_2]$ and s = [NADH] are as follows:

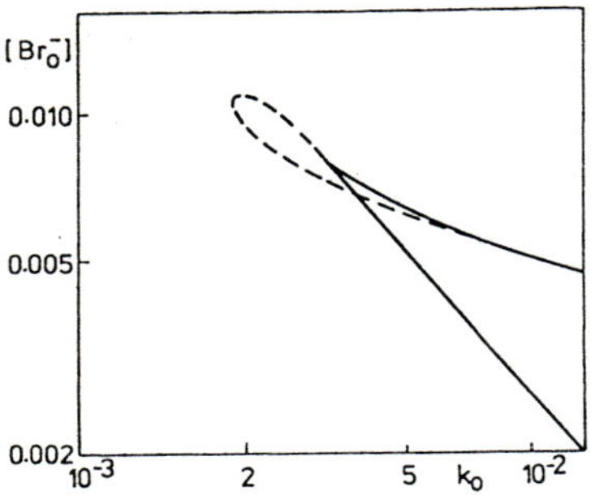

Fig. 4. Bifurcation diagram for the MB system showing the curves of Hopf bifurcation (dashed lines) and the curves of saddle-node bifurcation (full lines) in the k_0-y_0 parameter plane.

$$
\begin{aligned}
\frac{dx}{dt} &= 2r_{2+3} - r_4 - r_6 + r_7 - 2r_8 - k_0 x, \\
\frac{dy}{dt} &= -r_6 + k_t(y^* - y) - k_0 y, \\
\frac{dz}{dt} &= -r_4 + r_5 - k_0 z, \\
\frac{du}{dt} &= r_1 - r_{2+3} + r_4 - k_0 u, \\
\frac{dv}{dt} &= -r_1 + r_{2+3} - r_5 + k_0(v_0 - v), \\
\frac{dw}{dt} &= -r_5 + r_6 - r_7 - k_0 w, \\
\frac{dp}{dt} &= -r_1 + r_7 - k_0 p, \\
\frac{ds}{dt} &= -2r_{2+3} + k_0(s_0 - s),
\end{aligned}
\qquad (10)
$$

where the reaction rates are ([H^+] is included in the rate coefficients)

$$
\begin{aligned}
r_1 &= k_1 vp, & r_5 &= k_5 vw, & r_8 &= k_8 x^2, \\
r_{2+3} &= k_2 su, & r_6 &= k_6 xy, & & \\
r_4 &= k_4 xz, & r_7 &= k_7 ws & &
\end{aligned}
$$

The stoichiometric matrix ν implies one constraint binding the three enzyme species,

$$z + u + v = v_0. \qquad (11)$$

The interface exchange of oxygen is described by the transport coefficient k_t and the equilibrium concentration $y^* = [O_2^*]$.

The network analysis indicates the existence of one unstable pathway involving reactions (PO2)–(PO6) combined with the inflow of oxygen and NADH. The instability requires that x, y and u be sufficiently small. As in the earlier two examples, this suggests the following classification: x and u are autocatalytic species, y is an exit species. Since hydrogen peroxide is absent in the unstable pathway it is nonessential for oscillations. The rest of the species must be examined in more detail to determine their role; particularly, dynamics near a Hopf bifurcation has to be studied. This requires the values of rate coefficients. However, only very rough estimates are available for most of the constants [78].

Here we show how to derive a reasonable set of rate coefficients purely from the network stability considerations. Of course, there is quite a wide range of rate coefficients that may be chosen to display a Hopf bifurcation. Fortunately, the CSTR arrangement narrows that range considerably. By estimating experimentally feasible values of the steady state concentrations so that the small concentration condition is satisfied, $x \approx 0.01\mu$M, $y \approx 1\mu$M, $z \approx 1\mu$M, $u \approx 0.01\mu$M, $v \approx 1\mu$M, $w \approx 1\mu$M, $p \approx 0.1\mu$M and $s \approx 60\mu$M and by setting $k_0 \approx 0.001$ s^{-1} and $k_t \approx 0.015$ s^{-1}, [70, 74] there is an unambiguous choice of a reaction rate vector v_s in the steady state cone so that the Hopf bifurcation with a reasonable period of oscillations $T \approx 100$ s is found. This implies the following choice of rate coefficients: $k_1 = 2.3 \times 10^5$ M^{-2}s^{-1}, $k_2 = 2.8 \times 10^5$ M^{-2}s^{-1}, $k_4 = 7.1 \times 10^6$ M^{-2}s^{-1}, $k_5 = 1.4 \times 10^6$ M^{-2}s^{-1}, $k_6 = 8.3 \times 10^6$ M^{-2}s^{-1}, $k_7 = 3.8 \times 10^3$ M^{-2}s^{-1}, $k_8 = 5.6 \times 10^7$ M^{-2}s^{-1} and feed concentrations: $s_0 = 410$ μM, $z_0 = 2$ μM.

Thus we are left with three free parameters, k_0, k_t and y^*. A Hopf bifurcation is found, for example, at $k_0 = 0.001$ s^{-1}, $k_t = 0.0151$ s^{-1} and $y^* = 1.5 \times 10^{-5}$ M. Allowing k_t or k_0 to vary, the bifurcation is supercritical to the left and small amplitude sinusoidal oscillations develop. This situation is very different from that in both the BZ and MB systems. Fig. 5 shows the waveform of concentrations of the species NAD·, O_2, coIII, coI and O_2^- that were found to be essential by having large relative amplitudes. The species NAD· and coI are approximately in phase and advanced by coIII and O_2^-. O_2 is in opposite phase with coI. This confirms that NAD· and coI are the autocatalytic species, O_2 is the exit species and coIII and O_2^- are the feedback species. Therefore the category is 1C, the same as that of the MB oscillator.

To see the difference in the bifurcation behaviour between the two oscillators more clearly we allow k_0 and y^* to vary and examine the bifurcation diagram in Fig. 6. The parameter y^* controls the inflow of the exit species O_2 and plays a similar role as y_0 in the MB system even though the analogy is not perfect since the inflow of oxygen does not vanish with vanishing k_0 because of the interface transport. We immediately see that the 'cross shaped phase diagram' is no longer present. Instead, the Hopf curve forms a closed loop without self-intersections. Consequently, a subcritical Hopf bifurcation and excitability is not likely (even if possible) to occur. Indeed, a search for subcritical appearance of oscillations accompanied by excitability failed. However, we found complex dynamics coexisting with the small amplitude oscillations. When different initial concentrations are chosen, large amplitude relaxational oscillations occur for the same parameter values as

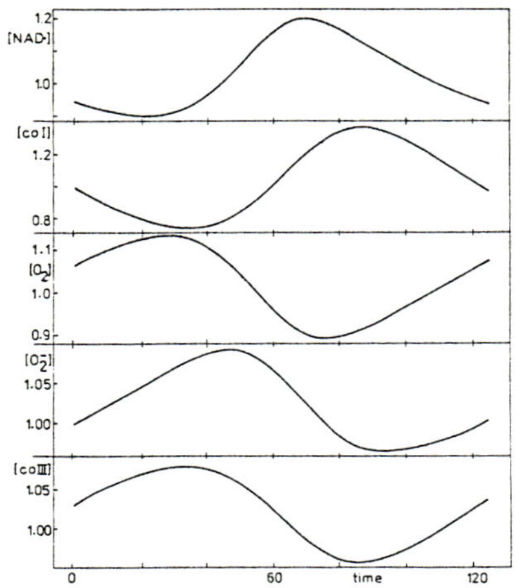

Fig. 5. Periodic oscillations of the dynamical variables x = [NAD·], u = [coI], y = $[O_2]$, w = $[O_2^-]$ and z = [coIII] relative to their average values for the PO system; $k_t = 1.5 \times 10^{-2}\ s^{-1}$, see text for other parameter values.

in Fig. 5, see Fig. 7. The new dynamical regime corresponds to mixed-mode oscillations with one large peak followed by a number of small amplitude oscillations. Such oscillations were found experimentally [74, 79] and in an abstract model [79]. Only recently, however, have they been observed in more realistic models of the PO system [80].

We conclude that even if the MB and PO systems are of the same category, they differ in their bifurcation and dynamical behaviour; the likely reason is a more complex topology of the PO reaction network, which has several irreversible cycles in addition to the basic autocatalytic one, particularly the one between NAD· and O_2^-, which compete and cause the mixed-mode dynamics.

3 Effects of Periodic Forcing and Coupling

In biological systems, oscillatory or excitatory cells are frequently coupled in one-, two- or three-dimensional arrays constituting a tissue that can mediate information in the form of wave trains of (electro)biochemical activity. Excitable or oscillatory chemical systems, if coupled by diffusion and periodically perturbed [59, 75, 81–83] at certain locations, display analogous behaviour, the chemical waves. Here we examine the basic properties of the simplest linear arrays of reaction cells (treated as CSTR's) coupled linearly to their neighbours.

A linear N-array of identical oscillatory or excitable units coupled linearly (i.e. diffusion-like) may be described by

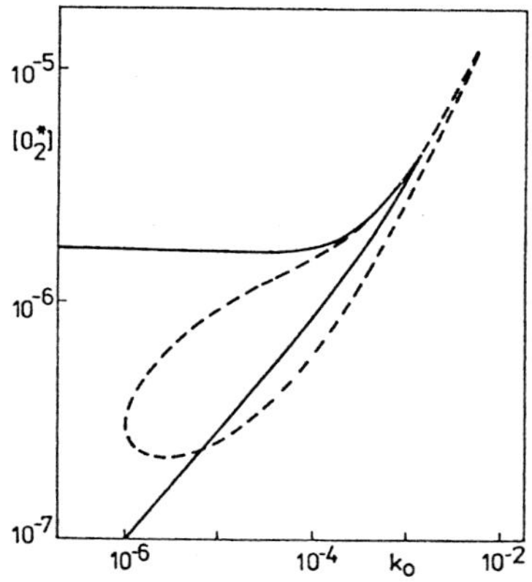

Fig. 6. Bifurcation diagram for the PO system showing the curves of Hopf bifurcation (dashed lines) and the curves of saddle-node bifurcation (full lines) in the k_0-y_0 parameter plane.

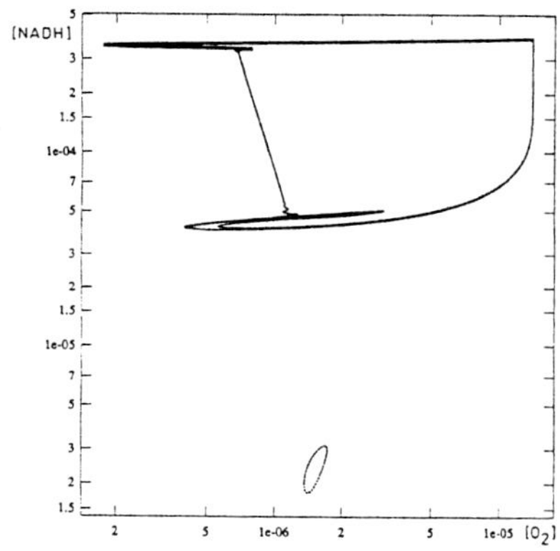

Fig. 7. Projection of two coexisting oscillatory regimes for the PO system to the $[O_2]$-[NADH] plane.

$$\frac{d\mathbf{x}_i}{dt} = \mathbf{f}(\mathbf{x}_i) + d\,(\mathbf{x}_{i+1} - 2\mathbf{x}_i + \mathbf{x}_{i-1}),\ i = 1, \cdots, N,\ \mathbf{x}_i \in R^n, \tag{12}$$

supplemented by a set of boundary conditions, for example no flux at the ends, $\mathbf{x}_0 = \mathbf{x}_1$, $\mathbf{x}_{N+1} = \mathbf{x}_N$. The local dynamics of each unit is given by the function $\mathbf{f}(\mathbf{x})$, d is the transport coefficient. A pulse perturbation of the first unit may cause its firing and the excitation can be propagated through the array. If the pulse is applied periodically, the conditions for propagation will depend on the input frequency, the amplitude of the pulse and the transport coefficient.

Our goal is to reduce the description of the dynamics in each cell to a single variable (which may be called the phase) and to find a way of coupling the units, that would correspond to the original system (12). The reduction of (12) to phase equations assumes that there is a one-dimensional set Λ in R^n such that, after a rapid transient, the local dynamics in each unit is confined to this set. Λ is a *limit cycle* for an oscillator and an *excitable cycle* for an excitator. Whereas the limit cycle is a single closed trajectory γ , the excitable cycle is formed by two trajectories γ_1, γ_2, both approaching a stable stationary state $\mathbf{x}_s$ from opposite directions as $t \to \infty$. It should be mentioned here that there is an additional characteristic mode of dynamics frequently occurring in biological as well as chemical systems. This takes the form of bursters, units responding to a superthreshold perturbation by repeated large amplitude oscillations during the transition back to the state of quiescence. Such dynamics require a non-planar topology in a state space of dimension ≥ 3. Bursters wil not be treated here and therefore let us assume that Λ can be embedded in a two-dimensional manifold.

There are two basic situations in state space corresponding to two basic modes of excitability: a) there is an additional (saddle) stationary point that γ_1 and γ_2 approach as $t \to -\infty$, and b) there is no such point; γ_1 and γ_2 come close to each other in reverse time, eventually creating a narrow gap, but never join up. Case (a) uniquely defines the excitable cycle Λ, whereas case (b) does not; an additional assumption of strong stability of Λ has to be made. In both cases, there exists a slow manifold, the *threshold set* [29], separating two kinds of trajectories that correspond to excitation events and no excitations. When perturbed at or near to $\mathbf{x}_s$ so that the state of the system gets beyond the threshold set, the system undergoes an excitation event; otherwise no excitation occurs. The threshold set may be a surface or a layer of finite thickness, locally separating the state space.

Once identified, Λ can be parametrized by a coordinate ϕ called the phase. A scaled time may be a good choice for an oscillator but a different parametrization, such as the arclength along Λ (positive time orientation is assumed), is more convenient for an excitator due to the presence of stationary state(s) in Λ. While the phase in the case of periodic oscillators and type (a) excitators is a cyclic variable, this is not the case with the type (b) excitators.

When T-periodically perturbed by pulse-like perturbations, the dynamics on a limit (or excitable) cycle defines a Poincaré mapping P carrying a phase ϕ_k at time kT to a phase ϕ_{k+1} at time $(k+1)T$ (cf. PRC and PEC maps discussed in the Introduction). The map P is a circle map for the oscillator or type (a) excitator, or a map of an interval for type (b) excitator. Two basic parameters of a periodic sequence of pulses are the amplitude A of a pulse and the period T with which the perturbations are repeated.

In case the perturbation initiates an autocatalytic process (the easiest way of achieving this is to decrease the concentration of an exit species) the response of the system is a rapid growth of the concentration of all autocatalytic species until the feedback species is exhausted. This stops the autocatalytic growth and makes it possible for the exit species to regain the control of the autocatalysis and remove the accumulated autocatalytic species. Therefore a pulse applied at phases before the autocatalytic growth shifts the phase rapidly whereas a pulse applied during the phase of deceleration of the autocatalytic process changes the phase slowly. As a result, the Poincaré mapping $P(\phi; A, T)$ has a steep portion (decreasing or increasing according to the size of the amplitude A) and a portion with a moderate slope. Due to the autonomous periodicity of an oscillator, P is periodic in T with the autonomous period T_a. In the case of an excitator, P becomes flat with increasing T, since all perturbations are eventually damped to the steady state. However, for moderate values of the forcing period, P has a similar shape for both cases and consequently the dynamics is very similar. The steep portion of P indicates a threshold phenomenon; the system undergoes a rapid response to a perturbation (a firing) if P maps a phase above the threshold to a subthreshold one. Unlike with the excitators, the threshold is crossed spontaneously in oscillators in the absence of perturbations.

The dynamics of P is well characterized by a firing number ξ which is an asymptotic average number of initiated autocatalytic processes per one forcing period. If $\xi = p/q$ (p,q integer) then there is a phase-locked periodic regime with p firing events per q perturbations. The firing number is typically a nondecreasing stepwise function of T which may be continuous (a devil's staircase) or may have discontinuous jumps (a broken devil's staircase); ξ vanishes at $T = 0$ and grows indefinitely with increasing T for oscillators, because the threshold is crossed spontaneously, but reaches a maximum value of 1 for excitators.

Examples of the Poincaré mappings for an oscillator and an excitator are shown in Fig. 8 and Fig. 9, respectively. P was extracted from the MB system at two values of k_0 near the Hopf bifurcation at $k_0^* = 2.76 \times 10^{-3}$ s^{-1} discussed in Sect. 2. The perturbation is mediated by silver ions added to the system. Ag^+ acts by removing Br^- via a fast equilibrium reaction Ag^+ + $Br^- \rightleftharpoons AgBr$; considering the solubility product S of this reaction, the amount δy of Br^- removed within the pulse is [25]

$$\delta y = (S/y + A + y - [(S/y + A - y)^2 + 4S]^{\frac{1}{2}})/2.$$

The map P is constructed numerically by taking many points on Λ (which is either a limit cycle or an excitable cycle), subtracting δy from the current value of $y \equiv [Br^-]$ and integrating Eqs. (7). The perturbation relaxes back to Λ within a short time (< 1 s) and after choosing a parametrization by ϕ on Λ, P is obtained, see Figs. 8, 9. The location of the sudden drop that corresponds to the threshold of autocatalysis is independent of T and is shifted to the left with increasing A (not shown in Figures). In both cases, curve a) corresponds to a small firing number since the iterations of the map must pass through a narrow channel before they pass the threshold value; the firing number is smaller for the oscillator since the channel is narrower. Case b) still reflects the similarity of a perturbed oscillator and excitator but the dynamics for case c) is different. For this value of T, the excitator comes close to the steady state (the curve is flat) whereas the oscillator begins a new cycle.

Now what happens when two (or more) units are coupled in a linear array and exposed to a periodic forcing at one of the ends. For simplicity, let us consider two units [81]. The Poincaré mapping is two dimensional, with

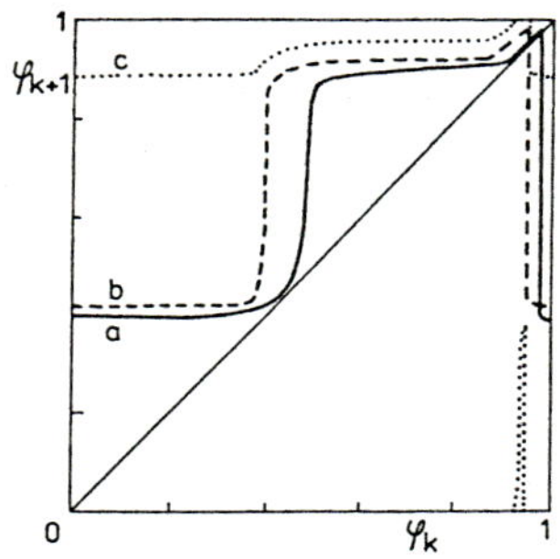

Fig. 8. The phase mapping P extracted from the MB system in an oscillatory regime; $k_0 = 2.7 \times 10^{-3}$ s^{-1}; the forcing period T is a) 100 s, b) 200 s, c) 600 s.

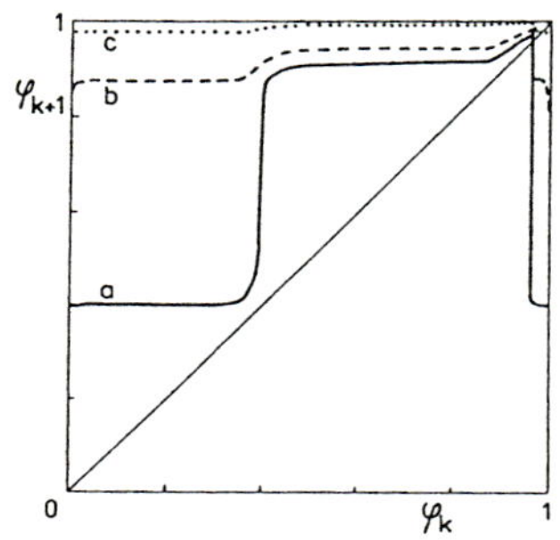

Fig. 9. The phase mapping P extracted from the MB system in an excitable regime; $k_0 = 2.8 \times 10^{-3}$ s^{-1}; the forcing period T is a) 100 s, b) 200 s, c) 600 s.

two phases ϕ_1, ϕ_2. A superthreshold perturbation of the first unit does not necessarily imply an initiation of the autocatalytic process because of a possible suppression by diffusion. Nonetheless, the firing numbers ξ_1, ξ_2 defined in the same fashion as for one cell accommodate the diffusion suppressed (or induced) firings. The dynamical regime is called a *complete propagation* when $\xi_1 = \xi_2 > 0$, a *partial propagation failure* when $\xi_1 > \xi_2 > 0$ and a *complete propagation failure* when $\xi_1 > \xi_2 = 0$.

The occurrence of the propagation failure phenomenon depends heavily on the coupling strength. If the cells are coupled strongly, all excitations propagate to the second cell; a weak coupling implies a complete propagation failure whereas at intermediate coupling strengths a partial propagation occurs. A dependence of ξ_1 and ξ_2 on T, experimentally determined for an excitable BZ reaction [59] at an intermediate coupling strength, is shown in Fig. 10. The complete propagation occurs in the right portion of each of the $1/q$-steps, q=1, 2, 3 in the first cell and breaks down due to the propagation failure. The complete propagation is repeatedly restored as T is decreased. The reason for the propagation failure behaviour is that the threshold value of ϕ_2 is generally larger than that of ϕ_1 due to a delay in transferring the excitation to the second cell. Extending this argument to N-arrays, wave trains are obtained, which have either the same or lower output frequency than the waves generated at the beginning. The frequency of these again is the same or lower than the forcing frequency.

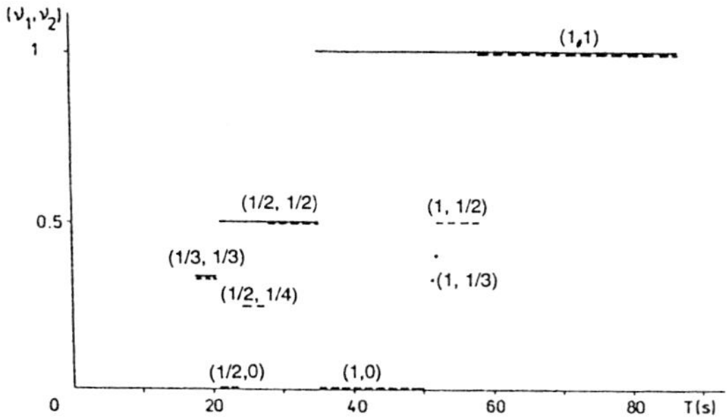

Fig. 10. The plot of (ξ_1, ξ_2) versus T constructed from experiments [59] ; $A \equiv c(Ag^+) = 7.5 \times 10^{-5}$ mol/l, the transport coefficient $z = 10$ mm; full line – first (externally driven) cell, dashed line – second cell.

Conclusions

Using three examples from chemical kinetics we have shown how stoichiometric network theory combined with bifurcation theory can be applied to reduction of models and estimation of parameter regions where oscillations, excitability and multiple steady states exist. We have used numerical continuation techniques to construct global bifurcation diagrams and to show how phase mappings describing the response to stimulations can be obtained. The stoichiometric network approach may be conveniently used in the description of biological systems provided that the power law formalism applies [7, 8, 13, 14].

Construction of models for networks of excitable units with a more complex topology (linear and cyclic arrays, various planar structures) based on coupled phase maps can proceed as a straightforward extension of the approach outlined here. Agreement between the excitation diagrams constructed from experiments on one hand [26, 27, 59] and calculated from simple phase models [81] on the other gives support to further development of the method. At the same time these results open wide possibilities of applications to biochemical and biological systems; for example those directly involved in signal transmission through synapses, such as the acetylcholine system. Once the phase map for a particular dynamics of an excitable unit has been constructed (either directly from experiments or from a detailed physico-chemical model), a model of a network is directly available. A detailed knowledge of the dynamics of such a coupled array would help to bridge the gap between the rather formal neural networks approach and the accumulating experimental knowledge about functioning of particular neurophysiological systems [84].

Even if the above described procedures have led to efficient ways of locating ranges of parameter values for oscillations, multiple steady states and excitability, the actual dynamic behaviour will most often depend on the detailed form of model equations. The ways of an improvement of predictions rest particularly in the numerical parts of the procedure. For example, the search for a subcritical Hopf bifurcation connected with excitability can be based on local small amplitude expansions in the neighbourhood of Hopf bifurcation points, fixing the relations among free parameters. Likewise, conditions for Takens-Bogdanov bifurcations points where a Hopf curve merges with a saddle-node curve can be used for location of parameter values near which complex oscillations may be expected [24, 51]. Also, when the unstable pathway leads to a small number of equations, it should be possible to relate the bifurcation structure to normal form equations. These approaches can be used not only in the case of mechanistic models with stoichiometric and charge balance constraints, but also in treating more formal models without such explicit constraints, that are mostly used for description of biological oscillatory or excitable media.

Recently, attempts to develop more systematic ways of model reduction for oscillatory and excitatory dynamics of neuron models based on Hodgkin-Huxley equations were described [85]. Modelling of driven oscillators by maps is the subject of a broader interest [86], as well as the phenomenon of propagation failure [87]. On one hand, the model chemical systems are being increasingly used to construct analogues of signal transmissions in neural systems [88], on the other hand models of chemical synaptic transmission and effects of neurotransmitters on function of neurons are being developed [89, 90]. We believe that the outlined systematic way of studying chemically sound models will find its applications in models of concrete biological oscillatory and excitatory systems in a near future.

References

1 I. Atwater, M. Dawson, A. Scott, G. Eddlestone, E. Rojas: In *Biochemistry and Biophysics of the Pancreatic Beta Cell*, G. Thieme, Ed., Springer-Verlag, New York 1980, pp. 100-107.
2 P. Rorsman, G. Trube: *J. Physiol.* **375**, 531-550 (1986).
3 S. D. Hughes, J. H. Johnson, C. Quaade, C. B. Newgard: *Proc. Natl. Acad. Sci.* **89**, 692 (1992).
4 J. E. Bailey: *Science* **252**, 1668-1675 (1991).
5 D. C. Cameron, I. T. Tong : *Applied Biochem. & Biotech.* **38**, 105-140 (1993).
6 R. M. Nerem *Ann. Biomed. Eng.* **19**, 529-545 (1991).
7 M. A. Savageau: *J. Theor. Biol.* **25**, 365-369 (1969); **25**, 370-379 (1969); **26**, 215-226,(1970); **154**, 131-136 (1992).
8 K. Hayashi, N. Sakamoto *Dynamic Analysis of Enzyme Systems*, Japan Scientific Societies Press and Springer-Verlag, Tokyo 1986.
9 H. Kacser, J. A. Burns: *Symp. Soc. Exp. Biol.* **27**, 65-104 (1973).
10 R. Heinrich, S. M. Rapoport, T. A. Rapoport: *Prog. Biophys. Molec. Biol.* **32**, 1-82 (1977).
11 W. J. Freeman: *Int. J. Bif. Chaos* **2** 451-483 (1992).
12 F. Crick: *Nature* **337** 129-132 (1989).
13 L. F. Abbot, G. LeMasson: *Neural Computation* **5**, 823-842 (1993).

14 A. Destexhe, Z. F. Mainen, T. J. Sejnowski: *Neural Computation* **6**, 14-18 (1994).
15 W. Gerstner, R. Ritz, J. L. van Hemmen: *Biol. Cybern.* **68**, 363-374 (1993).
16 J. J. Collins, I. Stewart: *Biol. Cybern.* **68**, 287-298 (1993).
17 A. J. Mandell, K. A. Selz: *SPIE* **2036**, Chaos in Biology and Medicine 86-99 (1993).
18 A. R. Brailove: *Int. J. of Bifurcations and Chaos*, **2**, 341-352 (1992).
19 T. Kondo, C. A. Strayer, R. D. Kulkarni, W. Taylor, M. Ishiura, S. S. Golden, C. A. Johnson: *Proc. Nat. Acad. Sci. USA* **90**, 5672-5676 (1993).
20 A. T. Winfree: *When time breaks down*, Princeton University Press, Princeton 1977.
21 L. Glass, M. C. Mackey, *From clocks to chaos. The rhythms of life.*, Princeton University Press, Princeton 1988.
22 L. Glass, P. Hunter, A. McCulloch: *Theory of heart*, Springer-Verlag, New York 1991.
23 M. Dolník, I. Schreiber, M. Marek: *Physica D* **21**, 78 (1986).
24 M. Marek, I. Schreiber: *Chaotic Behaviour of Deterministic Dissipative Systems*, Cambridge University Press, Cambridge 1991.
25 M. Dolník, J. Finkeová, I. Schreiber, M. Marek: *J. Phys. Chem.* **93**, 2764 (1989).
26 J. Finkeová, M. Dolník, B. Hrudka, M. Marek: *J. Phys. Chem.* **94**, 4100 (1990).
27 M. Dolník, M. Marek: *J. Phys. Chem.* **95**, 7267 (1991).
28 M. Dolník, M. Marek, I. Epstein: *J. Phys. Chem.* **99**, 3218 (1992).
29 J. C. Alexander, E. J. Doedel, H. G. Othmer: *Siam. J. Appl. Math* **50**, 1373-1418 (1990).
30 N. Takahashi, Y. Hanyu, T. Musha, R. Kubo, G. Matsumoto: *Physica D* **43**, 318 (1990).
31 K. Kopecz, G. Schöner, F. Spengler, H. R. Dinse: *Biol. Cybern.* **69**, 463-473 (1993).
32 R. F. Gilmour, M. Watanabe, D. R. Chialvo: *SPIE* **2036**, Chaos in Biology and Medicine, 2-9 (1993).
33 D. Cai, Y. Lai, R. L. Winslow: *SPIE* **2036**, Chaos in Biology and Medicine, 10-21 (1993).
34 K. Tanaka: in *Current Opinion in Neurobiology* **2**, 502-505 (1992).
35 I. Fujita, K. Tanaka, M. Ito, K. Cheng: *Nature* **360**, 343-346 (1992).
36 M. Hammer: *Nature* **366**, 59-63 (1993).
37 I. Schreiber, M. Marek: in *Chaos in Chemical and Biological Systems*, R. J. Field, Z. Györgyi, Eds., World Scientific, Singapore 1993.
38 M. Yoshimoto, K. Yoshikawa, Y. Mori: *Phys. Rew.* **47**, 864-874 (1993).
39 K. P. Zeyer, R. Holz, F. W. Schneider: *Ber. Bunsenges. Phys. Chem.* **97**, 1112-1119 (1993).
40 M. Dolník, J. Kosek, V. Votrubová, M. Marek: *J. Phys. Chem.* **98**, 3707-3715 (1994).
41 K. Kaneko: in *Formation, Dynamics and Statistics of Patterns*, Vol. 1, ed. by K. Kawasaki, M. Suzuki, A. Onuki, World Scientific, Singapore 1990.
42 T. Chawanya, T. Aoyagi, I. Nishikawa, K. Okuda, Y. Kuramoto: *Biol. Cybern.* **68**, 483-490 (1993).
43 A. V. Holden, M. Markus, H. G. Othmer, Eds.: *Nonlinear Wave Processes in Excitable Media*, Plenum Press, New York 1991.
44 D. Barkley: *Physica D* **49**, 61-70 (1991).
45 B. L. Clarke: *Adv. Chem. Phys.* **43**, 1 (1980).
46 M. Feinberg: *Arch. Ration. Mech. Anal.* **49**, 187 (1973).
47 M. Feinberg, F. Horn: *Chem. Eng. Sci.* **29**, 775 (1974).
48 B. G. Johnson, P. L. Corio: *J. Phys. Chem.* **97**, 12100-12105 (1993).

49 M. Eiswirth, A. Freund, J. Ross: *Adv. Chem. Phys.* **90**, 127 (1991); *J. Phys. Chem.* **95**, 1294 (1991).
50 T. Chevalier, I. Schreiber, J. Ross: *J. Phys. Chem.* **97**, 6776 (1993).
51 M. Kubíček, M. Marek: *Computational Methods in Bifurcation Theory and Dissipative Structures*, Springer-Verlag, New York 1983.
52 R. J. Field, E. Körös, R. M. Noyes: *J. Am. Chem. Soc.* **94**, 8649 (1972).
53 R. J. Field and M. Burger, Eds.: *Oscillations and Traveling Waves in Chemical Systems*, Wiley, New York 1985.
54 R. J. Field, R. M. Noyes: *J. Chem. Phys.* **60**, 1877 (1974).
55 A. M. Zhabotinsky, F. Buchholz, A. B. Kiyatkin, Epstein, I. *J. Phys. Chem.* **97**, 7578-7584 (1993).
56 H.-D. Fösterling, L. Stuk, A. Barr, W. D. McCormick: *J. Phys. Chem.* **97**, 2623 (1993).
57 R. J. Field, H.-D. Fösterling: *J. Phys. Chem.* **90**, 5400 (1986).
58 Z. Nagy-Ungvarai, J. J. Tyson, B. Hess: *J. Phys. Chem.* **93**, 2760 (1989).
59 J. Kosek, M. Marek: *J. Phys. Chem.* **97**, 120-127 (1993).
60 P. De Kepper, A. Pacault: *Compt. Rend. Sc. Acad. Sci., Ser. C* **286**, 437 (1978); P. De Kepper, J. Boissonade: In *Oscillations and Traveling Waves in Chemical Systems*, R. J. Field and M. Burger, Eds., Wiley, New York, 1985, p.223.
61 J. Guckenheimer: *Physica D* **20**, 1 (1986).
62 M.-L. Smoes: In *Dynamics of Synergetic Systems*, Haken, H., Ed., Springer-Verlag, Berlin, 1980, p. 80.
63 M. Marek, P. Kaštánek, S. C. Müller: *J. Phys. Chem.* (1994), submitted.
64 P. Kaštánek, D. Šnita, J. Kosek, I. Schreiber, I., M. Marek: *Physica D* (1994), submitted.
65 R. M. Noyes, R. J. Field, R. C. Thompson: *J. Am. Chem. Soc.* **93**, 7315 (1971).
66 K. Bar-Eli, R. J. Field: *J. Phys. Chem* **94**, 3660 (1990).
67 I. Yamazaki, K. Yokota: *Biochim. Biophys. Acta* **132**, 310 (1967).
68 K. Yokota, I. Yamazaki: *Biochemistry* **16**, 1913 (1977).
69 L. F. Olsen, H. Degn: *Nature* **267**, 177 (1977).
70 B. D. Aguda, L.-L. Hofmann Frisch, L. F. Olsen: *J. Am. Chem. Soc.* **112**, 6652 (1990).
71 J. Lazar, J. Ross: *Science* **247**, 189 (1990).
72 M. Samples, Y.-F. Hung, J. Ross: *J. of Phys. Chem.* **96**, 7338-7342 (1992).
73 T. Geest, C. G. Steinmetz, R. Larter, L. F. Olsen: *J. Phys. Chem.* **96**, 5678-5680 (1992).
74 T. Hauck, F. W. Schneider: *J. Phys. Chem.* **97**, 391 (1993).
75 A. Förster, T. Hauck, F. W. Schneider: *J. Phys. Chem.* **98**, 184-189 (1994).
76 A. Scheeline, D. L. Olson: *SPIE* **2036**, Chaos in Biology and Medicine, 253-255 (1993).
77 B. D. Aguda, B. L. Clarke: *J. of Chem. Phys.* **87** (6), 3461-3470 (1987).
78 B. Aguda, R. Larter: *J. Am. Chem. Soc.* **112**, 2167 (1990).
79 H. Degn, L. F. Olsen, J. Perram: *Ann. N. Y. Acad. Sci.* **316**, 623 (1979).
80 Y.-F. Hung, I. Schreiber, J. Ross, J., in preparation.
81 J. Kosek, I. Schreiber, M. Marek: *Trans. Roy. Soc. A*, in press.
82 M. Marek, I. Schreiber: Chapter 4 in *Chaos in Chemical and Biological Systems*, R. Field and Z. Györgi, Eds., World Scientific, Singapore (1993).
83 I. Kuramoto: *Physica D* **50**, 15-30 (1991).
84 M. Steriade, D. A. McCormick, T. J. Sejnowski: *Science* **262**, 679-685 (1993).
85 D. Golomb, J. Guckenheimer, S. Gueron: *Biol. Cybern.* **69**, 129-137 (1993).
86 U. Parlitz, C. Scheffczyk, T. Kurz, T., W. Lauterborn: *Int. J. of Bifurcations and Chaos* **1**, 261-264 (1993).

87 V. Perez-Munuzuri, V. Perez-Villar, L. O. Chua: *Int. J. of Bifurcations and Chaos* **2**, 403-406 (1992).
88 A. Toth, V. Gaspar, K. Showalter: *J. Phys. Chem.* **98**, 522-531 (1994).
89 D. S. Melkonian: *Biol. Cybern.* **68**, 341-350 (1993).
90 R. Bertram: *Biol. Cybern.* **69**, 257-267 (1993); ibid **70**, 359-368 (1994).

Localized Turing and Turing-Hopf Patterns

P. Borckmans, O. Jensen, V.O. Pannbacker, E. Mosekilde, G. Dewel, A. De Wit

Abstract

In systems driven away from thermodynamic equilibrium, patchiness often arises through the occurrence of symmetry breaking bifurcations. Diffusive instabilities resulting from differential diffusion processes acting in the presence of some autocatalytic kinetic scheme enter that class of phenomena to produce stationary space periodic (Turing) or spatiotemporal (Hopf) patterns. Turing patterns have at last recently been obtained when the isothermal Chlorite-Iodide-Malonic Acid (CIMA) reaction takes place in a continuously fed gel reactor in the presence of starch. On varying the malonic acid or starch concentrations a transition from stationary Turing structures to Hopf wavy patterns occurs. In the transition region, where both instabilities interact, a host of interesting behaviours may occur. Among these, new intrinsically localized patterns may form that give rise to new types of patchinesses. All of these structures may be accounted for through the study of the bifurcation behaviour of simple theoretical reaction-diffusion models.

1 Introduction

The propensity of systems driven away from thermodynamic equilibrium, by fluxes of matter and/or energy, to spontaneously exhibit organized spatiotemporal behaviour on a macroscopic level can be observed in an ever increasing variety of fields. Classical examples arise in convective and centrifugal flows [1], liquid crystals (electro-) hydrodynamics [2], non linear optics [3], solidification processes [4], chemistry [5], ... Although this phenomenon has generated scores of studies, especially in the last three decades, its analysis remains as fascinating as ever. It is now clear that many such organisations arise through the occurrence of *symmetry breaking instabilities*.

In the realm of chemistry, in which we are interested here, the breaking of time translation symmetry, leading to regular oscillations in time of the concentrations of the reacting species, has been known since the discovery of the celebrated Belousov-Zhabotinsky reaction [6]. Numerous other autocatalytic chemical reactions, i.e. with antagonistic feedbacks, have since been shown to exhibit that same characteristic behaviour. They have been studied quantitatively in continuous stirred tank reactors [7] with such reproducibility and precision that these experiments allowed for the discovery of genuine deterministic chaotic chemical dynamics [8]. When carried out in unstirred reactors, these oscillating reactions may then give rise to a variety of waves, among which the much studied spirals, that have been shown to result solely from the interaction of nonlinear chemistry and diffusion processes [9, 10]. Spiral waves may also occur when these reactions are operating under excitable, rather than self-oscillating, conditions [10]. Such waves seem to play an important role in the transmission of information in certain biological systems (CAMP waves, waves of activity on the myocardium, slime molds aggregation, ...). Let us furthermore notice that numerous wave phenomena have also been observed for heterogenous reactions taking place on well defined surfaces under ultra high vacuum conditions [11].

On the other hand, already in 1952, A.Turing proposed the complementary, space translation symmetry breaking, as a "Chemical Basis for Morphogenesis" [12]. This instability leads to stationary, space-periodic patterns of concentration, dissipative structures [13], that arise because the species involved in stable non-linear chemical interactions may organize spontaneously in space as a result of the presence of differential diffusive processes.

To sustain symmetry breaking instabilities, the system has to present some type of positive feedback loop initiated by an activator species that reinforces its own changes and that are controlled by an inhibitor. Structures can form when the inhibition is of longer range than the activation process. This allows the local growth of the activator while the lateral inhibition prevents the spreading of the activated centre. Such a process will repeat itself, finally leading to a stationary periodic pattern. Its wavelength is a function solely of the values of the diffusion coefficients and the kinetic parameters and not of some geometrical length characteristic of the reactor. The beauty of Turing's idea lies in this counterintuitive organizing role of diffusion that usually tends to smear out any concentration inhomogeneity.

Numerous applications of reaction-diffusion instabilities to biological patterns formation have been developed [14, 15]. However the idea remains to be fully vindicated because there is as yet no confirmed example of a Turing instability playing a role in any biological pattern forming process. In particular in the problem of differentiation, that was directly addressed by Turing, only little success has been achieved in finding and identifying to so-called morphogenic substances which are supposed to be involved [16, 17].

As such structures for many years also remained elusive in the field of nonlinear chemistry that, on the other hand, produced multistability, oscillations, waves (and even chaos) for rather general conditions, one was led to believe that Turing structures were an appealing theoretical idea with no utility in practice.

The difficulties mainly arose from two sources. First, to maintain the system at a controlled distance from equilibrium it is necessary to supply fresh reactants to – and extract the products from – the system continuously. This is to be carried out without generating hydrodynamic currents that would disrupt the all important diffusion processes. Furthermore, convection currents, that may be generated by the reactions because of non-isothermicity or local density variations, must also be avoided. Secondly, the possible occurrence of Turing instabilities relies on the need for the activator, i.e. the species responsible for the positive feedback, to diffuse much slower than the inhibitor. This has presented a major obstacle since, in the aqueous media used to study non-linear chemical phenomena, all the simple molecules and ions diffuse more or less at the same rate $D \approx 10^{-5}\mathrm{cm}^2/\mathrm{s}$.

2 Turing Structures in the CIMA Reaction

Finally, using the chlorite-iodide-malonic acid (CIMA) reaction, which for other conditions is known to produce multistability, oscillations and travelling waves [18], the first laboratory observation of a genuine Turing pattern was made in December 1989 by P. De Kepper and his coworkers in Bordeaux [19], and then later on in Austin (TX) [20].

The first of the above problems was solved by using new, *unstirred continuously fed reactors* [21] (Fig. 1). The core of these reactors consists of a block of transparent, chemically inert, *hydrogel* (polyacrylamide, polyvinylalcohol or agarose) that quenches convective currents while allowing the small soluble species of the CIMA reaction to diffuse. The temperature is held constant by a waterjacket. Two opposite sides of this block are in contact with open tanks containing non reactive subsets of the reagents that are continuously stirred and refreshed by pumps. Malonic acid, in acetic acid solution, is introduced only in tank I and chlorite, in basic solution, only in tank II. Sodium hydroxyde stabilizes chlorite while acetic acid enhances the iodine solubility and acts as a pH buffer. Iodide is introduced symmetrically in both tanks. The other sides are in contact with impermeable barriers. Triiodide, I_3^-, that is produced during the reaction, is used to visualize the structures as it may form an optically detectable complex with macromolecular substances such as starch (Thiodene) or polyvinylalcohol of dark blue or reddish-brown color, respectively. A video CCD camera is used to monitor the experiments.

The concentrations of the reagents are thus maintained and controlled at the two feeding planes. The species can then permeate through the gel by diffusion and react when they meet. If no spatial symmetry breaking instability occurs, concentration profiles establish naturally into isoconcentration planes parallel to the feed surfaces. High iodide concentrations are typically found along tank I. This region is dark due to the formation of the color indicator - I_3^- complex. On the opposite side, iodide and iodine are rapidly oxidized to iodate and the gel remains clear. A monotonous profile of I_3^- is then established under such conditions. Beyond critical conditions, Turing structures appear that lead to more complex profiles or eventually the breaking of the symmetry due to the feeding.

Unexpectedly, the experimental set up also solved the second difficulty, that of the necessary difference of diffusion coefficients. Indeed, because of their mere size, the color indicator molecules are unable to diffuse in the gel. (They must be preloaded into it when the gel is prepared). As a result their reversible complex with I_3^-, is immobile in the gel matrix. This phenomenon then leads to a separation of the time scales of the dynamics of the activator (the iodine species) and the inhibitor (chlorite) proportional to the complexant concentration [22, 23]. This mechanism modifies the stability properties of the uniform steady reference state. The (Hopf) instability that may give rise to oscillations in the absence of complexant is shifted proportionally to the concentration of this complexing agent. Thereby the Turing diffusion induced instability (that is not affected by the process) may now occur before the oscillations at parameters values that were inside the oscillatory region of the complexant-free system [24].

In order to compare with the theoretical analysis, it is of some interest to try to characterize the extension in space (= dimensionality) of the structures obtained experimentally. To fix the ideas let us consider (Fig. 1) a block of gel of lengths L_x (in the direction of the feeding gradients or ramps), L_y and L_z. The lower limits of these lengths are primarily set by the difficulties of manipulation of the gels due to their brittleness. Let us recall the important property that the Turing structures have an intrinsic wavelength λ. In the experiments with the CIMA reaction, it is of the order of 0.20 mm.

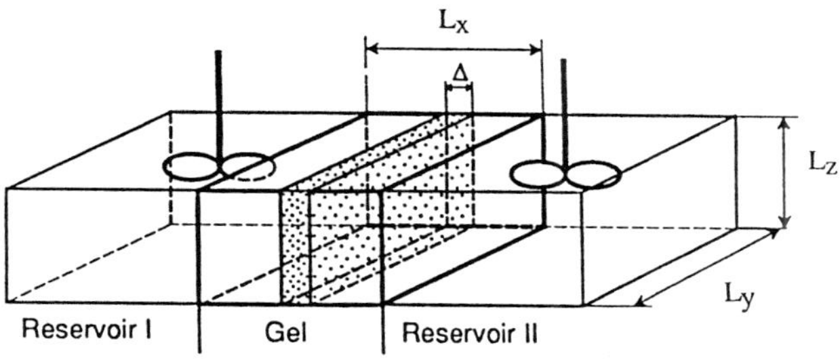

Fig. 1. Sketch of the open spatial reactors. See description in the text.

Now in the x direction, the pattern will form in the region, of length Λ, where the concentration profiles of the chemical species allow the system to be above the instability threshold. The size of Λ ($\lambda \leq \Lambda \leq L_x$) is determined by the values of the concentrations maintained at the boundaries, the diffusion constants, the kinetic constants (hence the temperature) and the width L_x of the reactor itself as it also enters in the control of the steepness of the profiles. Parameter Λ is thus not easy to control, and it is important to state that in full generality the Turing patterns obtained in such gel reactors are three dimensional (3D) objects [25] localized in space along the x-direction by the feeding ramps. The characteristics of the Turing patterns therefore vary along this x-direction (ramps localized structures). Patterns, free from the influence of these feeding gradients may be obtained, but then only transitorily, in batch reactors [26]. However, by careful conception and manipulation, patterns of lower dimensionality, some of which under quasi-uniform conditions, may be obtained.

For instance if the parameters are such that $\Lambda \leq \lambda$ and if the gel block has L_y (or L_z) $\leq \lambda$ while the third length is arbitrary (Type a), a uniform quasi-1D pattern results. It can be analyzed along the x or y (z) directions. Also if $\Lambda \leq \lambda$, but both other lengths are arbitrarily large (Type b) then the result is a quasi-2D pattern in a "plane" orthogonal to the direction of the feedings. It is then studied looking along the x-direction. This has probably been the most used experimental configuration with a circular geometry of the gel block in the yz plane (the so-called "disc reactor").

Ramps localized patterns of lower dimensions can also be produced. If $\Lambda > \lambda$, but $L_y \approx L_z \leq \lambda$, we have a quasi-1D ramps localized structure. This is similar to the Turing structures that have been obtained in a capillary [27] to show that the gel is not a prerequisite for their obtention. The situation where $\Lambda > \lambda$ whereas L_y (or L_z) $\leq \lambda$ while the third length is arbitrary, that would lead to quasi-2D ramps localized structures have not been used so far.

The first observations of Turing patterns [18, 19] sparked off subsequent experiments to determine the bifurcation diagrams, the role of the gel matrix and the starch [25, 28, 29]. New theoretical studies aimed at the uncovering of the different pattern modes and understanding of the role of the feeding concentration ramps [25, 30].

3 Theoretical Analysis of the Global Turing Patterns

The basic equations for the study of Turing structures are the *reaction-diffusion equations*

$$\frac{\partial \mathbf{C}}{\partial \mathrm{t}} = \mathbf{F}(\mathbf{C}; \mathrm{B}) + \mathbf{D}\nabla^2\mathbf{C} \tag{1}$$

where $\mathbf{C}$ is the concentration vector, the components of which are the concentrations of all chemical species participating in the reaction scheme. Then $\mathbf{F}(\mathbf{C};\, \mathrm{B})$ represents the reaction kinetics itself, and B is a control parameter. The Laplacian term represents the diffusion processes characterized by a usually diagonal *diffusion* matrix $\mathbf{D}$.

Let us remark that theoretical work relies heavily on the use of nonlinear kinetic models with a limited number of chemical species, typically two or three. These models stand as a compromise between a minimum of chemical realism and mathematical tractability. Indeed the kinetic schemes of the reactions put to work experimentally are very complex and bring a large amount (usually more than 10) of species into play, and their schemes are for the most not yet unequivocally determined.

The results we discuss here have been obtained on two such models, involving only two competing chemical species that we denote X and Y:
– The Brusselator (BX) that has been at the forefront of the studies of Turing patterns since the early days [13] and the kinetics of which is of the form

$$\mathbf{F}_{\mathrm{BX}} = \begin{bmatrix} \mathrm{A} - (\mathrm{B}+1)\mathrm{X} + \mathrm{X}^2\mathrm{Y} \\ \\ \sigma(\mathrm{BX} - \mathrm{X}^2\mathrm{Y}) \end{bmatrix} \tag{2}$$

– The Lengyel-Epstein (LE) model that has been derived [31] from the experimental kinetic studies of the chlorine dioxide-iodide-malonic acid reaction closely related to the CIMA reaction. Its rate is

$$\mathbf{F}_{\mathrm{LE}} = \begin{bmatrix} \mathrm{A} - \mathrm{X} - 4\mathrm{XY}/(1+\mathrm{X}^2) \\ \\ \sigma\left(\mathrm{BX} - \mathrm{BXY}/(1+\mathrm{X}^2)\right) \end{bmatrix} \tag{3}$$

where $\mathrm{X} = [\ \mathrm{I}^-\ ]$ and $\mathrm{Y} = [\ \mathrm{ClO}_2^-\ ]$.

In both models, X and Y are respectively the activator and inhibitor while σ describes the influence of the complexation process [22, 23] ($\sigma = 1$ if no complexant is present). The parameters A and B are related to feed concentrations of the species and the rate constants. In the following B is taken as the control (bifurcation) parameter. Note however that A, D_x, D_y or σ are equally valid bifurcation parameters.

Both systems possess a uniform steady reference state (uss). If D_y is sufficiently larger than D_x (in absence of complexant) or if the concentration of complexant is sufficient (when D_x and D_y are of the same order of magnitude), this reference state may become diffusively unstable at a finite distance from thermodynamic equilibrium, when the driving parameter B reaches a characteristic value $\mathrm{B_c}$. This Turing instability sets in with modes possessing a finite wave number $\mathrm{k_c}$. Both critical parameters may be determined by the *linear stability analysis* of the reference state.

To characterise the behaviour beyond this threshold of instability ($\mathrm{B} > \mathrm{B_c}$ for BX while $\mathrm{B} < \mathrm{B_c}$ for LE), mathematical methods of bifurcation theory [1, 32, 33] may be brought to bear to extract various quantitative features. The outcome of such analytical work may then hopefully serve to organize the results that may be obtained by straightforward numerical integration of the reaction-diffusion equations, before the results are compared with experiments.

These theoretical methods lead to the derivation of nonlinear partial differential equations for the amplitudes of the concentration patterns describing their dynamical properties. The concentration field is approximated by a linear superposition of active modes:

$$\mathbf{C} = \mathbf{C_o} + \mathbf{E_T} \sum_{i=1}^{M} \left[\mathrm{A_i e^{i\mathbf{k}_i \mathbf{r}}} + \mathrm{c.c}\right] \tag{4}$$

where $\mathbf{C}_o$ is the reference concentration (uss). Then the amplitude equations read

$$\frac{\partial \mathrm{A_i}}{\partial \mathrm{t}} = \mu \mathrm{A_i} + \mathrm{G_i}\left(\{\mathrm{A_j}\}\right) + \xi_o^2 \Box^2 \mathrm{A_i} \tag{5}$$

where μ is the reduced distance to the bifurcation point, $\mathrm{G_i}\left(\{\mathrm{A_j}\}\right)$ is a nonlinear polynomial in the active amplitudes, and $\Box$ a spatial operator describing the modulations of the patterns.

The experimental characteristic wavelength ($\approx k_c^{-1} \approx 0.2$mm) of the patterns is small compared to the size of the reactor. For such large aspect ratio systems, the boundaries play a minor role and the underlying rotational invariance of the system then leads to a large multiplicity in the number of solutions that emerge from the bifurcation points. Pattern selection is then the study, with the aid of the above mentioned theoretical and numerical tools, of the relative stability of, and the competition among, the spatial modes that grow as the result of these Turing symmetry-breaking bifurcations. More precisely, its aim lies in the determination, for given parametric conditions, of the possible structures – geometrical aspect, orientation in the reactor, wavelength – and their stability properties. These aspects can be compared with the determination of equilibrium phase diagrams in the field of phase transitions. This information is summarized in so-called bifurcation diagrams showing, for instance, the change in the amplitudes of the various allowed patterns as the control parameter B is varied. An extensive discussion is given in ref. [30].

Here, we merely show illustrative diagrams, respectively for the 2D BX [30, 34] (Fig. 2a) and LE [35] (Fig. 2b) models together with the aspects of the realized Turing patterns. As in numerous 2D situations in other fields (convection, nonlinear optics, solidification, ...) the competition here also takes place between stripes (structures periodic in one direction) and patterns of hexagonal symmetry. The hexagons are the first to appear in a subcritical way. They are followed by stripes that may be either supercritical or subcritical as illustrated respectively for the BX and LE models for the chosen values of the parameters. Let us also note the existence of a reentrant hexagonal phase [36] for higher values of B for the BX model while the competing oscillating patterns originating from the linear Hopf bifurcation at $\mathrm{B} = 21.25$, are not yet apparent for B = 30. On the contrary, for the LE model, the time dependent behaviour is visible while no reentrant hexagons appear.

This general competition near B_c, that is also present for other pattern forming chemical models [37], between hexagons and stripes is also the norm in the experiments [18, 19, 25, 28] (Fig. 3). It follows from the fact that the amplitude equations possess universal features as their structure is determined solely by symmetry arguments. Only the values of the coefficients in the equations take care of the specificity of a particular problem or model and may for instance transform a supercritical bifurcation into a subcritical one. It is nevertheless important to keep in mind that universality is soon lost as a result of secondary, tertiary, ..., and higher order bifurcations. This is already clear in comparing the illustrated behaviours of BX and LE models.

One is then led to consider scenarios eventually leading to the spatio-temporal complex phenomena that arise in these large aspect ratio systems with loss of various kinds of correlations and proliferation of defects as was done earlier for the routes to chaos in small systems. Defects, similar to those in the experiments are also apparent on our numerical simulations (Fig. 4). They play an important role at many levels of the dynamics of large aspect ratio systems. Indeed, in the experiments or in the numerical simulations, the particular orientation of the patterns has to be chosen by some external or initial bias, because of the orientational degeneracy. Often in the experiments – true or numerical – different biases are present in different parts of the reactor and structures, with diverse orientations, start to grow in various parts of the system. The way in which these orientationally competing patches form compatible patterns is the subject of much present interest [1, 33]. It is not unusual for these patterns to remain time dependent over very long times and indeed never settle down at all. The resulting mismatch leads to the formation of defects (dislocations, disclinations, grain boundaries) that play an important role in pattern rearrangement. The spatial operators of the amplitude equations allow the treatment of the modulational effects of the patterns resulting from the presence of boundaries or defects.

Such spatial operators are also important for reactor conditions where one has to take into account the influence of the feeding profiles on the pattern selection [34]. In this case we have the unfolding in space of the bifurcation diagram: Each pattern develops in the region of the reactor where the local value of the bifurcation parameter allows it to be stable. Numerical analysis shows that this selection becomes very complicated, even in the presence of the simplest forms of profiles. The form of the profiles, the boundaries, and the domain walls separating patterns of different symmetries, all compete to determine the selected orientation of the patterns.

Pattern selection may also be studied for 3D systems [38, 30]. In this case a body centered cubic lattice, hexagonal packed cylinders or lamellae are in competition. However the general 3D experimental reactor, as noted before, is always under the influence of the feeding ramps. This complex problem is

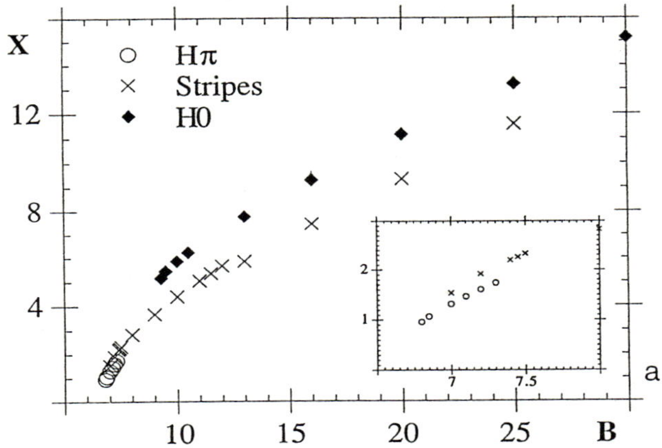

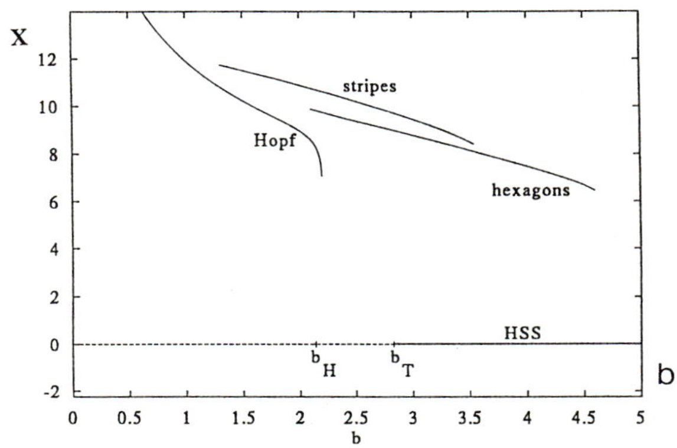

Fig. 2. (a) Bifurcation diagram for the concentration of the activator species (X) for the BX model where the bifurcation parameter B is varied while the other parameters are held fixed at respectively A=4.5 and $D_y/D_x = 8$, while $\sigma = 1$. $[X_{max} - A]$ is represented. It exhibits the standard hexagons-stripes competition with hysteresis loop (inset), near the primary Turing bifurcation ($B_c = 6.71$). On increasing B, a pattern of hexagonal symmetry is the first to appear subcritically ($B < B_c$). A striped structure then appears supercritically ($B > B_c$). Reentrant patterns of hexagonal symmetry become stable again at higher values of the bifurcation parameter and, in their own competition with the stripes, give rise to another hysteresis loop. (b) Bifurcation diagram for the concentration of species X for the LE model where the bifurcation parameter b (=B) is varied while the other parameters are taken as A=30, $D_y/D_x = 1.5$ and $\sigma = 8$. Here on decreasing B hexagons are also the first to appear subcritically ($B > B_T$), but for these conditions the stripes also form in the subcritical region. The competing Hopf bifurcation (B_H) leading to a branch of oscillating solutions is also visible.

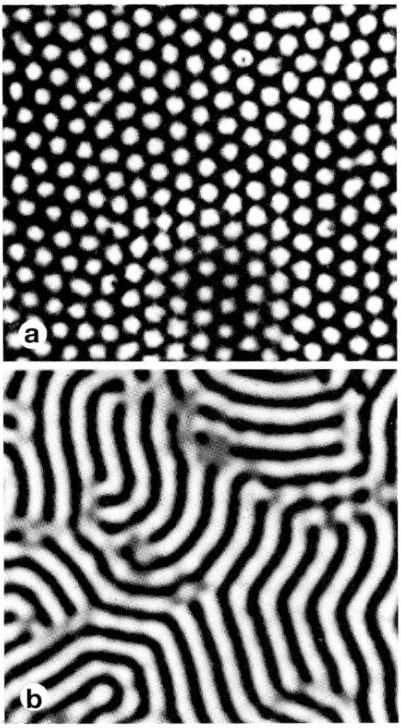

Fig. 3. Sustained Turing concentration patterns in the disc reactor of diameter ≈ 20 mm (type b). Contrast enhanced images of the central portion of the reactor. Dark and clear regions respectively correspond to reduced or oxidized iodine states. Both pictures are at a scale corresponding to a view of 2.8 mm × 2.8 mm. Experimental conditions: $T^o = 6\ ^oC$, residence time in the feeding tanks=10 min., Thiodene = 25g/l of gel, $[I^-]_o = 2.9\times10^{-3}$ M, $[ClO_2^-]_o^{II} = 2.0\times10^{-2}$ M, $[NaOH]_o^{II} = 8\times10^{-3}$ M, $[\text{Acetic Acid}]_o^I = 2.3$ M. (a) Hexagonal pattern: $[\text{Malonic Acid}]_o^I = 3.5\times10^{-3}$ M. (b) Striped structure: $[\text{Malonic Acid}]_o^I = 5\times10^{-3}$ M. $[X]_o^i$ is the feeding concentration of species X in tank i.

only now being studied theoretically and experimentally. In the latter case difficult 3D visualization problems are encountered [27, 25].

4 Stability and Intrinsic Localized Turing Structures

To investigate the stability of the nonlinear branches that emerge from the bifurcation points, and thereby detect eventual secondary, tertiary, ..., n-ary bifurcations, one again resorts to linear stability analysis, now, however, on the new emerging solutions. For each state, and there lies the difficulty, one has to test with respect to all possible types of "small" perturbations : amplitude, modules and orientation of the wavevector and resonant perturbations

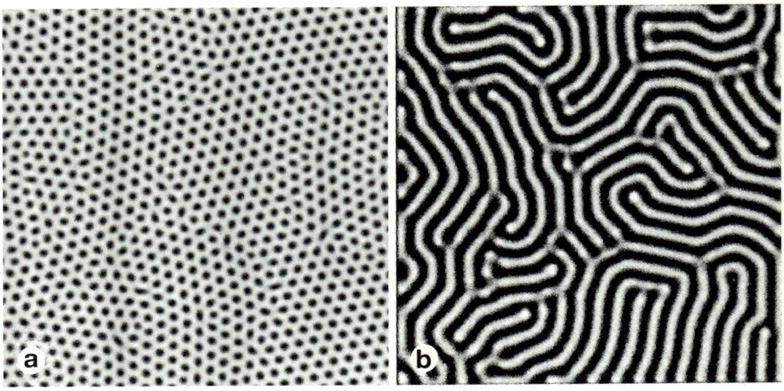

Fig. 4. Large aspect ratio Turing patterns for species X in the BX model with hexagonal (a) and stripe (b) symmetries and exhibiting defects and grain boundaries. The gray scale corresponds to the concentrations lying between the absolute minimum (black) and maximum (white); it thus measures the concentration relative to that of the uniform reference state.

to structures with a different symmetry. As it is already the case for the primary Turing bifurcation, this means of action does not preclude the possible existence of isolated branches that can only be reached and detected through the application of perturbations of finite size.

Near the primary bifurcation ($\mu = 0$), in the domain where the amplitude equations are valid (weakly nonlinear theory) another means of investigation of the stability is available. Indeed the amplitude equations then have a relaxational character and can thus be derived from a *Lyapunov functional* $L[\mathrm{A}, \mathrm{A}^*]$ such that [1,33]

$$\frac{\partial \mathrm{A_i}}{\partial t} = \frac{\delta L[\mathrm{A}, \mathrm{A}^*]}{\delta \mathrm{A_i^*}} \tag{6}$$

where $L[A, A^*]$ itself decreases monotonously in time. Therefore the globally stable pattern corresponds, for a given value of μ, to the absolute minimum of $L[A, A^*]$, whereas the relative minima represent metastable structures that arise when multistability is present. The maxima naturally correspond to the unstable states. This functional therefore plays a role similar to a free energy in the determination of the equilibrium phase diagram.

Such an organizing principle based on some extremal property to determine the stability of solutions is, however, far from general under the nonequilibrium situations considered here. In fact it constitutes the exception as the full reaction-diffusion equations in general possess a non-variational form. This then allows for solutions to the problem, of which oscillations and waves are examples, that have no equivalent in the equilibrium phase diagram to arise as we will illustrate below. Also in contrast to equilibrium phase transitions, the fluctuations generally play a very minor role in these chemical instabilities. The hysteresis regions in the experimental bifurcation diagrams are quite large, indicating that the transitions between the different states indeed take place at the marginal stability points via the decay of the unstable modes: traditional nucleation phenomena are absent. The systems we consider therefore in general exhibit the phenomenon of multistability and one may observe theoretically and experimentally more than one stable state for a given value of the control parameter even though some states are only metastable according to the value of $L[A, A^*]$.

Metastability is better apprehended if one considers the velocity, v, of a front (or domain boundary or interface) connecting two solutions (say A_1 and A_2) of the amplitude equations that are bistable in the interval $\mu_1 \leq \mu \leq \mu_2$. When a Lyapunov functional may be defined, there exists a single value μ_{co} in this interval for which the front is stationary ($v = 0$) and the solutions A_1 and A_2 may coexist in space. For all other values in the interval, one solution will dominate the other and invade the whole system. This state is then, in that sense, the most stable. As in classical nucleation theory, droplets of one state imbedded in the other are then always unstable. Those with an extension smaller than some critical radius will shrink away while those with a size larger than critical will fill the whole system.

This seemingly simple organization is lost as soon as *non-variational* effects come into play. These result from contributions that cannot be incorporated in the Lyapunov functional. Then for a given value of the parameters one can have fronts moving in either directions depending on the initial conditions [39]. These effects, typical of systems far from equilibrium, may furthermore give rise to interactions between fronts that may then lead to the stabilization of droplets of finite size [1, 40, 41] that constitute new intrinsic localized solutions of the pattern formation problem at hand. Examples are discussed in the following.

This is still not the complete picture. Supplementary new effects arise for the general class of problems where a stable periodic pattern grows, beyond a moving front, at the expense of another, metastable, state. This is of direct interest for our discussion.

The simplest example of a new effect is provided, in 1D, by a steady *front* separating a state of Turing stripes from the uniform reference steady state (uss). Such a front (Fig. 5a) may be constructed whenever the two states are bistable, i.e. when the stripes appear subcritically [35]. It is found numerically that such a stable stationary interface exists for a whole band (*locking band*) of values of the control parameter and not only for the single value μ_{co} as before (Fig. 5b).

It has been conjectured that the origin of the existence of this band lies in the pinning of the large scale front by the underlying small characteristic scale (k_c^{-1}) of the pattern: here the array of 1D stripes [42]. Physically this phenomenon arises from the difference in this interaction when the center of the front sits on top or between the stripes. The two length scales involved

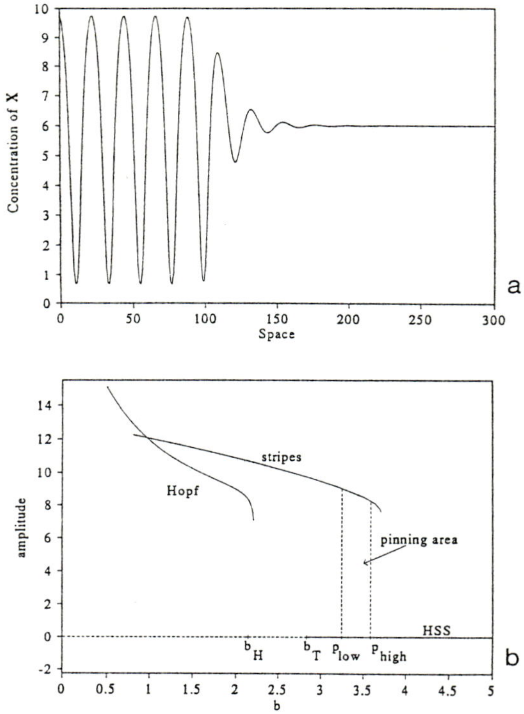

Fig. 5. (a) Pinned front of concentration of X, for the LE model, between the uniform reference state and a 1D Turing structure for b(=B)=3.4. All the other parameters have uniform values in the system that are as in Fig. 2b. Such locked front exists in all the $p_{low} \leq b(=B) \leq p_{high}$ region (see below). (b) Bifurcation diagram for a 1D LE system corresponding to that of Fig.2b (2D) exhibiting the span of parameter b (=B) where pinning of the front (and of the other localized patterns) occurs.

couple through so-called non-adiabatic contributions that do not appear in the amplitude equations [43].

As soon as the bifurcation parameter leaves the locking band, the disequilibrium between the states in presence becomes sufficient to overcome the pinning effect. So one of the states may again start to invade the other (depinning transition). One may then distinguish a freezing front (stripes invading uss) from a melting front (stripes being erased by uss). However, the motions of these fronts are still affected by the effects that gave rise to pinning and the more so as one gets closer to the locking band edges. Indeed, outside the locking band, the velocity of the front is not uniform. The structure is created (or destroyed) by "jumps" of one wavelength. This then leads to the staircase aspect of a space-time plot of the front position (Fig. 6). Because the underlying structure is periodic, the position of the front with respect to the small scale structure can be mapped to a phase variable. Its dynamical behaviour is then similar to the phase slippage staircase of forced oscillators [44]. As one nears the edges of the locking band, the mean velocity of the front goes to zero exhibiting a characteristic critical behaviour as the time between the phase slippages becomes longer and longer and finally diverges at the edge creating the observed locking. Such an interference between a kink and a structure is frequent in condensed matter physics [45] where they are known to dominate the dynamics of dislocations, charge density waves or domains.

Freezing and melting fronts may be used as building blocks for the construction of coherent structures formed by *droplets* of one state embedded in the other. Such intrinsic localized (as opposed to localization by a profile in space of the bifurcation parameter) Turing states, composed of a structured core, consisting of N wavelengths, connected on both sides by a front to the uniform reference state, have been obtained [35, 46] in our numerical simulations. These coherent states are stabilized by the pinning of both fronts. This mechanism therefore leads to multistability between states composed of a different number of spikes in their core. The number N naturally depends on the initial conditions as the width of the locking band is the same for all values of N as for the simple front. Random distribution of such localized domains, in the extreme case single-spiked, akin to glassy structures may also be obtained [47]. Outside the pinning band, the boundary fronts unlock and, depending of the edge considered, the structure either invades the whole system or disappears.

The complementary coherent structures where a bubble of the uniform reference state exists inside a Turing background is also seen in the simulations [35, 46].

Similarly coherent structures may also exist in higher space dimensions [35, 46, 48]. Here, however, two new effects come into play. First one has to take into account the existence of patterns of different symmetries (e.g. stripes and hexagons in 2D). So that not only can we now also have localized Turing patterns in the background of the uniform reference state when the bifurcation to the global structured states are subcritical (in 2D, although it is usually the case for hexagons, it is not always true for the stripes), but we may also construct droplets of a structure of a given symmetry in the background of a structure with a different symmetry whenever the global structured states are bistable (as it is usually the case for stripes and hexagons in 2D). Secondly, the relative orientations of the domain boundaries and of the wavevectors characterizing the patterns constitute important parameters. Pinning is favored when one of the wavevectors is perpendicular to the front [49]. Already in 2D a large variety of localized structures may be formed. The depinning behaviours are also more complex as is already exhibited by the dynamical behavior (freezing) of a localized state of hexagonal symmetry invading the uniform reference state. This growth occurs by adding new

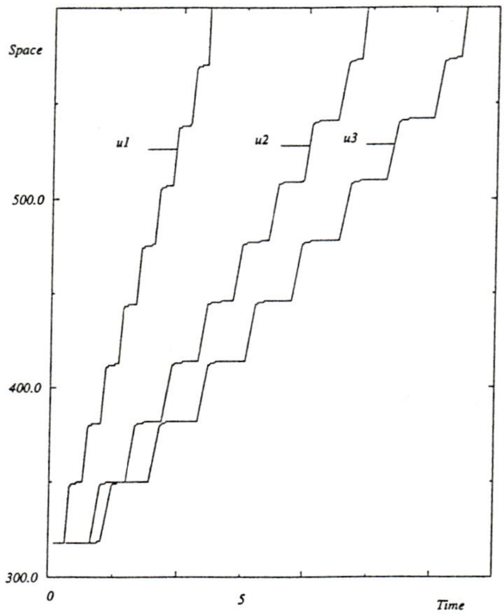

Fig. 6. Evolution in time of the position of a freezing front (LE model) that at time t=0 was pinned $p_{low} \leq b(= B) \leq p_{high}$ at the center of the 1D system. The bifurcation parameter is then lowered respectively to b=u1, u2 or u3, such that u1 < u2 < u3 < p_{low} outside the pinning band. The Turing structure invades the domain of the uniform reference states with a nonuniform velocity. The movement proceeds by 'jumps' of one wavelength. The time between each jump increases while the mean velocity of the front decreasse as the bifurcation parameter increases towards the pinning band edge. At this edge the time between jumps diverges so that pinning occurs.

"spots" to systematically form new triangles of spots along the boundaries of the growing structure [50].

5 Turing and Hopf Bifurcations Interactions

As predicted by the complexation and immobilization mechanism invoked in Section 2, a transition between stationary periodic (Turing modes) and propagating wavelike (Hopf modes) patterns are experimentally observed by decreasing the color indicator concentration [27, 25]. At very low concentration only waves can be observed. Let us mention that this is not true for the polyacrylamide gel as the CIMA reaction may modify the structure of the gel matrix [51].

Because it has to be preloaded in the gel, the color indicator concentration is not an easily tunable parameter. However a similar transition was discovered to exist at constant, but sufficiently low, color indicator concentration by increasing instead the malonic acid concentration [52]. In such conditions, trains of plane waves resulting from a Hopf bifurcation of the uniform reference state are observed experimentally [27, 25, 52](Fig. 7). Their wavelength is different from that of the Turing patterns.

In the transition region, where a Turing and a Hopf bifurcation are near each other and interact, a wealth of new spatio-temporal behaviors is observed [53].

Near such a codimension 2 bifurcation point, as suggested by the possibility of tuning the two bifurcations by varying two independant parameters, color indicator and malonic acid concentrations, new amplitude equations may be derived [54, 52]. This is done by expressing the concentration field $\mathbf{C}$(r,t) in terms of Turing ($\mathrm{A_T}$) and Hopf ($\mathrm{A_H}$) amplitudes. For instance in a 1D system one writes:

$$\mathbf{C}(\mathrm{x,t}) = \mathbf{C_o} + (\mathbf{E_T}\mathrm{A_T}\mathrm{e}^{\mathrm{i}k_c x} + \mathbf{E_H}\mathrm{A_H}\mathrm{e}^{\mathrm{i}\Omega_c t} + \mathrm{c.c}) \tag{7}$$

where $\mathrm{k_c}$ and Ω_c are respectively the critical wavenumber of the structure and the critical frequency around the degenerate bifurcation point One gets

$$\begin{aligned} \frac{dA_T}{dt} &= \mu_T A_T - g_{TT}|A_T|^2 A_T - g_{TH}|A_H|^2 A_T + D_T \frac{\partial^2 A_T}{\partial x^2} \\ \frac{dA_H}{dt} &= \mu_H A_H - (\beta_r + i\beta_i)|A_H|^2 A_H - (\delta_r + i\delta_i)|A_T|^2 A_H + \\ &\quad (D_{H,r} + iD_{H,i}) \frac{\partial^2 A_H}{\partial x^2} \end{aligned} \tag{8}$$

where μ_T and μ_H are the reduced distances from the Turing and Hopf bifurcation points in the unfolding of the degenerate bifurcation.

Near this bifurcation, three types of global patterns may compete: (i) a band of pure Turing structures ($\mathrm{A_T} \neq 0, \mathrm{A_H} = 0$), (ii) a one parameter family of plane waves ($\mathrm{A_T} = 0, \mathrm{A_H} \neq 0$) and (iii) a two-parameter family of mixed states ($\mathrm{A_T} \neq 0, \mathrm{A_H} \neq 0$) that corresponds to pulsating Turing structures. Numerous bifurcation scenarios are therefore possible depending on which of the Turing and Hopf bifurcation arises first on varying the bifurcation parameter, or on the possibility for each bifurcation to be super- or subcritical. However, the competition between the modes boils down to determining the stability of the mixed mode [55]. If it is unstable ($\beta_\mathrm{r}\mathrm{g_{TT}} - \delta_\mathrm{r}\mathrm{g_{TH}} < 0$) then there exists a region of bistability between the Turing and Hopf modes in the bifurcation diagram. On the contrary when ($\beta_\mathrm{r}\mathrm{g_{TT}} - \delta_\mathrm{r}\mathrm{g_{TH}} > 0$) the mixed mode is stable and necessarily connects the regions of bifurcation parameter values where the Turing and Hopf modes are individually stable.

In 2D, patterns of hexagonal symmetry enter the competition with the stripes and waves. Undoubtedly the number of possible bifurcation scenarios increases again and this problem has by no means been analyzed systematically yet.

Fig. 7. Sustained travelling concentration waves in the disc reactor (type b). Such a pattern occurs in the region of parameters where the Hopf bifurcation preceeds the Turing instability. Dimensions are as in Fig. 3. Experimental conditions: $T^o = 6°C$, residence time in the feeding tanks = 10min., Polyvinylalcohol = 5g/l of gel, $[I^-]_o = 2.8 \times 10^{-3}$M, $[ClO_2^-]_o^{II} = 2.5 \times 10^{-2}$M, $[NaOH]_o^{II} = 8 \times 10^{-3}$M, $[\text{Acetic Acid}]_o^{I} = 2.3$M, $[\text{Malonic Acid}]_o^{I} = 9.8 \times 10^{-3}$M.

For 1D systems, in the conditions where the mixed mode is stable a new scenario to spatio-temporal chaos has been characterized [56]. It is not known at present if this behaviour is related in some way to the spatio-temporal chaos observed experimentally within the Turing-Hopf interaction region in the disc reactor [53].

In the following section we will concentrate on conditions where the mixed mode is unstable and where new structures emerge that are relevant to the interpretation of some experimental results.

6 Intrinsic Localized Turing-Hopf Structures

An unstable mixed mode implies a region of bistability between the global Turing and Hopf modes. Therefore, in this zone, one may also enquire about the possible existence of coherent localized solutions, extending those discussed already in Section 3.

Once more the simplest corresponds to a *front*. However, it now connects a Turing pattern to a train of plane phase waves, the wavenumber and the period of which are selected by the front among the family of possible Hopf modes (Fig. 8a) [52, 55]. Depending on the competition between the nonlinear (β_i) and spatial (D_i) dispersions, these waves may be either emitted by the front or absorbed by it.

Again this front may be pinned, through its interaction with the underlying periodical Turing structure, for a range of values of the bifurcation parameter giving rise to a locking band [55].

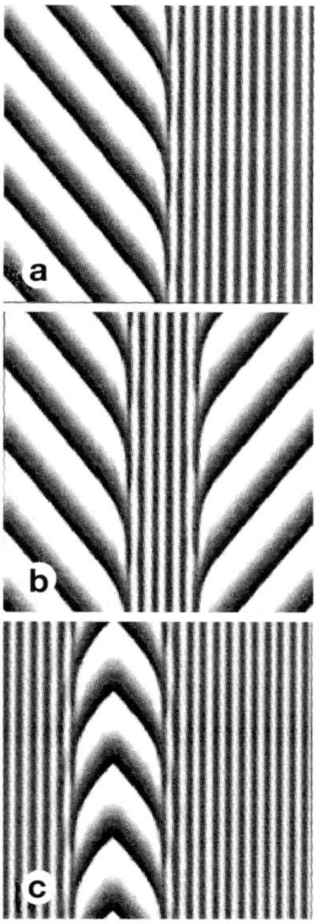

Fig. 8. Space-time maps (time running upwards) of stable coherent 1D Turing-Hopf structures for the BX model ($A = 2.5, D_x = 4.11, D_y = 9.73$) when the Turing and Hopf bifurcations interact. The color coding is as previously. (a) Turing-Hopf pinned front (B=10) (b) Turing droplet (N=5) emitting waves to both sides (B=10) (c) Hopf 'bubble' (of sink type) resulting from the interaction of two Turing-Hopf fronts (B=9.5).

The depinning of such Turing-Hopf front when the control parameter exits the locking band is much more complex than in the case previously considered in Section 3.

A relatively simple situation [55] is provided by the conditions of Fig. 9a. Starting from a locked front at B=9 (BX model), one decreases the parameter to 8.82. There we see that, behind a freezing front, the size of the 1D Turing pattern increases by one wavelength every period of oscillation of the receeding Hopf domain. This again leads to the staircase aspect if one plots the position of the front versus time. If one looks carefully it takes 19 periods of oscillation for the front to progress by 19 wavelengths towards the boundary of the system. Thus we seem to be in the presence of some kind of frequency locking phenomena. The two periods here are the period of oscillation of the Hopf mode and the time needed to advance the front by one wavelength of the Turing structure. That frequency locking is at play is further emphasized by the possibility for the front of moving p wavelengths over q periods of oscillations (p and q integers). In fact the adaptativity of the system is such that one even finds random mixtures in time of such events as shown, for instance, on Fig. 9b for a melting front when the locked front at B=13.4 is unlocked at B=13.7. But a still finer matching mechanism may come into play. Its trace is already apparent on Fig. 9a for B=8.82. The two "blobs" lying above the front on the space-time map represent oscillations in time of the width of the newly created Turing cells (for about two periods). This corresponds to the existence of some kind of transient mixed-mode in the front that freezes up as the front moves on. However, as one modifies the distance in B to the locking band edge, not only may the lifetime of such mixed-mode change but its wavelength (or amplitude of oscillation of width of the new Turing cell) may also be modified giving rise to space-time patterns such as those represented on Fig. 9c for B=8.8. Ultimately one may even have the temporary creation of a Hopf zone temporarily embedded (see below) in the Turing structure being created. On this last figure the front moves by 19 wavelengths in 16 periods of oscillations. Hence its mean velocity is larger than for B=8.8 as should be expected because one is further from the edge of the locking band. All these adjustment procedures also lead to a large multiplicity of transient behaviours.

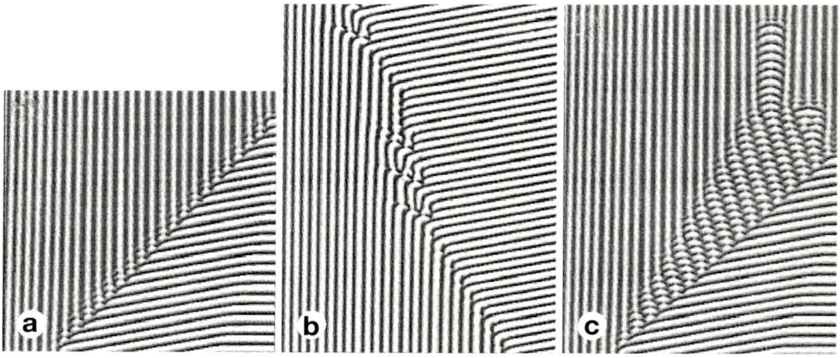

Fig. 9. (a) Space-time map (time running upwards) exhibiting a freezing front behind which a 1D Turing pattern invades a Hopf domain after it was unlocked by lowering B from 9 (inside the locking band) to 8.82 (outside this band). Initially each state (Turing and Hopf) occupy half the system. A space-time plot of the position of the front would present a staircase aspect (as in Fig.6), the front advancing by 'steps' of one wavelength for every period of the oscillation. (b) Space-time plot showing a melting front behind which the Hopf domain now absorbs the Turing pattern. At B=13.4, the front is pinned and each state (Turing and Hopf) coexist in half the box. When it is unlocked at B=13.7, the front moves in the complex way shown. (c) Space-time map exhibiting a situation similar to (a), but when B is lowered to 8.8. Localized oscillating states in the vicinity of the front help the system to adapt (increase in the present case) its mean velocity.

Further studies are necessary to unravel the complete picture. Some of the ideas of the simpler, but somewhat related problem of front propagation in a chain of coupled bistable oscillators [57], may be useful. Such Turing-Hopf fronts have also been studied in an array of resistively coupled LC oscillators [58].

As before such fronts may be combined to form droplets of one of the two states embedded in the other. Numerically we have indeed produced [52, 55] stable states with a Turing structure, truncated to N wavelengths in the core, emitting plane waves both to the left and the right (Fig. 8b). The complementary structure, where Hopf modes are embedded in a Turing structure has also been obtained (Fig. 8c). The center of the Hopf bubble acts as a sink or a source depending on whether the fronts emit or absorb the waves. For the conditions we have analyzed, both Turing and Hopf bifurcations are supercritical and in the bistable Turing-Hopf region both the uniform reference state and the mixed mode are unstable. Therefore the presence of the Turing mode inhibits the Hopf mode and vice-versa. Equivalently, the amplitude of the Hopf mode goes to zero whenever the Turing mode is present and thus maximum, while the Turing mode vanishes in the regions where the Hopf mode is present and maximum.

It is again found [55] that all localized Turing states where N is sufficiently large (larger than 5 for the parameters that were used in our simulations for conditions such that the Turing instability preceeds the linear Hopf instability point) have the same domain of stability as that of the simple front. Their stability may then also be ascribed to the pinning effects. However when N is smaller the stability domain of the structured bubbles is increased in the direction where the Hopf mode becomes dominant (the region where the high N localized structures are destroyed by the propagation of two melting fronts). The locking band therefore becomes wider as N progressively decreases (Fig. 10). Ultimately one is left with a single Turing "spot" that acts as an anti-synchronous wave source (or sink) that was therefore called a flip-flop [52, 55, 46]. Its stability must now depend on the interaction between the fronts because there is no spatial periodicity involved anymore. It results from two competing effects. As we are in a region where the Hopf mode is dominant, the fronts tend to attract each other and to fill the Hopf hole in the core of this wave source. On the other hand, the fact that the two fronts emit waves with a phase shift of π induces an effective repulsion between them (non variational effects). The balance between these two effects gives rise to a stable bound state between the two fronts that delineate the core of the flip-flop. As N increases these effects become weaker as the fronts separate from one another while the pinning effects increase due to the progressive formation of a spatial periodicity. The stability of the complementary Turing-Hopf structures may be discussed along the same lines.

Using a thin strip reactor of Type a (see Section 2), experimental endogeneous, quasi-1D, antisynchronous wave sources are readily obtained [52]. One starts the experiment with a pattern consisting of a single row of spots (quasi-1D Turing structure) oriented perpendicularly to the feeding gradients (see Fig. 2a of reference 52). A large increase of the input concentration of malonic acid leads to the development of oscillatory modes that gradually erase the Turing spots. These oscillations then organize to waves (see Fig. 7

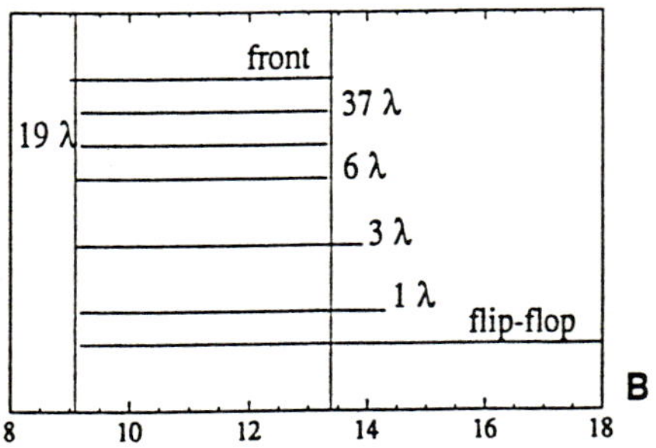

Fig. 10. Numerically obtained regions of stability for the Turing-Hopf front and the Turing embedded in Hopf structures. The number nλ refers to the number of Turing wavelengths present in the core of such localized patterns. The parameters are the same as for Fig. 8.

of reference 27) that propagate along the direction where the Turing spots existed before, i.e. parallel to the feed boundaries. Characteristic of the Hopf domain of parameters, these plane waves are the quasi-1D equivalent of the phase waves illustrated in Fig. 7 for the 2D disc reactor. When the increase in malonic acid is weaker, so that one enters the parameter region where Turing and Hopf modes interact, some isolated Turing spots remain (Fig. 11a) and act as antisynchronous sources of waves. A time-averaged image, cancelling out the periodic time dependant phenomena then clearly shows the stationary core (Fig. 11b). The experimental dynamics has been shown to be effectively 1D and thus this chemical flip-flop acts as a kind of 1D "spiral". The dynamics is best represented by a space-time plot that bears strong resemblance with that obtained numerically (Fig. 12) [52].

In the quasi-2D disc reactor, one may also study the competition of Turing and Hopf modes in the region between that giving rise to Turing structures as in Fig. 3 and that exhibiting characteristic Hopf phase wave modes as in Fig. 7. In this intermediary region, for parameters similar to those used in the experiments described in the preceding paragraph, a host of structures emitting waves were discovered [53, 59] but most of them were unstable. However a stable pattern consisted of a spiral source (Fig. 13). The Turing mode is then absent except in the core of this new kind of spiral. The Turing mode amplitude that can only be guessed at the tip of the spiral in the snapshot may again be revealed as a bright spot on a grey background by time-averaging. In the core, the amplitude of oscillations drops to zero while that of the Turing mode grows to its maximum. 2D computer simulations yield a similar result [46].

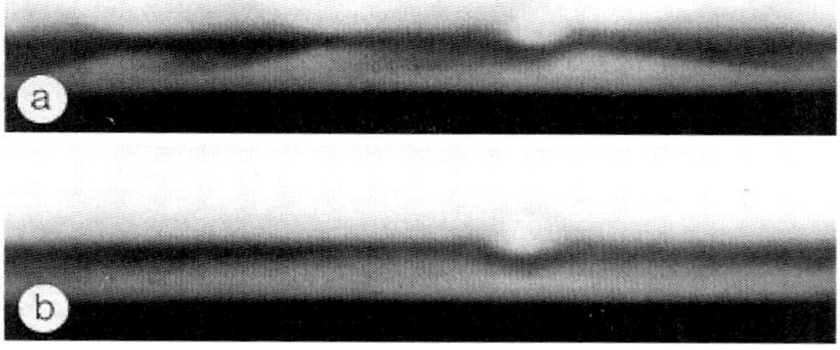

Fig. 11. Chemical flip-flop in a thin strip gel-reactor (type a): $L_x = 3$mm, $L_y = 10$mm and $L_z = 0.14$mm. Experimental conditions: $T^o = 3°C$, residence time in the feeding tanks = 2.0min, Thiodene = 20 g/l of gel, $[I^-]_o = 2.5 \times 10^{-3}$M, $[ClO_2^-]_o^{II} = 2.4 \times 10^{-2}$M, $[NaOH]_o^{II} = 2 \times 10^{-3}$M, [Acetic Acid]$_o^I = 2.3$M. One first forms a stationary pattern consisting of a single row of spots (1D Turing structure: see Fig. 2a in reference [52]) for [Malonic Acid]$_o^I = 5.0 \times 10^{-3}$M. If this last concentration is doubled the spots disappear and are replaced by waves characteristic of the Hopf domain. However, for weaker changes the system falls in a region where Turing and Hopf modes compete: (a) Antisynchronous wave source obtained when [Malonic Acid]$_o^I = 7.5 \times 10^{-3}$M. The arrowhead patterns corresponds to travelling waves emitted antisynchronously from the remaining 'Turing spot'. (b) Time average of the previous image enhancing the Turing spot emitter.

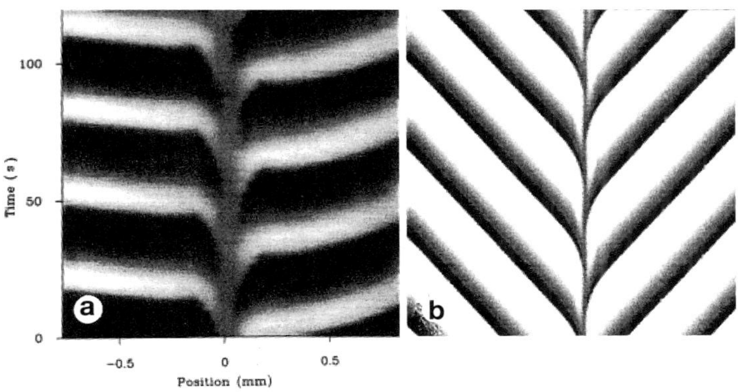

Fig. 12. (a) Space-time plot of the light intensity across the experimental flip-flop wave source (Fig. 11), parallel to the feed boundaries. (b) Space-time plot, for conditions similar to those of Fig.5, for the Turing droplet with the smallest core (flip-flop).

7 Conclusions

We have reviewed some recent developments of experiments and theory on Turing structures, stationary space periodic concentration patterns that arise through diffusive instabilities in chemical systems with suitable feedbacks.

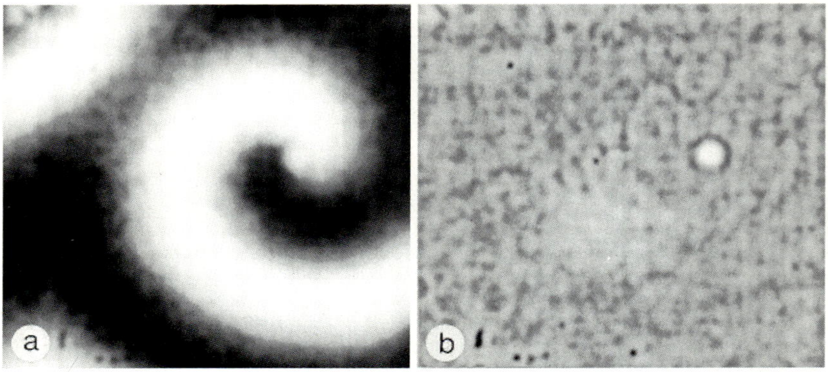

Fig. 13. Turing/Hopf spiral obtained in the disc reactor (type b). Dimensions as in Fig. 3. Experimental conditions as in Fig. 6 but with $[\text{MalonicAcid}]_o^I = 9.5 \times 10^{-3}$M (a) snapshot, (b) time averaged image.

These patterns were proposed some time ago as a possible origin of patchiness in biological problems. They should however not be thought of as perfect objects as these structures may support defects that allow for some adaptation to boundaries or non-uniform environments.

We have also illustrated that for appropriate conditions other types of organization in space are possible.

For situations of multistability, the pinning of the boundary of a Turing domain to its underlying structure may lead to clumps of patches where islets of Turing order of a given symmetry coexist with other states : uniform or of a different symmetry.

In the case of the interaction with a Hopf bifurcation that may always loom in the background because of the necessary condition for differential diffusion coefficients, such Turing islands are seen to emit waves that "explore" the inter islet space.

Recently a different type of structure, that relies on front interaction because of the long range inhibitory processes, have been obtained [60] in the bistable iodate-ferrocyanide-sulfite (the so-called EOE) reaction [61]. For low ferrocyanide concentrations labyrinthine-type patterns appear spontaneously through a standard Turing bifurcation. However for higher concentration they can only be created by applying a finite perturbation to the uniform reference state. For some such latter conditions the structure is generated by self-replication of "spots" [62]. This mechanism has also been observed to come into play with the CIMA reaction [53,59].

It therefore seems that far from equilibrium conditions provide us with a never ending list of rich and complex principles of organization. Their relevance to the realm of biological problems, however, still remains to be proven.

Acknowledgments

We would like to thank Professor G. Nicolis for his interest in this work. We are very grateful to P. De Kepper and J.J. Perraud for providing us with their experimental iconography. Quasi continuous discussions with them and other members of the CRPP group, ever since the discovery of the structures, have been both stimulating and enlightning. P.B. and G.D. are Research Associates with the FNRS (Belgium). This work was supported by Erasmus Program B 1028/13 and sponsored by Twinning contract SCI*-CT91-0706 of the EU Science Program.

References

1 M.C. Cross and P.C. Hohenberg, Rev. Mod. Phys. **65**, 854 (1993)
2 S. Kai, *Physics of Pattern Formation in Complex Dissipative Systems* (World Scientific, Singapore, 1992)
3 A.C. Newell and J.V. Moloney, *Nonlinear Optics* (Addison-Wesley, Redwood City, 1992)
4 J.M. Flesselles, A.J. Simon and A. Libchaber, Adv. Phys. **40**, 1 (1991)
5 R. Kapral and K. Showalter, Eds, *Chemical Waves and Patterns* (Kluwer, Amsterdam,to appear 1994)
6 A.M. Zhabotinskii, in *Oscillations and Travelling Waves in Chemical Systems* (R.J. Field and M. Burger, Eds., Wiley, New York, 1985)
7 I.R. Epstein, J. Phys. Chem. **88**, 187 (1984)
8 see in P. Bergé, Y. Pomeau and C. Vidal, *Order within Chaos* (Wiley, New York, 1987)
9 Y. Kuramoto, *Chemical Oscillations, Waves, and Turbulence* (Springer, Berlin, 1984)
10 J. Ross, S.C. Müller and C. Vidal, Science **240**, 460 (1988)
11 G. Ertl, Science **254**, 1750 (1991)
12 A. Turing, Philos. Trans. R. Soc. Lond. B **237**, 37 (1952)
13 G. Nicolis and I. Prigogine, *Self-Organization in Nonequilibrium Systems* (Wiley, New York, 1977)
14 H. Meinhardt, *Models of Biological Pattern Formation* (Academic Press, New York, 1982)
15 J.D. Murray, *Mathematical Biology* (Springer, Berlin, 1989)
16 G.M. Edelman, *Topobiology: An Introduction to Molecular Biology* (Basic Books, New York, 1988)
17 L.G. Harrison, *Kinetic Theory of Living Pattern* (Cambridge University Press: Developmental and Cell Biology Series 28, New York, 1993)
18 P. De Kepper, J. Boissonade and I.R. Epstein, J. Phys. Chem. **94**, 6525 (1990)
19 V. Castets, E. Dulos, J. Boissonade and P. De Kepper, Phys. Rev. Lett. **64**, 2953 (1990)
20 Q. Ouyang and H.L. Swinney, Nature **352**, 610 (1991)
21 Z. Noszticzius, W. Horsthemke, W.D. McCormick, H.L. Swinney and W.Y. Tam, Nature **329**, 619 (1987)
22 A. Hunding and P.G. Sørensen, J. Math. Biol. **26**, 27 (1988)
23 I. Lengyel and I.R. Epstein, Proc. Natl. Acad. Sci. USA **89**, 3977 (1992)
24 J.E. Pearson and W.J. Bruno, Chaos **2**, 513 (1992)

25 J. Boissonade, E. Dulos and P. De Kepper in ref. 5
26 I. Lengyel, S. Kadar and I.R. Epstein, Science **259**, 493 (1993)
27 J.J. Perraud, K. Agladze, E. Dulos and P. De Kepper, Physica A **188**, 1 (1992)
28 Q. Ouyang and H.L. Swinney in ref. 5
29 I. Lengyel and I.R. Epstein in ref. 5
30 P. Borckmans, G. Dewel, A. De Wit and D. Walgraef in ref. 5
31 I. Lengyel and I.R. Epstein, Science **251**, 650 (1991)
32 P. Manneville, *Dissipative Structures and Weak Turbulence* (Academic Press, San Diego, 1990)
33 A.C. Newell, T. Passot and J. Lega, Annu. Rev. Fluid Mech. **25**, 399 (1993)
34 P. Borckmans, A. De Wit and G. Dewel, Physica A **188**, 137 (1992)
35 O. Jensen, V.O. Pannbacker, G. Dewel and P. Borckmans, Phys. Lett. A **179**, 91 (1993)
36 J. Verdasca, A. De Wit, G. Dewel and P. Borckmans, Phys. Lett. A **168**, 194 (1992)
37 V. Dufiet and J. Boissonade, Physica A **188**, 158 (1992)
38 A. De Wit, G. Dewel, P. Borckmans and D. Walgraef, Physica D **61**, 289 (1993)
39 P. Ortoleva and J. Ross, J. Chem. Phys. **63**, 3398 (1975)
40 S. Koga and Y. Kuramoto, Prog. Theor. Phys. **63**, 106 (1980)
41 O. Thual and S. Fauve, J. Phys. (Paris) **49**, 1829 (1988)
42 Y. Pomeau, Physica D **23**, 3 (1986)
43 D. Bensimon, B.I. Shraiman and V. Croquette, Phys. Rev. A **38**, 5461 (1988)
44 P. Rhemus and J. Ross, J. Chem. Phys. **78**, 3747 (1983)
45 M. Peyrard and M.D. Kruskal, Physica D **14**, 88 (1984)
46 V.O. Pannbacker, O. Jensen, E. Mosekilde, G. Dewel and P. Borckmans, in *Spatio-Temporal Patterns in Nonequilibrium Complex Systems* (P. Palffy-Muhoray and P. Cladis, Eds., Kluwer, Amsterdam, to appear 1994)
47 M.I. Rabinovich and A.L. Fabrikant in *Future Directions of Nonlinear Dynamics in Physical and Biological Systems* (P.L. Christiansen et al., Eds, Plenum Press, New York, 1993)
48 I.S. Aranson, K.A. Gorshkov, A.S. Lomov and M.I. Rabinovich, Physica D **43**, 435 (1990)
49 B.A. Malomed, A.A. Nepomnyashchii and M.I. Tribelskii, Phys. Rev. A **42**, 7244 (1990)
50 O. Jensen, V.O. Pannbacker, E. Mosekilde, G. Dewel and P. Borckmans, Phys. Rev. E (submitted)
51 K.J. Lee, W.D. McCormick and H.L. Swinney, J. Chem. Phys. **96**, 4048 (1992)
52 J.J. Perraud, A. De Wit, E. Dulos, P. De Kepper, G. Dewel and P. Borckmans, Phys. Rev. Lett. **71**, 1272 (1993)
53 P. De Kepper, J.-J. Perraud, B. Rudovics and E. Dulos, 'Experimental Study of Stationary Turing Patterns and their Interaction with Travelling Waves in a Chemical System', Int. J. Bifurcation and Chaos (to appear)
54 H. Kidachi, Progr. Theor. Phys. **63**, 1152 (1980)
55 A. De Wit, 'Brisure de Symétrie Spatiale et Dynamique Spatio-temporelle dans les Systèmes Réaction-Diffusion', Ph. D. Thesis, Université Libre de Bruxelles, 1993
56 A. De Wit, G. Dewel and P. Borckmans, Phys. Rev. E **48**, R4191 (1993)
57 A.-D. Defontaines, Y. Pomeau and B. Rostan, Physica D **46**, 201 (1990)
58 G. Heidemann, M. Bode and H.G. Purwins, Phys. Lett. A **177**, 225 (1993)
59 B. Rudovics, J.-J. Perraud, P. De Kepper and E. Dulos in *Far-from-equilibrium Dynamics of Chemical Systems* (J. Gorecki, A.S. Cukrowski, A.L. Kawczynski and B. Nowakowski, Eds., World Scientific, Singapore, to appear 1994)

60 K.J. Lee, W.D. McCormick, Q. Ouyang and H.L. Swinney, Science **261**, 192 (1993)
61 E.C. Edblom, M. Orban and I.R. Epstein, J. Am. Chem. Soc. **108**, 2826 (1986)
62 K.J. Lee, W.D. McCormick, J. Pearson and H.L. Swinney, Nature **369**, 215 (1994)

Part II

Biological Patterns

Domains and Patterns in Biological Membranes

Ole G. Mouritsen, Kent Jørgensen, John Hjort Ipsen and M. M. Sperotto

Abstract

Computer simulations of the Monte Carlo importance-sampling type have been applied to a series of statistical mechanical lattice models of lipid bilayers with a view to establishing local and global membrane structures, in particular the dynamic heterogeneous states, domains, and patterns that are formed due to cooperative modes. Results obtained from such simulations are reported in the case of single-component bilayers and binary mixtures. The simulations presented in this chapter also illustrate how membrane agents like cholesterol, polypeptides, as well as various drugs, have a strong influence on the dynamic membrane heterogeneity, which in turn relates to their function in the membrane.

1 Introduction

A large number of physiological processes associated with cell membranes are controlled by the structure and thermodynamics of the fluid lipid-bilayer component of the membrane (Gennis 1989, Bloom et al. 1991). The lipid bilayer is a many-component mixture of a considerable complexity governed by the global phase structure of the mixture. The cooperative nature of the membrane assembly furthermore imparts the bilayer with a considerable degree of dynamic heterogeneity in terms of lipid domains and lateral patterns which are believed to influence the functioning of membrane-bound proteins and enzymes (Mouritsen and Jørgensen 1992, Jacobsen and Vaz 1992, Mouritsen and Biltonen, 1993). These local structures, which persist on a nanoscale of typically 10–1000Å are due to lateral density and compositional fluctuations and are of a dynamic origin.

Whereas the strong hydrophobic effect leads to the formation of thermodynamically stable lipid bilayer aggregates, it is the much weaker intra-layer molecular interactions which are responsible for the very special dynamic and static physical properties of the lipid bilayer. The pseudo-two-dimensional character of the bilayer, being only 50-80Å thick, implies a condition of extreme mechanical anisotropy. The cooperative behavior of the ensemble of lipid molecules determines the thermodynamic phase of the bilayer and hence its static physical properties, including average bulk macroscopic properties such as specific heat, compressibility, and bilayer thickness as well as area. The cooperative nature of the interactions also manifests itself in specific dynamic events involving large numbers of molecules, specifically phase transitions, protein aggregation, and the formation of microscopic lipid domains and special interfacial regions in the bilayer. The latter type of processes may be induced by thermal density fluctuations accompanying phase transitions or by compositional fluctuations leading to heterogeneous bilayer structures and to differentiated regions in biological membranes. Pure lipid bilayers display phase transitions (Mouritsen 1991) and have a very rich phase structure. We shall here mainly be concerned with the so-called main transition of the lipid bilayers which occurs at a temperature T_m that depends on the lipid species in question. The main transition takes the bilayer from a low-temperature solid, chain-conformationally ordered phase to a high-temperature fluid, chain-conformationally disordered phase. The main transition can be driven thermally or by electric fields, pH, lateral or hydrostatic pressures, or bilayer composition. The phase transitions of a one-component lipid bilayer become smeared in the presence of other bilayer components, such as other lipid species, cholesterol, proteins, or drugs. These other components often induce lateral phase separation in the bilayer and can in some cases lead to critical demixing. In those cases, it is the phase equilibria of the mixed system which govern the physical properties of the bilayer.

2 Models and Simulation Techniques

The common strategy underlying computer simulation studies of bilayer membrane systems involves basically two steps: (i) Formulation of a microscopic interaction model which, in terms of a chosen set of variables (e.g. translational coordinates or conformational states), accounts for the forces between the constituents of the system. (ii) Numerical solution of the statistical mechanical problem posed by (i) using either Monte Carlo or Molecular Dynamics methods (Mouritsen 1990a). Step (i) always involves approximations of some type. These approximations are dictated by computational feasibility as well as choice of properties to be calculated. Most simulations take as their starting point the existence of the lipid bilayer aggregate and only model the secondary effects of the lipid-water interactions.

The formulation of a useful theoretical model of a phospholipid bilayer is a balance between physical realism and computational feasibility (Mouritsen 1991). The fundamental microscopic variables have to be identified and the coupling between them described by an appropriate Hamiltonian. The Hamiltonian includes the proper interaction potentials, possibly parameterized in terms of some interaction constants (which are often unknown). The microscopic model calls for a full statistical mechanical treatment of the many-body problem. In the case of lipid bilayer models the translational variables are often frozen out by adopting a lattice approximation in which each acyl chain is positioned on a triangular two-dimensional lattice. The multi-state Pink model (Pink et al. 1980, Mouritsen 1990a) describes the interactions within such a lattice approximation. The Pink model assumes that the conformational properties of a single acyl chain can be described by a small number of selected conformational states corresponding to the mapping of the three-dimensional acyl-chain conformations upon a finite, discrete set of projected two-dimensional coarse-grained variables. These variables assign to each chain a cross-sectional area, a conformational energy, and a single-chain degeneracy. The number of states included in a multi-state model depends on the level of detail required. Most often ten different states are chosen on the basis of optimal packing properties and low conformational energy. In the Pink lattice model the conformational chain variables are coupled by hydrophobic anisotropic van der Waals interactions in the spirit of anisotropic liquid crystals. The interactions between the lipid-acyl chains and various other molecular compounds interacting with membranes, such as sterols, polypeptides, and drugs, can be taken into account each in their own way by allowing for their specific interaction with the different lipid-acyl-chain conformational states.

In this chapter we report some results obtained from Monte Carlo computer-simulation calculations on membrane models and their phase transitions with particular emphasis on heterogeneous membrane structures and patterns. For the purpose of revealing such structures, computer-simulation methods are unique since they allow for a full treatment of fluctuations and permit inspection and analysis of the system on the molecular level.

2.1 Model of a One-Component Lipid Bilayer

The Pink model allows for a series of ten conformational states of the acyl chains of the lipid molecules. The ten–state model provides a reasonably accurate description of the phase behavior of one-component lipid bilayers and the associated density fluctuations since it accounts for the most important conformational acyl-chain states of the acyl chains as well as their mutual interactions and statistics. In the Pink model the bilayer is considered as composed of two monolayer sheets which are independent of each other. Each monolayer is represented by a triangular lattice. The model is therefore a pseudo–two–dimensional lattice model which neglects the translational modes of the lipid molecules and focusses on the conformational degrees of freedom of the acyl chains. Since we will be extending the formalism to include several molecular species, we label the lipid variables corresponding to a particular lipid species, A. Each acyl chain can take on one of ten conformational states m, each of which is characterized by an internal energy E_m^{A}, a hydrocarbon chain length d_m^{A} (corresponding to half a bilayer, $2d_m^{\mathrm{A}} \sim d_{\mathrm{L}}$), and a degeneracy D_m^{A}, which accounts for the number of conformations that have the same area A_m^{A} and the same energy E_m^{A}, where $m = 1, 2, \ldots, 10$. The ten states can be derived from the all–*trans* state in terms of *trans*–*gauche* isomerism. The state $m = 1$ is the non–degenerate gel–like ground state, representing the all–*trans* conformation, while the state $m = 10$ is a highly degenerate excited state characteristic of the melted or fluid phase. The eight intermediate states are gel–like states containing kink and jogs excitations satisfying the requirement of low conformational energy and optimal packing. The conformational energies E_m^{A} are obtained from the energy required for a gauche rotation (0.45×10^{-13} erg) relative to the all–*trans* conformation. The values of D_m^{A} are determined by combinatorial considerations (Pink et al. 1980). The chain cross-sectional areas, A_m^{A}, are trivially related to the values of d_m^{A} since the volume of an acyl chain varies only slightly under temperature changes. The saturated hydrocarbon chains are coupled by nearest-neighbor anisotropic forces which represent both van der Waals and steric interactions. These interactions are formulated in terms of products of shape–dependent nematic factors. The lattice approximation automatically accounts for the excluded volume effects and to some extent for that part of the interaction with the aqueous medium which allows for bilayer integrity. An effective lateral pressure, Π, is included in the model to assure bilayer stability.

The Hamiltonian for the one-component lipid bilayer can then be written

$$\mathcal{H}^{\mathrm{A}} = \sum_i \sum_{m=1}^{10} (E_m^{\mathrm{A}} + \Pi A_m^{\mathrm{A}})\, \mathcal{L}_{mi}^{\mathrm{A}} - \frac{J_{\mathrm{A}}}{2} \sum_{<i,j>} \sum_{m,n=1}^{10} I_m^{\mathrm{A}} I_n^{\mathrm{A}} \mathcal{L}_{mi}^{\mathrm{A}} \mathcal{L}_{nj}^{\mathrm{A}}, \qquad (1)$$

where J_{A} is the strength of the van der Waals interaction between neighboring chains, and $I_m^{\mathrm{A}} I_n^{\mathrm{A}}$ is an interaction matrix which involves both distance and shape dependence (Pink et al. 1980). $\langle i, j \rangle$ denotes nearest-neighbor indices on the triangular lattice. The Hamiltonian is expressed in terms of site occupation variables $\mathcal{L}_{mi}^{\mathrm{A}} : \mathcal{L}_{mi}^{\mathrm{A}} = 1$ if the chain on site i is in state m, otherwise $\mathcal{L}_{mi}^{\mathrm{A}} = 0$. The model parameters J_{A} and Π are chosen so as to reproduce the transition temperature and transition enthalpy for a one-component $\mathrm{DC_{16}PC}$[1] bilayer (Mouritsen 1990a). The values for other phospholipids are then determined by simple scaling (Jørgensen et al. 1993, Risbo et al. 1994).

[1] Abbreviations used: $\mathrm{DC}_n\mathrm{PC}$, saturated di-acyl phosphatidylcholine with n carbon atoms in each acyl chain

2.2 Model of Binary Lipid Mixtures

The ten-state Pink model for a one-component lipid bilayer has been extended to binary lipid mixtures (Sperotto and Mouritsen 1993; Jørgensen et al. 1993, Risbo et al. 1994) by explicitly incorporating a mismatch term which accounts for the incompatibility of acyl chains of different hydrophobic lengths of the two species. The Hamiltonian for a binary mixture of the two lipid species A and B is written

$$\mathcal{H} = \mathcal{H}^{\mathrm{A}} + \mathcal{H}^{\mathrm{B}} + \mathcal{H}^{\mathrm{AB}}, \tag{2}$$

where the two first terms describe the interaction between like species and the last term the interaction between different species. The composition of the mixture is given by $x_{\mathrm{B}} = \sum_m \langle \mathcal{L}^{\mathrm{B}}_{im} \rangle = 1 - x_{\mathrm{A}}$. The interaction between different lipid species is described by the Hamiltonian

$$\begin{aligned}\mathcal{H}^{\mathrm{AB}} =& \frac{-J_{\mathrm{AB}}}{2} \sum_{\langle i,j \rangle} \sum_{m,n=1}^{10} (I^{\mathrm{A}}_m I^{\mathrm{B}}_n \mathcal{L}^{\mathrm{A}}_{im} \mathcal{L}^{\mathrm{B}}_{jn} + I^{\mathrm{B}}_m I^{\mathrm{A}}_n \mathcal{L}^{B}_{im} \mathcal{L}^{\mathrm{A}}_{jn}) \\ &+ \frac{\Gamma_{\mathrm{AB}}}{2} \sum_{\langle i,j \rangle} \sum_{m,n=1}^{10} (|d^{\mathrm{A}}_{im} - d^{\mathrm{B}}_{jn}| \mathcal{L}^{\mathrm{A}}_{im} \mathcal{L}^{\mathrm{B}}_{jn} + |d^{\mathrm{B}}_{im} - d^{\mathrm{A}}_{jn}| \mathcal{L}^{\mathrm{B}}_{im} \mathcal{L}^{\mathrm{A}}_{jn}).\end{aligned} \tag{3}$$

The first term in $\mathcal{H}^{\mathrm{AB}}$ describes the direct van der Waals hydrophobic contact interaction between different acyl chains. The corresponding interaction constant is taken to be the geometric average $J_{\mathrm{AB}} = \sqrt{J_{\mathrm{A}} J_{\mathrm{B}}}$. Γ_{AB} in the second term of Eq. (3) represents the mismatch interaction and has been shown to be 'universal' in the sense that its value does not depend on the lipid species for binary mixtures characterized by a non-ideal phase behavior. The value of the mismatch parameter used in the simulation approach to the statistical mechanics of the model in Eq. (3) was found to be $\Gamma_{\mathrm{AB}} = 0.038$ erg/Å (Jørgensen et al. 1993).

A special case of a binary lipid mixture is that of lipid-cholesterol mixtures. The model in Eq. (3) can readily be transcribed to account for cholesterol by considering component B as a stiff molecule with no internal flexibility (Cruzeiro-Hansson et al. 1989, Zuckermann et al. 1993). Cholesterol substitutes on lipid sites of the triangular lattice and the Hamiltonian is formally the same as in Eq. (3) except that the nematic factor, I, for cholesterol is a constant. The cross-sectional area of cholesterol is taken to be 32Å^2.

2.3 Model of Lipid-Protein Bilayers: the Mattress Model

Several extensions of the ten-state Pink model have been proposed in order to account for the lipid-protein interactions in a specific fashion which depends on the conformational states of the lipid chains (for references, see Mouritsen and Sperotto 1993). Here we consider a particular model in which the lipid–protein interactions have been incorporated into the microscopic Pink model by identifying part of the interaction parameters in terms of hydrophobic matching between the hydrophobic length of the lipid chain and hydrophobic protein length. This is implemented in the spirit of the phenomenological mattress model of lipid–protein interactions (Mouritsen and Bloom 1984, 1993, Fattal and Ben-Shaul 1993). The lipid–protein interactions were included in the model by assuming that the hydrophobic membrane–spanning part of the protein or polypeptide molecule is a stiff, rod–like, and hydrophobically smooth object with no appreciable internal flexibility. In this way the protein is characterized geometrically only by a cross sectional area, A_{P} (or circumference ρ_{P}), and a hydrophobic length, d_{P}. The protein can occupy one or more sites of the lipid lattice depending on its actual hydrophobic circumference.

The Hamiltonian of the microscopic version of the mattress model for a lipid bilayer of species A interacting with a protein P is now written (Sperotto and Mouritsen 1991a, 1991b)

$$\begin{aligned}
\mathcal{H}^{\mathrm{AP}} =& \Pi A_{\mathrm{P}} \sum_i \mathcal{L}_{\mathrm{P}i} \\
&+ \frac{\Gamma_{\mathrm{AP}}}{4}\left(\frac{\rho_{\mathrm{P}}}{z}\right) \sum_{<i,j>} \sum_{m=1}^{10} \left(|2d_{im}^{\mathrm{A}} - d_{\mathrm{P}}|\mathcal{L}_{im}^{\mathrm{A}}\mathcal{L}_{\mathrm{P}j} + \right. \\
&\left. |2d_{jm}^{\mathrm{A}} - d_{\mathrm{P}}|\mathcal{L}_{jm}^{\mathrm{A}}\mathcal{L}_{\mathrm{P}i}\right) \\
&- \frac{J_{\mathrm{AP}}}{4}\left(\frac{\rho_{\mathrm{P}}}{z}\right) \sum_{<i,j>} \sum_{m=1}^{10} \left(\min(2d_{im}^{\mathrm{A}}, d_{\mathrm{P}})\mathcal{L}_{im}^{\mathrm{A}}\mathcal{L}_{\mathrm{P}j} + \right. \\
&\left. \min(2d_{jm}^{\mathrm{A}}, d_{\mathrm{P}})\mathcal{L}_{jm}^{\mathrm{A}}\mathcal{L}_{\mathrm{P}i}\right) .
\end{aligned} \tag{4}$$

The parameter J_{AP} is related to the direct lipid-protein van der Waals–like interaction which is associated with the interfacial hydrophobic contact of the two molecules, while the parameter Γ_{AP} is related to the hydrophobic effect. $\mathcal{L}_{\mathrm{P}i} = 0, 1$ is the protein occupation variable. $\mathcal{L}_{im}^{\mathrm{A}}$ and $\mathcal{L}_{\mathrm{P}i}$ satisfy a completeness relation at each lattice site, $\sum_m \mathcal{L}_{im}^{\mathrm{A}} + \mathcal{L}_{\mathrm{P}i} = 1$. The model in Eq. (4) can readily be extended to binary lipid mixtures using Eqs. (2) and (3). The appropriate values of the lipid–protein interaction parameters, Γ_{AP} and J_{AP}, are discussed in Mouritsen et al. (1994).

2.4 Model of Lipid Bilayers Containing Drugs

The Hamiltonian for the microscopic interaction model used to describe the effects of drugs, such as anaesthetics, as well as insecticides, on lipid bilayer properties is described in the work by Jørgensen at al. (1991,1993). The model consists of three terms describing the pure lipid bilayer interactions, the lipid–drug molecule interactions, and the interactions between neighboring drug molecules, respectively. The first term is given by Eq. (1). The second term in the model Hamiltonian is based on an extension of the hypothesis proposed by de Verteuil et al. (1981). This extended hypothesis states that those foreign molecules, which do not change the transition enthalpy of the bilayer, intercalate between either the flexible acyl chains of the membrane or the polar heads of the lipid molecules and that their contribution to the exluded volume effect can be ignored. The adsorbed foreign molecules can therefore be considered as interstitial impurities whose available locations are the centers of the triangles formed by three neighboring lipid acyl chains. The sites available to the foreign molecules therefore form a honeycomb lattice embedded in the triangular lattice formed by the acyl chains. The requirement that the lipid chains and the foreign molecules occupy separate lattices and hence cannot mix, results in the absence of an entropy of mixing. It is assumed that the interstitial impurities interact with the three neighboring lipid acyl chains as well as with other impurities occupying neighboring interstitial sites on the honeycomb lattice. The third term of the model Hamiltonian describes direct interactions between neighboring impurities on occupied interstitial sites.

The total Hamiltonian for a drug molecules (D) in a lipid bilayer (of type A) is written as follows

$$\mathcal{H} = \mathcal{H}^{\mathrm{A}} - \sum_{i}\sum_{m}\sum_{\ell} J_{\mathrm{AD}}^{m} I_{m}^{\mathrm{A}} \mathcal{L}_{im}^{\mathrm{A}} \mathcal{L}_{\ell}^{\mathrm{D}} - \frac{J_{\mathrm{DD}}}{2} \sum_{\ell,k} \mathcal{L}_{\ell}^{\mathrm{D}} \mathcal{L}_{k}^{\mathrm{D}} . \tag{5}$$

$\mathcal{L}_{\ell}^{\mathrm{D}} = 0, 1$ is an interstitial site occupation variable. J_{AD}^{m}, and J_{DD} are interaction constants for the lipid-drug and the drug-drug interactions, respectively. The lipid-drug molecule interaction is taken to be specific, i.e. J_{AD}^{m} depends explicitly on the lipid-acyl-chain conformational state m (Jørgensen et al. 1991).

The concentration, x, of drug molecules in the membrane is not conserved since partitioning between the membrane and the aqueous phase is assumed to occur. The use of a grand canonical ensemble is therefore appropriate and x is controlled by a chemical potential, μ. The effective Hamiltonian then becomes

$$\mathcal{H}_{\mathrm{grand}} = \mathcal{H} - \mu \sum_{\ell} \mathcal{L}_{\ell}^{\mathrm{D}} . \tag{6}$$

The model for lipid-drug interactions can readily be extended to account for a ternary mixture of a drug interacting with a binary lipid mixture.

2.5 Monte Carlo Computer-Simulation Techniques

A full description of the implementation of standard Monte Carlo simulations will not be given, but we refer the reader to general references on this subject (Mouritsen 1984, 1990). We shall, however, mention several new techniques which are extremely powerful for the characterization of phase equilibria in lipid bilayers and lipid-protein systems when combined with Monte Carlo simulations.

The fundamental problem in numerical investigations of cooperative phenomena is associated with an unambiguous assessment of the nature of phase transitions as well as a determination of the precise location of the phase boundaries in a phase diagram. Numerical simulations share this problem with laboratory experiments. The root of the problem is that most approaches do not function on the level of the free energy but rather in terms of derivatives of the free energy, such as densities and order parameters (first derivatives) or response functions (second derivatives). In the case of one-component systems these derivatives are seldom known with sufficient accuracy to discern between, for example, a fluctuation-dominated first-order phase transition and a continuous transition (critical point). In the case of phase coexistence in, e.g. a two–component system, there is no direct way of relating features in the response functions to the precise location of equilibrium phase boundaries (Risbo et al. 1994). However, by use of novel techniques (Zhang et al. 1993a) it is possible numerically to circumvent this problem and obtain access to that part of the free energy which is necessary to locate the phase equilibria. These techniques involve calculation of distribution functions (histograms) (Ferrenberg and Swendsen 1988) of thermodynamic functions, e.g. order parameters (composition) or internal energy, thermodynamic reweighting of the distribution functions in order to locate the phase transition or the phase equilibria, and then a subsequent analysis of the size dependence of the reweighted distribution functions by means of finite-size scaling theory (Lee and Kosterlitz 1991).

3 Density Fluctuations and Dynamic Heterogeneity

Monte Carlo computer-simulation studies of the ten-state Pink model of the main bilayer transition, Eq. (1) indicate that the main transition is close to a critical point and associated with strong lateral density fluctuations of a thermal nature (Ipsen et al. 1990). In Fig. 1 are shown snapshots of microconfigurations typical of the main transition region obtained from a simulation on Eq. (1) for $DC_{14}PC$, $DC_{16}PC$, and $DC_{18}PC$ bilayers in their respective transition region. The thermal density fluctuations are manifested as a kind of cluster-formation process leading to dynamic bilayer heterogeneity on the nanoscale. The heterogeneity, which becomes more pronounced the shorter the acyl chains are, implies the existence of a network of interfaces within the bilayer. The interfaces formed in the transition region have very special packing properties, they are soft and characterized by excited acyl-chain conformations, and they act as sinks for other bilayer-bound components. The interfacial tension of these interfaces is very low at the transition since it is close to a critical point. The interfaces and the lipid domains are dynamic entities

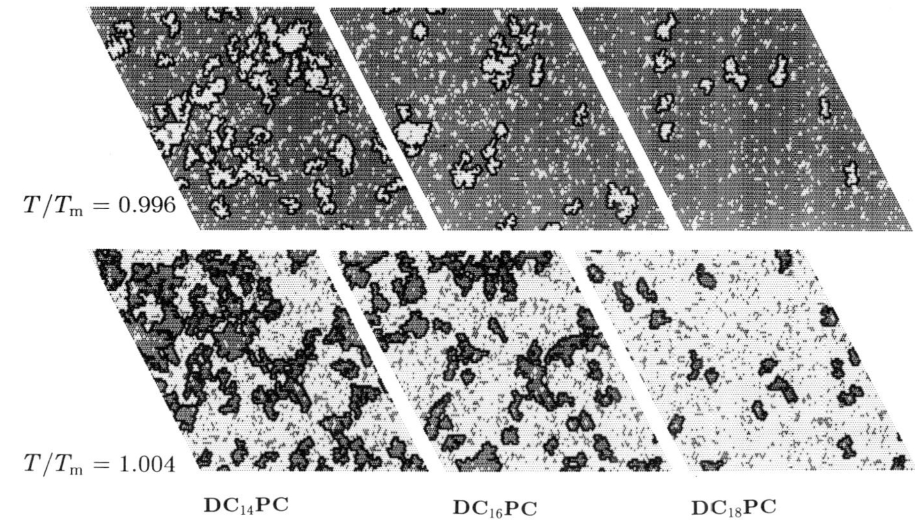

Fig. 1. Lipid-domain formation and dynamic heterogeneity in the transition region of $DC_{14}PC$, $DC_{16}PC$, and $DC_{18}PC$ bilayers. The transition temperature of the lipid in question is is T_m. The pictures illustrate configurations obtained from computer-simulation calculations on a system with 5000 lipid molecules using the model in Eq. (1). Gel and fluid regions are denoted by grey and light regions, and the interfaces of the lipid domains are highlighted in black.

which fluctuate in time. However, they are described by average equilibrium quantities, e.g. a lipid-domain size-distribution function, $\langle \ell \rangle$, which is shown in Fig. 2 for five different lipids. The fluctuating nature of the bilayer hence entails a non-uniform lateral organization of the bilayer components which may be termed *dynamic membrane heterogeneity* (Mouritsen and Jørgensen 1992, Mouritsen and Biltonen 1993). Macroscopically, the fluctuations are signalled by strong peaks in the specific heat and the lateral compressibility. These observations are in close agreement with experimental observations obtained using a variety of techniques (Biltonen 1990).

4 Compositional Fluctuations and Dynamic Heterogeneity

In the case of binary lipid bilayers, a great variety of phase diagrams can result, depending on the difference between the mixed species, i.e. the degree of non-ideality of the mixture. The two-component systems give rise to different types of heterogeneous bilayer states. Apart from the obvious phase-separated states, compositional fluctuations arise in the mixed fluid state which turns out to have a substantial measure of local structure and compositional fluc-

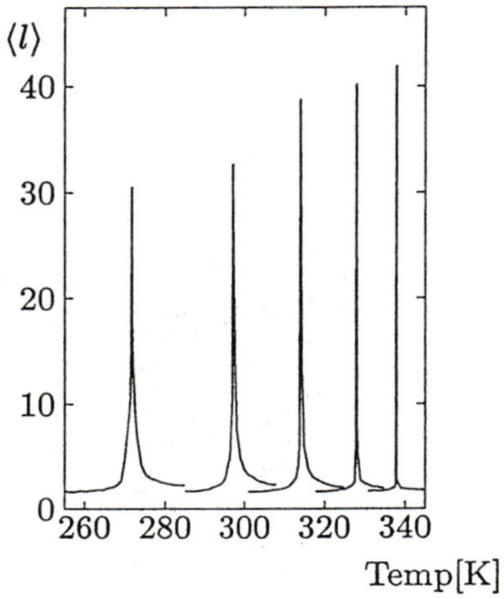

Fig. 2. Average lipid domain size, $\langle \ell \rangle$ (in units of number of lipid molecules) for $DC_{12}PC$, $DC_{14}PC$, $DC_{16}PC$, $DC_{18}PC$, and $DC_{20}PC$ (from left to right) lipid bilayers in the transition region. Data as obtained from computer simulation calculations on the model in Eq. (1) are shown as a function of temperature.

tuations (Jørgensen et al. 1993, Mouritsen and Jørgensen 1994a,b). As an example, typical microconfigurations of three different equimolar lipid mixtures in the fluid phase as described by the model in Eqs. (2) and (3) are shown in Fig. 3. The figure shows that under isothermal conditions the lateral dynamic heterogeneity is most pronounced for the most non-ideal mixtures, i.e. those mixtures of lipids which differ most in acyl-chain length.

A more quantitative description of the strongly fluctuating fluid mixture is given in Fig. 4 which shows the fluid-state pair correlation function for the long-chain lipid in the $DC_{12}PC$-$DC_{18}PC$ mixture.

$$g_{\mathrm{ff}}(\mathrm{DC_{18}PC})(R) = \langle \mathcal{L}_{i10}^{\mathrm{DC_{18}PC}}(R_i)\mathcal{L}_{j10}^{\mathrm{DC_{18}PC}}(R_j)\rangle - \langle \mathcal{L}_{i10}^{\mathrm{DC_{18}PC}}\rangle^2. \tag{7}$$

The second term in Eq. (7) corrects for the background correlation, i.e. the global concentration. Hence the correlation function is a measure of the correlation of both thermal and compositional fluctuations, thereby providing a complete description of short-range order effects and local structure. There are several features of this function that are interesting. Firstly it is seen that the correlation is fairly long-ranged and that the local structure at any given time is considerably different from that of the average global structure given by the composition. Secondly, there is a broad peak around equimolar concentration reflecting that the compositional fluctuations are larger in this range. Finally, there is another peak at the axis of the high-melting lipid ($DC_{18}PC$). This peak reflects the density fluctuations near the transition temperature of $DC_{18}PC$. A similar peak at the other axis is not visible because the transition of $DC_{12}PC$ is too far below the current temperature to have any effect on the correlation function. The coherence length of the compositional fluctuations and the local structure has a strong temperature dependence. As the gel-fluid coexistence region is approached, the correlation becomes more long-ranged.

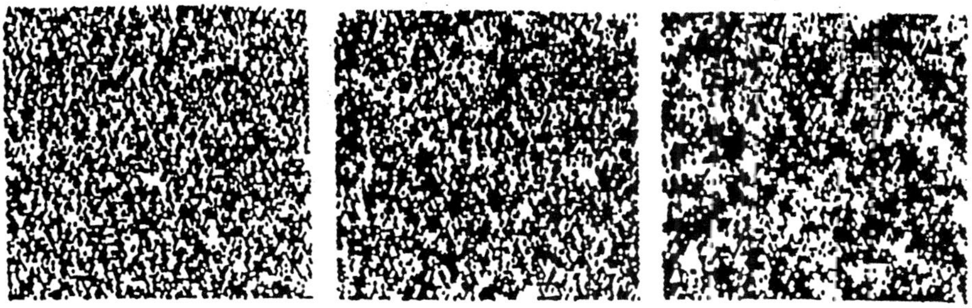

$DC_{14}PC$ - $DC_{18}PC$ (fluid) $DC_{12}PC$ - $DC_{18}PC$ (fluid) $DC_{12}PC$ - $DC_{20}PC$ (fluid)

Fig. 3. Local structure and pattern formation in three different fluid lipid mixtures, $DC_{14}PC$-$DC_{18}PC$, $DC_{12}PC$-$DC_{18}PC$, and $DC_{12}PC$-$DC_{20}PC$, at the same temperature, $T = 338$K, and the same composition $x = 0.5$. The two lipid species in each snapshot are indicated by white and grey tones. The configurations are derived from computer-simulation calculations on systems with 5000 lipid molecules using the model in Eqs. (2) and (3).

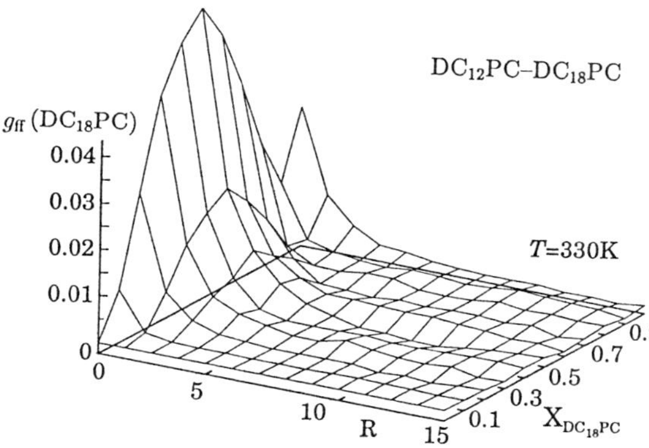

Fig. 4. Pair correlation function, $g_{ff}(DC_{18}PC)$ in Eq. (7), for fluid lipid states of $DC_{18}PC$ in the $DC_{12}PC$-$DC_{18}PC$ mixture at $T = 330$K. Data are given as a function of composition, $x_{DC_{18}PC}$, and acyl-chain separation, R (in units of lattice spacings). The data have been derived from computer-simulation calculations on the model in Eqs. (2) and (3).

The compositional fluctuations are clearly revealed and the bilayer mixture is best described as a highly correlated fluid.

5 Spinodal Decomposition and Phase Separation Dynamics

Mixtures of lipids can undergo dynamic phase-separation phenomena resulting in coexisting phases due to the non-ideal mixing properties of the lipids, as described above. The coexistence of different thermodynamic phases in the membrane plane can be induced by e.g. a change in composition or temperature as described by the phase diagram. The two-dimensional heterogeneous membrane structure within the phase-separated regions induced by either a change in composition or temperature has been investigated by several experimental techniques (for a list of references, se Mouritsen and Jørgensen 1994a). The outcome from such experiments has provided new insight into the lateral membrane organization within the phase-separated regions, and emphasized the importance of understanding the properties that control two-dimensional membrane organization, e.g. in relation to enzymatic reactions that take place in restricted (fractal) geometries (Melo et al. 1992).

Results from computer-simulation studies on the model in Eqs. (2) and (3) of the ordering process of an equimolar binary mixture of $DC_{12}PC$-$DC_{18}PC$ quenched from a temperature in the thermodynamic one-phase fluid region to a temperature within the gel-fluid coexistence region are given in Fig. 5. The mixture is first equilibrated in the fluid phase and an instantaneous thermal quench is then performed. The relaxation of the energy and the order parameters is not exponential but rather algebraic, as expected from general arguments for non-equilibrium interface-driven ordering processes in condensed matter (Mouritsen 1990b). Figure 5 show characteristic time-dependent microconfigurations of the system together with the initial ($t \leq 0$) fluid and final ($t = \infty$) phase-separated gel-fluid equilibrium configurations. The series of snapshots show how order evolves via formation of a time-dependent percolative structure characterized by a single length scale that increases in time. It can be shown that the time dependence of this length scale is described by an algebraic law (K. Jørgensen, unpublished work). It is clear that the dynamic ordering process has a strong influence on the nano- and mesocopic heterogeneous lateral membrane structure as well as on the microscopic local order of the acyl chains. This is the process known as spinodal decomposition at critical concentrations (equimolar mixture).

The snapshots in Fig. 5 also reveal that the percolative structure of the membrane sustains structure on a smaller length scale within the phase-separated regions similar to that found in equilibrium. In particular, the low-melting lipid ($DC_{12}PC$) in its conformationally ordered acyl-chain state predominantly accumulates and wets the gel-fluid interfaces. In the early stage of the phase-separation process, the low-melting lipid covers nearly all of the gel-fluid interfaces, a phenomena which becomes less pronounced as the fluid phase separates out.

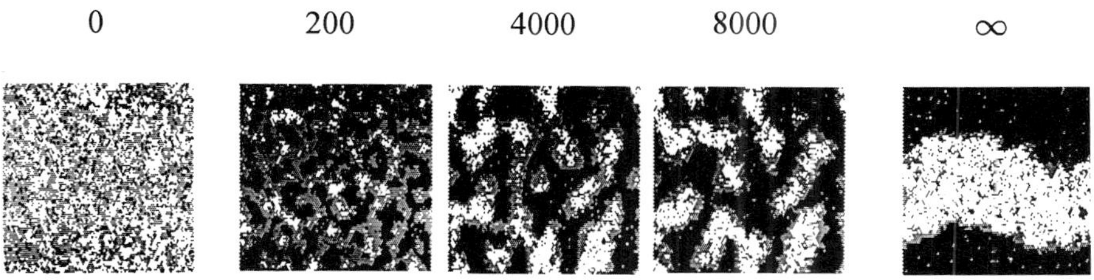

Fig. 5. Non-equilibrium dynamics of the phase-separation process in a $DC_{12}PC$-$DC_{18}PC$ mixture at equimolar composition that has been thermally quenched from a temperature, $T = 330K$, in the fluid phase to a temperature, $T = 278K$, within the gel-fluid coexistence region. The phase-separation process is monitored in time (in units of Monte Carlo step per acyl chain). The mixture is first equilibrated in the fluid phase and the instantaneous quench is performed at time $t = 0$. The results are obtained from a computer-simulation calculation on a system with 5000 lipid molecules using the model in Eqs. (2) and (3). Microconfigurations typical of an increasing series of times are shown along with the initial ($t \leq 0$) fluid and final ($t = \infty$) phase-separated gel-fluid equilibrium configurations.

6 Effects of Cholesterol on Membrane Heterogeneity

The lipid-cholesterol mixture is not only the most important binary lipid system but it is probably also the system which has proved to be one of the most difficult to study in relation to phase equilibria (Zuckermann et al. 1993). The phase diagram has been explained theoretically using a model which appreciates the subtleties of the coupling between the cholesterol molecule and the different degrees of freedom of the lipid acyl chains. We shall here only focus on the effect of cholesterol on the conformational degrees of freedom of the lipids. The description only applies at low cholesterol contents. A Monte Carlo computer-simulation study (Cruzeiro-Hansson et al. 1989) of the lipid-cholesterol system has been performed on a simplified version of the model of lipid-cholesterol interactions by focussing on the acyl-chain conformational variables alone.

Snapshots obtained from Monte Carlo calculations on the ten-state Pink model showing the influence of cholesterol are presented in Fig. 6 for low cholesterol concentrations, 0–7.5%. It is seen that the presence of cholesterol leads to a more heterogeneous membrane structure with larger and more ramified domains. A quantitative analysis of the domain distributions shows that cholesterol at low concentrations decreases the average domain size at the transition but increases it away from the transition. A closer inspection of the snapshots of microconfigurations in Fig. 6 indicates that the cholesterol molecules are not randomly distributed in the plane of the bilayer but accumulate at the lipid-domain interfaces. Cholesterol tends to increase the amount of interface at temperatures in the neighborhood of the transition but not at the transition point itself. This effect may be rationalized in phys-

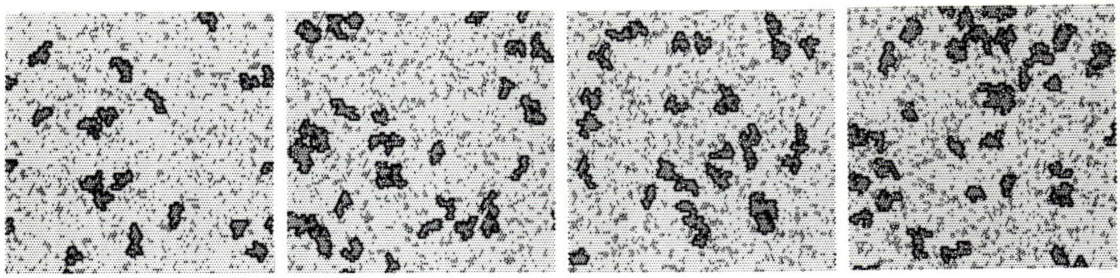

Fig. 6. Typical lipid-domain configurations in the fluid phase of $DC_{16}PC$-cholesterol mixtures at a temperature, $T = 319K$, in the fluid phase. Gel and fluid regions are denoted by grey and light regions, and the interfaces of the lipid domains are highlighted in black. The results are obtained from computer-simulation calculations on mixed systems with varying amounts of cholesterol. The cholesterol molecules are denoted by $\circ$.

ical terms by remarking that cholesterol causes a decrease in the interfacial tension between the lipid domains and the bulk. One of the macroscopic consequences of the cholesterol-induced enhanced density fluctuations near the main transition is a concomitant enhancement of the passive permeability which in fact is observed experimentally at low cholesterol concentrations (Corvera et al. 1992).

7 Effects of Drugs on Membrane Heterogeneity

The partitioning of amphiphilic foreign molecules into lipid bilayer membranes is of importance for a range of biological processes. The ability of these compounds to perturb living systems has resulted in their widespread use in medicine and biology. Well-known examples are anaesthetics, alcohols, halucinogens, and a range of drugs and pesticides. The underlying pharmacological mechanisms for the action of most of these compounds are still unknown. We shall here describe modelling of the way in which drugs affect the main transition characteristics and the dynamic bilayer heterogeneity.

We study the dynamic pattern formation and heterogeneity in mixtures of lipids and drugs via the model in Eqs. (5) and (6) (Jørgensen et al. 1991). This model is based on a characterization of drugs as interstitial impurities, in which case they intercalate between the lipid acyl chains. The most important feature of the model is its ability to distinguish between certain drugs whose concentration in the bilayer is fixed and others that partition between the bilayer and various external gaseous and/or aqueous media. Monte Carlo computer simulations have been carried out on this model in the case where the interactions between the lipid chains and the drugs are specific in the sense that the lipid-foreign molecule interactions depend explicitly on the excited gel-like conformational states of the acyl chains. The principal result

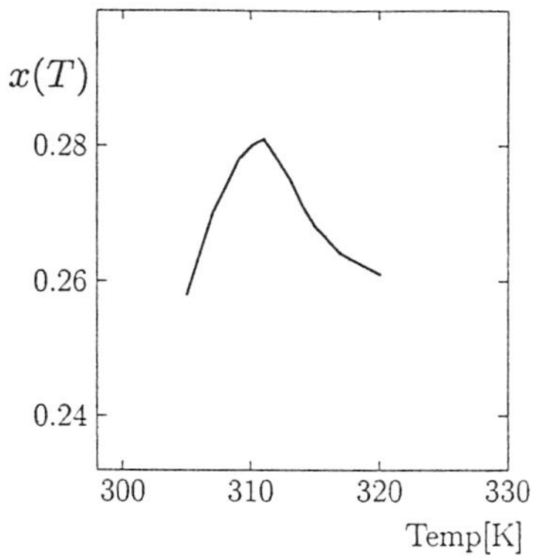

Fig. 7. Partition coefficient, $x(T)$, for a water-soluble anaesthetic like halothane in a $DC_{16}PC$ bilayer in the region around the transition temperature, $T_m = 314K$. The data are obtained from computer-simulation calculations on the model in Eqs. (5) and (6).

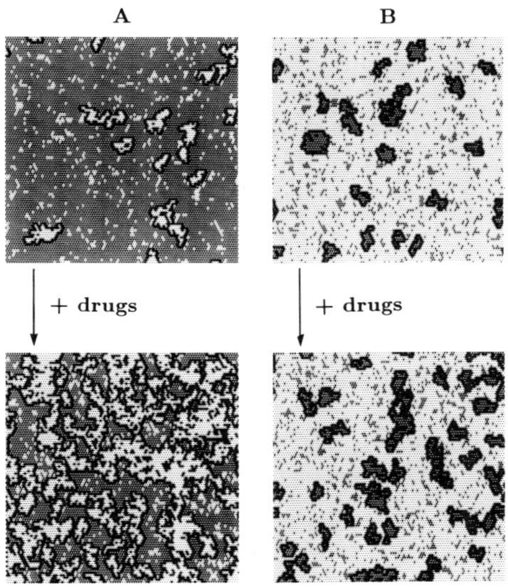

Fig. 8. Effects on lipid-bilayer dynamic heterogeneity due to a water-soluble drug, like halothane. Results are shown for a $DC_{16}PC$ bilayer at temperatures slightly below (A) and slightly above (B) the transition temperature of the pure lipid bilayer, $T_m = 314K$. The amount of drug dissolved in the bilayer varies significantly in the transition region. The configurations are obtained from computer-simulation calculations on a system with 5000 lipid molecules using the model in Eqs. (5) and (6). Gel and fluid regions are denoted by grey and light regions, and the interfaces of the lipid domains are highlighted in black.

of the simulation is the discovery that foreign molecules can have a dramatic effect on the fluctuations in the transition region. As shown in Fig. 7, the partition coefficient is found to exhibit a definite fluctuation-induced peak below the phase transition temperature. The specific heat peak is simultaneously broadened and shifted to lower temperatures. The transition enthalpy is nearly independent of the chemical potential.

It is necessary to examine the system at the microscopic level in order to understand the origin of the peak in the partition coefficient. Fig. 8 shows the effect on the dynamic membrane organization for a lipid bilayer model doped with drugs which place themselves interstitially in the lipid matrix and which interact in a specific manner with excited lipid acyl-chain conformations. The figure shows that the drug molecules lead to a dramatic enhancement of the membrane heterogeneity by inducing more interfacial regions. Moreover, it is found that the local concentration of the foreign molecules in the interfaces can easily be an order of magnitude larger than the global concentration, indicating a significant dynamic accumulation of the molecules in the interfaces. This observation may be of relevance in connection with the evaluation of 'clinical' concentrations of drugs. Hence, the drug molecules modulate the dynamic membrane heterogeneity, leading in turn to a lateral organization of the bilayer components which is very heterogeneous. There are several macroscopic consequences of these microscopic and mesoscopic events. Firstly, the doping of the bilayer leads to enhanced passive membrane permeability. Secondly, the partition coefficient, as mentioned, displays a peak in the gel-to-fluid transition region.

Several pharmacological compounds have been found experimentally to lead to enhancement of bilayer fluctuations and a shift of transition temperature in consistency with the behavior described above (Jørgensen 1993). Examples include volatile general aneasthetics like halothane, local anaesthetics like cocaine derivatives, and non-steroidal anti-inflammatory agents like indomethacin. Moreover, it is interesting to note that a completely different class of compounds interacting with bilayers, the organochlorine insecticides lindane, DDT, parathion, and malathion, have similar effects on the physical properties of lipid bilayers (Jørgensen et al. 1991).

The remarkable ability of drugs to modify global order in lipid bilayers is illustrated in Fig. 9 in the case of a binary mixture of $DC_{14}PC$ and $DC_{18}PC$ in the gel-fluid phase-separation region. The figure shows that the drug under isothermal conditions completely abolishes the phase separation and induces a fluid phase with a considerable degree of local order. The former macroscopic gel phase is emulsified and appears as gel domains in a very loosely connected network.

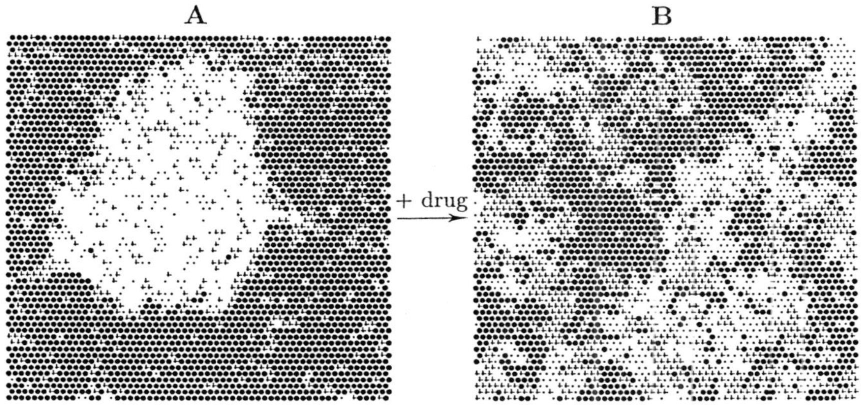

DC$_{14}$PC–DC$_{18}$PC (gel + fluid)

Fig. 9. Effect on lateral organization in the gel-fluid phase-coexistence region of DC$_{14}$PC-DC$_{18}$PC due to the presence of a water-soluble drug like halothane. The configurations are obtained from computer-simulation calculations on a system with 1800 lipid molecules using the model in Eqs. (5) and (6). Gel (predominantly DC$_{18}$PC chains) and fluid (predominantly DC$_{14}$PC chains) regions are denoted by dark and light regions, respectively.

8 Local Lipid Structure and Demixing Near Integral Proteins

The mattress model (Mouritsen and Sperotto 1993, Mouritsen and Bloom 1993) of lipid-protein interactions in membranes, cf. Eq. (4), represents an attempt to capture an essential part of lipid-protein or lipid-polypeptide interactions in membranes by focussing on the degree of hydrophobic mismatch between the protein and lipid-bilayer hydrophobic thicknesses. In addition to the mismatch energy terms, the mattress model accounts for the direct lipid-protein hydrophobic interactions as well as the elastic deformation energy of the bilayer. It has been found that the mismatch is a major determinant of the phase diagram of the lipid-protein system. However, whereas the hydrophobic mismatch controls the phase behavior, the aggregation state of the proteins within the individual phases and phase coexistence regions is predominantly determined by the direct lipid-protein hydrophobic interaction (Sperotto and Mouritsen 1991a,b). In Fig. 10 is shown an example of a heterogeneous bilayer configuration which is caused by a protein-induced fluid-gel phase separation in a lipid bilayer in the case of a rather small protein. Due to a better hydrophobic match, the proteins are accumulated in patches in the fluid phase.

A particular type of local lipid bilayer dynamic heterogeneity may be induced by integral membrane proteins in a fluid mixture of two lipid species

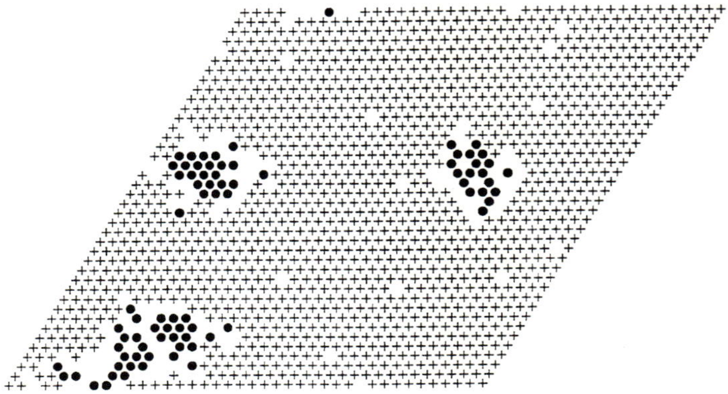

Fig. 10. Simulated microconfiguration of a $DC_{16}PC$ lipid bilayer incorporated with trans-membrane polypeptides (small proteins). The data are obtained from the model of lipid-protein intercations in Eq. (4). The polypeptides, which are most soluble in the fluid lipid phase due to a better hydrophobic matching, induce a gel-fluid phase separation. The proteins are indicated by dots, and gel and fluid lipid regions are denoted by grey and white areas, respectively.

which have different degrees of hydrophobic matching to the protein. This is illustrated in Fig. 11 which shows a distinct demixing of the binary mixture near the lipid-protein interface of a very large protein (a protein wall). This phenomenon is a special type of interfacial enrichment/depletion which supports a physical mechanism behind lipid specificy of certain proteins (Sperotto and Mouritsen 1993).

The simulations results of protein-induced lipid-demixing shown in Fig. 11 were obtained in the extreme limit of a very large protein and under the assumption that the protein is immobile. For smaller proteins which are laterally mobile, the overlap in compositional profiles from adjacent proteins results in an effective lipid-mediated protein-protein interaction which can give rise to an attractive force and hence protein aggregation. The mattress model for lipid-protein interactions, Eq. (4) has recently been used (Sperotto and Mouritsen 1994) in conjunction with the model for binary mixtures of phospholipids with different acyl-chain lengths, Eqs. (2) and (3), to study the thermodynamic properties as well as the local compositional structure in the case of $DC_{14}PC$-$DC_{18}PC$ mixtures incorporated with a protein of a size corresponding to that of bacteriorhodopsin. The model simulations have been inspired by recent experimental work on the effect of bacteriorhodopsin on lipid bilayers of different thickness (Piknova et al. 1994). The large proteins are mobile and each is assumed to be of hexagonal form occupying $n_P = 19$ sites of the lipid lattice.

Results from Monte Carlo simulations of this model of bacteriorhodopsin in equimolar $DC_{12}PC$-$DC_{18}PC$ mixture for a protein/lipid molar ratio of $P/L = 0.025$ (Sperotto and Mouritsen 1994) are shown Fig. 12. The lateral

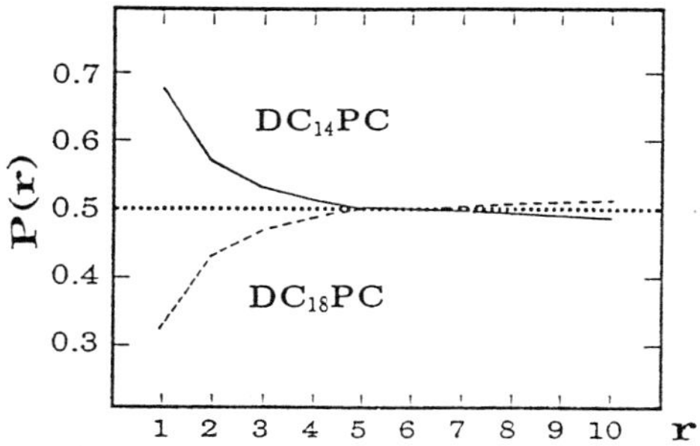

Fig. 11. Example of simulated protein-lipid interface enrichment and physical lipid specificity in a binary lipid mixture. Lipid concentration profiles, $P(r)$, for the two lipid species are shown as a function of distance, r, from a very large integral membrane protein. The data, which are obtained from simulations on the microscopic mattress model of lipid-protein interactions, Eq. (4), used together with the binary lipid mismatch model in Eqs. (2) and (3), refer to an equimolar binary mixture of $DC_{14}PC$ and $DC_{18}PC$ at a temperature well inside the fluid region. The protein hydrophobic thickness is close to twice the acyl–chain length in a fluid $DC_{14}PC$ lipid bilayer.

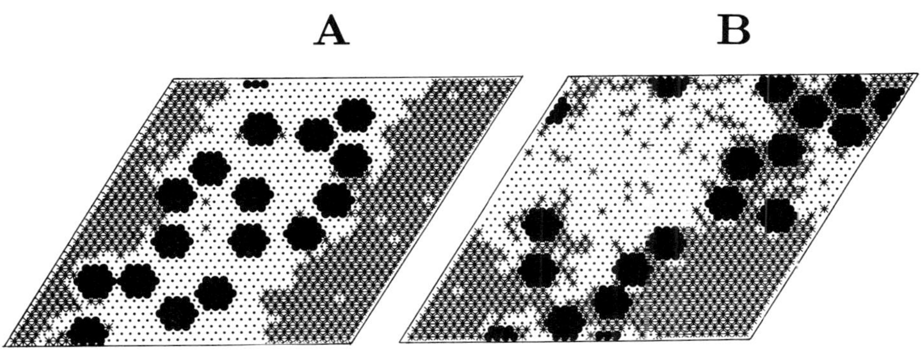

Fig. 12. (A): Typical microconfigurations of a $DC_{12}PC$-$DC_{18}PC$ mixture and with a dispersion of model bacteriorhodopsin in a molar protein-lipid ratio of 0.025 and at temperature $T = 270K$. The data are obtained from simulations on the microscopic mattress model of lipid-protein interactions, Eq. (4), used together with the binary lipid mismatch model in Eqs. (2) and (3). The proteins are shown as solid hexagons. Dark and white regions correspond to gel and fluid regions, respectively. (B): Typical microconfiguration at $T = 310K$.

configuration is shown corresponding to two cases, the gel-gel coexistence region (A) and the gel-fluid coexistence region (B). The higher solubility, caused by a better hydrophobic matching of bacteriorhodopsin in the gel phase to the lipid species with the shorter chain length, $DC_{12}PC$, is clearly seen in Fig. 12A. In the gel-fluid coexistence region, where bacteriorhodopsin again is expelled from the predominantly $DC_{18}PC$ gel phase due to poor matching, the protein in the fluid region prefers to be wetted by the well-matched fluid $DC_{18}PC$ lipids. The fluid $DC_{12}PC$ lipids, which are too short, are repelled by the protein.

The local lipid compositional structure around bacteriorhodopsin in the fluid phase is analyzed quantitively in Fig. 13. This figure clearly shows that there is a statistical dynamic lipid annulus for this mobile protein. The well-matched fluid $DC_{18}PC$ lipids are enriched in concentration near the protein, whereas there is a concomitant depletion of the poorly matched fluid $DC_{14}PC$ lipids. The figure also shows the protein-protein correlation function which has a highly non-trivial distance dependence due to the delicate demixing phenomenon in the lipid mixture near the protein surface. Hence, bacteriorhodopsin within this model has a distinct statistical preference for $DC_{18}PC$ lipids in the fluid phase. This type of physically controlled lipid selectivity of integral membrane proteins has yet to be investigated experimentally.

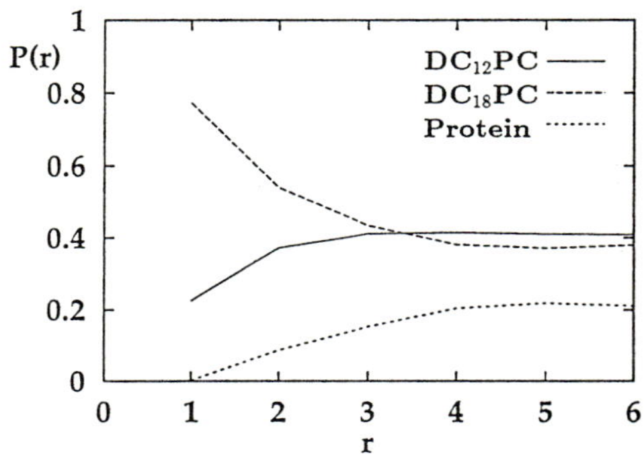

Fig. 13. Computer simulation data for the lipid composition profiles, $P(r)$, for $DC_{12}PC$ and $DC_{18}PC$ as a function of the distance, r, from the lipid–protein interface. The results refer to $T = 330K$ for an equimolar $DC_{12}PC$-$DC_{18}PC$ mixture. The system is in the fluid phase. The data are obtained from simulations on the mattress model of lipid-protein interactions, Eq. (4), used together with the binary lipid mismatch model in Eqs. (2) and (3). The protein-protein correlation function is also shown.

9 Dynamic Membrane Heterogeneity and Membrane Functions

From the point of view of membrane science, the finding from computer-simulation work of dynamically heterogeneous membrane states and pattern, cf. Figs. 1, 3, 5, 6, 8, 10, and 12, induced by strong lateral density fluctuations or compositional fluctuations is important since it leads to a proposal for a simple and general physical mechanism underlying diverse phenomena as transmembrane permeation, exchange of molecules between different membranes, and the activity of certain interfacially active enzymes. The basic idea behind this general mechanism is that the lipid-domain formation implies a concomitant formation of a fluctuating interfacial environment bounding the lipid domains and the bulk matrix. As an example of the predictive power of the simulation of the microscopic models dealt with in the present chapter, Fig. 14A shows the results of a simulation of the transmembrane permeability of Na^+ ions which is built on the assumption that the ions predominantly leak through the lipid domain interfaces (Cruzeiro-Hansson et al. 1989). A similarly striking agreement between the simulated bilayer heterogeneity and an active functional membrane property can be obtained in the case of phospholipase A_2 activity, as illustrated in Fig. 14B (Mouritsen and Jørgensen 1992).

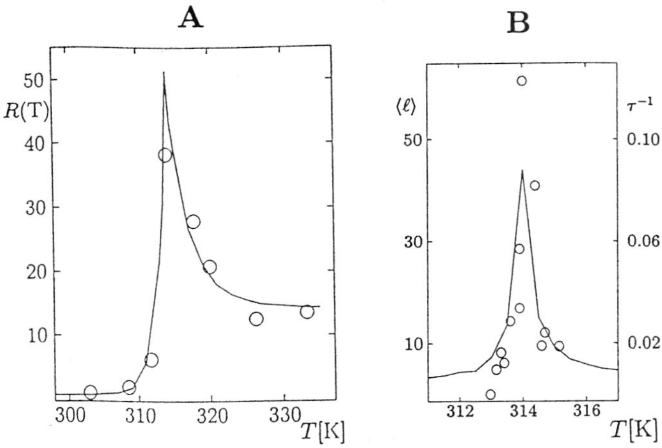

Fig. 14. Correlations between dynamic membrane heterogeneity and membrane function. **(A):** *Dynamic membrane heterogeneity enhances the passive transmembrane permeability:* Relative permeability, $R(T)$, of Na^+-ions in $DC_{16}PC$ liposomes as measured by radioactive ^{22}Na-techniques (○) (Papahadjopoulos et al. 1973) and as calculated from a computer simulation (——) of dynamic heterogeneity of a $DC_{16}PC$ bilayer membrane using the model in Eq. (1) together with a simple assumption for the transmission coefficients (Cruzeiro-Hansson et al. 1989). **(B):** *Dynamic membrane heterogeneity enhances active membrane functions:* Rate of activation (○) ($\sim$ inverse time, τ^{-1} [min^{-1}]), for hydrolysis of large unilamellar $DC_{18}PC$ vesicles by porcine pancreatic phospholipase A_2 in the neighborhood of the gel-to-fluid phase transition of the vesicles at 314K (Biltonen 1990). τ is the time required to reach half-maximum activity of the enzymatic process. Also shown are the results of a model simulation, cf. Eq. (1), of the average lipid-domain size (——), $\langle \ell \rangle$, in a $DC_{16}PC$ lipid bilayer. $\langle \ell \rangle$ is in units of number of lipid molecules.

Acknowledgments

This work was supported by the Danish Natural Science Research Council under grant 11-0065-1, by Jenny Vissings Fond, and by the Carlsberg Foundation.

References

Biltonen, R. L. 1990. A statistical-thermodynamic view of cooperative structural changes in phospholipid bilayer membranes: their potential role in biological function. J. Chem. Thermodyn. **22**, 1-19.

Bloom, M., Evans, E. and Mouritsen, O. G. 1991. Physical properties of the fluid-bilayer component of cell membranes: a perspective. Q. Rev. Biophys. **24**, 293-397.

Corvera, E., Mouritsen, O. G., Singer, M. A., and Zuckermann, M. J. 1992. The permeability and the effect of acyl chain length for phospholipid bilayers containing cholesterol: theory and experiment. Biochim. Biophys. Acta **1107**, 261-270.

Cruzeiro-Hansson, L., Ipsen, J. H., and Mouritsen, O. G. 1989. Intrinsic molecules in lipid membranes change the lipid-domain interfacial area: cholesterol at domain boundaries. Biochim. Biophys. Acta **979**, 166-176.

Fattal, D. R. and Ben-Shaul, A. 1993 A molecular model for the lipid-protein interaction in membranes: the role of hydrophobic mismatch. Biophys. J. **65**, 1795-1809.

Ferrenberg, A. M. and Swendsen, R. H. 1988. New Monte Carlo technique of studying phase transitions. Phys. Rev. Lett. **61**, 2635-2638.

Gennis, R. B. 1989. Biomembranes. Molecular Structure and Function. Springer-Verlag, London.

Ipsen, J. H., Jørgensen, K. and Mouritsen, O. G. 1990. Density fluctuations in saturated phospholipid bilayer increase as the acyl-chain length decreases. Biophys. J. **58**, 1099-1107.

Jacobsen, K. and Vaz, W. L. C. 1992. Domains in Biological Membranes. Comm. Mol. Cell. Biophys. **8**, 1-114.

Jørgensen, K., Ipsen, J. H., Mouritsen, O. G., Bennett, D. and Zuckermann, M. J. 1991. The effects of density fluctuations on the partitioning of foreign molecules into lipid bilayers: application to anesthetics and insecticides. Biochim. Biophys. Acta **1067**, 241-253.

Jørgensen, K., Ipsen, J. H., Mouritsen, O. G., and M. J. Zuckermann. 1992. The effect of anesthetics on the dynamic heterogeneity of lipid membranes. Chem. Phys. Lipids **65**, 205–216.

Jørgensen, K., Sperotto, M. M., Mouritsen, O. G., Ipsen, J. H., and Zuckermann, M. J. 1993. Phase equilibria and local structure in binary lipid bilayers. Biochim. Biophys. Acta. **1152**, 135–145.

Lee, J. and Kosterlitz, J. M. 1991. Finite–size scaling and Monte Carlo simulations of first order phase transitions. Phys. Rev. B **43**, 3265-3277.

Melo, E. C. C., Lourtie, I. M., Sankaram, M. B., Thompson, T. E. and Vaz, W. L. C. 1992. Effects of domain connection and disconnection on the yields of in-plane bimolecular reactions in membranes. Biophys. J. **63**, 1506–1512.

Mouritsen, O. G. 1990a. Computer simulation of cooperative phenomena in lipid membranes. *In* Molecular Description of Biological Membrane Components by Computer Aided Conformational Analysis, Vol. 1 (ed. R. Brasseur), pp. 3-83. CRC Press, Boca Raton, Florida.

Mouritsen, O. G. 1990b. Pattern formation in condensed matter. Mod. Phys. B **4**, 1925–1954.

Mouritsen, O. G. 1991. Theoretical models of of phospholipid phase transitions. Chem. Phys. Lipids **57**, 178-194.

Mouritsen, O. G. and Biltonen, R. L. 1993. Protein-lipid interactions and membrane heterogeneity. *In* New Comprehensive Biochemsitry. Protein-Lipid Interactions (ed. A. Watts), pp.1-39. Elsevier Scientific Press, Amsterdam.

Mouritsen, O. G. and Bloom, M. 1993. Models of lipid-protein interactions in membranes. Annu. Rev. Biophys. Biomol. Struct. **22**, 147-171.

Mouritsen, O. G. and Jørgensen, K. 1992. Dynamic lipid-bilayer heterogeneity: a mesoscopic vehicle for membrane function? BioEssays **14**, 129-136.

Mouritsen, O. G. and Jørgensen, K. 1994a. Dynamical order and disorder in lipid bilayers. Chem. Phys. Lipids (in press).

Mouritsen, O. G. and Jørgensen, K. 1994b. Micro-, nano-, and meso-scale heterogeneity of lipid bilayers and its influence on macroscopic membrane properties. Mol. Memb. Biol. (in press).

Mouritsen, O. G. and Sperotto, M. M. 1993. Thermodynamics of lipid-protein interactions in lipid membranes. *In* Thermodynamics of Membrane Receptors and Channels. (ed. M. B. Jackson). pp.127-181. CRC Press, Boca Raton, Florida.

Mouritsen, O. G., Sperotto, M. M., Risbo, J., Zhang, Z., and Zuckermann, M. J. 1994. Computational approach to lipid -protein interactions in membranes. In: H. O. Villar (ed.) Adv. Comp. Biol. JAI Press, Greenwich, Connecticut (in press).

Papahadjopoulos, D., Jacobson, K., Nir, S., and Isac, T. 1973. Phase transitions in phospholipid vesicles. Fluorescense polarization and permeability measurements concerning the effect of temperature and cholesterol. Biochim. Biophys. Acta **311**, 330-348.

Piknova, B., Perochon, E., and Tocanne, J.-F. 1993. Hydrophobic mismatch and long-range protein-lipid interactions in bacteriorhodopsin/phosphatidylcholine vesicles. Eur. J. Biochem. **218**, 385-396.

Pink, A., T. Green, and D. Chapman. 1980. Raman scattering in bilayers of saturated phospatidylcholine. Biochemistry **19**, 349-357.

Risbo, J., Sperotto, M. M., and Mouritsen, O. G. 1994. Theory of phase equilibria and critical mixing points in binary lipid bilayers: Free energy, enthalpy, specific heat, and interfacial tension. J. Chem. Phys. (submitted).

Sperotto, M. M. and Mouritsen, O. G. 1991a. Mean-field and Monte Carlo studies of the lateral distribution of proteins in membranes. Eur. Biophys. J. **19**, 157-168.

Sperotto, M. M. and Mouritsen, O. G. 1991b. Monte Carlo simulation studies of lipid order parameter profiles near integral membrane proteins. Biophys. J. **59**, 261–270.

Sperotto, M. M. and Mouritsen, O. G. 1993. Lipid enrichment and selectivity of integral membrane proteins in two-component lipid membranes. Biochim. Biophys. Acta. **22**, 323–328.

Sperotto, M. M. and Mouritsen, O. G. 1994. Lipid selectivity of integral membrane proteins in two-component lipid bilayers. Bacteriorhodopsin in DLPC-DSPC mixtures (in preparation).

de Verteuil, F., Pink, D. A., Vadas, E. B., and Zuckermann, M. J. 1981. Phase diagrams for impure lipid systems. Application to lipid/anaesthetics mixtures. Biochim. Biophys. Acta **640**, 207–222.

Zhang, Z., Mouritsen, O. G., and Zuckermann, M. J. 1993a. Detecting phase equilibria in models of thermotropic and lyotropic liquid crystals. Mod. Phys. Lett. B **7**, 217-232.

Zhang, Z., Sperotto, M. M., Mouritsen, O. G., and Zuckermann, M. J. 1993b. A microscopic model for lipid-protein bilayers with critical mixing. Biochim. Biophys. Acta **1147**, 154-160.

Zuckermann, M. J., Ipsen, J. H., and Mouritsen, O. G. 1993. Theoretical studies of the phase behavior of lipid membranes containing cholesterol. *In* Cholesterol and Membrane Models (ed. L. X. Finegold), pp. 223-257. CRC Press, Boca Raton, Florida.

Modelling Pattern Formation on Primate Visual Cortex

J. R. Thomson, Wm Cowan, K. R. Elder, Ph. Daviet, G. Soga, Z. Zhang, Martin Grant and Martin J. Zuckermann

Abstract

We begin this chapter by giving an overview of the types of theoretical model used to describe pattern formation related to ocular dominance on the visual cortex. We then present a new neural network model for the formation of ocular dominance stripes on primate visual cortex and examine the generic phase behavior and dynamics of the model. The dynamical equation of ocular dominance development can be identified with a class of Langevin equations with a non-conserved order parameter. We find that the phase diagram for our model comprises three phases: a striped phase, a hexagonal "bubble" phase and a uniform paramagnetic phase. We then examine the dynamics of the striped phase by solving the Langevin equation both numerically and by singular perturbation theory. Finally, we compare the results of the model with physiological data. The typical striped structure of the ocular dominance columns corresponds to the zero-field configurations of the model. We also use our model to simulate mononuclear deprivation.

1 Introduction

Spontaneous stripe formation occurs in many physical and biological systems; this includes Rayleigh-Benard instabilities [1], magnetic domain patterns [2, 3] and reaction-diffusion systems [4]. Many of these systems have been successfully modeled using the methods of statistical physics. One system of this type is the striped pattern of ocular dominance in layer IV, area 17 of primate visual cortex.

The ocular dominance columns of area 17 form a pattern which is striped when layer IV is observed in sections parallel to the surface of the cortex. The ocular dominance of a cortical neuron measures the difference in excitation of the neuron when each eye is stimulated separately. In primates ocular dominance begins to develop shortly before birth. When both eyes experience a normal visual environment or are deprived of pattern vision (binocular deprivation) ocular dominance forms in parallel stripes (see Fig. 1a and [5]). This development is disturbed, however, when a single eye is deprived of pattern vision (monocular deprivation) in which case the stripes are spaced less regularly. They also thin, or 'corrode' into patterns with broken stripes (see Fig. 1b and [6]). These ocular dominance patterns are symptoms of changes in the strength of synaptic connections between cortical neurons and the neurons in the lateral geniculate nucleus from which they receive their input.

In section 2.1 we briefly review the current theoretical models describing ocular dominace and then in section 2.2 we present a new theory of synaptic development for ocular dominance [7–10]. It consists of a neural net whose structure is based on known physiology of the primate visual system and whose synaptic strengths change according to the rules of Hebbian learning [11]. A novel feature of this theory is essential for stripe formation: correlations that are induced on the visual cortex, which are responsible for the creation of ocular dominance stripes, are determined by aspects of retinal connectivity that are well understood and easily measurable. This network is developed formally in section 2.2 in terms of correlations of visual inputs which are obtained from a Born-Oppenheimer approximation. The result is a dynamical equation which incorporates an intra-cortical interaction that depends on the known receptive field, $W(\mathbf{x} - \mathbf{x}')$, of the retinal ganglion cells. We show that, when both eyes receive the same visual input, this equation reduces to a Langevin equation for the difference in the cortical synaptic strength functions. An equivalent Ising model is used to examine the phase behavior of the model as a function of its parameters. The phase behavior is described in section 3, where we review our previous results and present new Monte Carlo data for anisotropic interactions. The dynamics of the model are studied in section 4 from both numerical and analytical solutions of the Langevin equation. This section ends with an analysis of monocular deprivation based on the inclusion of an external field which destroys eye–eye symmetry. Since this does not give the pattern of Fig. 1(b) observed for monocular deprivation, we present in section 5 a novel extension of our model based on a local field which gives the correct behavior for monocular deprivation and can be used to make predictions for pattern development in this case. These solutions are then discussed in section 6 in terms of their physical and physiological relevance.

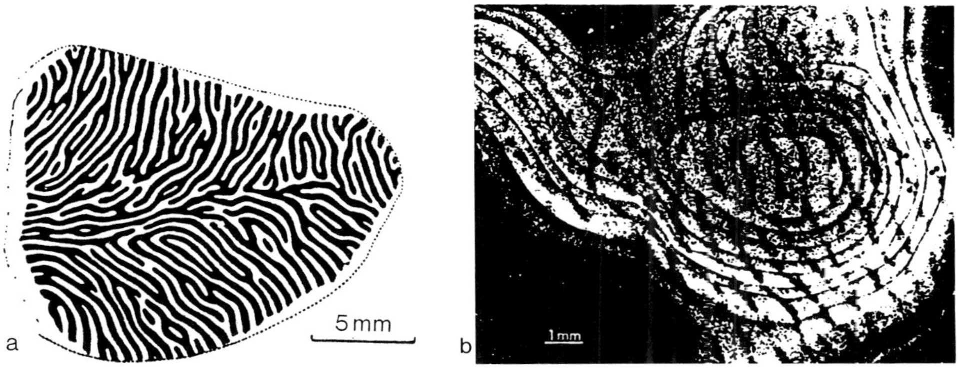

Fig. 1. Dark-field autographs of monkey striate cortex: (a) normal monkey (after [5]) and (b) deprived monkey (after [6]).

2 From Neurophysiology to Statistical Mechanics

2.1 Classifiaction of Theoretical Models for Occular Dominance

Most neurophysiological models for occular doninance are more concerned with the developmental process of the formation and segregation of ocular dominance columns than the complete pattern of Fig 1a. In these models, segregation is due to a reinforcement of afferents to the visual cortex from the same eye and a competition or repulsion between afferents from different eyes. This process determines the width of the ocular dominance columns. We give here a very brief overview of some types of models and we first consider topographical models. Our list is not meant to be exhaustive and the reader is encouraged to read and find new references in the cited literature.

One type of topographical model is based on the concept of chemospecific matching [12]. A model of this type proposed by von der Malsburg [13] considered the cortex as a passive area on which chemical messengers diffuse. These models are based on the concept of a coarse retinotopic map constructed from the first contacts of the afferents at the cortex. Neighboring afferents from the same eye are assumed to attract one another during this process. The coarse map is then refined as follows by the chemical markers which are carried by the afferents themselves. Afferents from the same eye are attracted to the same location on the cortex as they have the same markers while afferents from different eyes would be repelled at the same location due to different chemical markers. Whilshaw and von der Marlsburg [14] also proposed a model based on correlated neural activity on a pre-synaptic sheet. More recent theories which combine neural activity models with chemospecific matching are described in the excellent thesis of Goodhill [15] whose descriptions of the various models are summarised in this section. Most of these theories were generalised from models describing the retino–tectal system in frogs etc.

Many models tend to sacrifice some neurophysiological details for algorithms which are computationally fast. The best known of this class of model is due to Kohonen who very effectively simplified previous models by sacrificing biological detail. Other models, such as the elastic net algorithm, use an energy functional which can be optimised in a similar way to the well–known travelling salesman model. In the elastic net algorithm [16, 17] the two retinae are represented by two planar sheets with neural units lying in an abstract 'feature space'. The cortex is also a plane of cells which are connected to the units in each retinal plane and to neighbouring cortical units. This model is then minimized from a geometric point of veiw.

Several authors propose that pattern formation on the visual cortex is due to correlations between presynaptic sheets relative to the cortex. These models are concerned both with development of the pattern and the final pattern itself and are based on Hebbian learning between a pre-synaptic sheet (for example the lateral geniculate neucleus) and a post–synaptic sheet taken to be the visual cortex. Our model is of this type and is most similar to the model of Swindale [18]. However, as shown below, it has the advantage that the intra-cortical interaction can be related to physiological parameters. More recently, an elaborate model proposed by Miller et al. [19] describes synaptic development in cat visual cortex. Miller's model includes the dynamical development of detailed interactions in the lateral geniculate nucleus and visual cortex in addition to arborization of afferents from the lateral geniculate to the visual cortex. This model requires a priori knowledge of several interactions and has sufficient dynamical variables that simulations can be performed only on small systems. Within these limitations it shows that connections between lateral geniculate nucleus neurons and cortical cells evolve much faster than the cortical pattern. A later model extending Miller's ideas and including the effects of topography was proposed and analysed by Goodhill [15, 21], who placed considerable emphasis on the role of weight normalisation in competitive learning of this type. Lyons and Harrison [20] have proposed a class of reaction-diffusion mechanisms which preferentially select striped patterns. The configurations obtained using their method are very similar to those obtained using our model.

2.2 A New Model and the Associated Langevin Equation

We begin the description of our model by giving a detailed derivation of the associated Langevin equation which describes the evolution of dynamical variables which represent synaptic activity in the early visual system. This derivation starts with conventional ideas from the theory of neural networks wheras the resulting Langevin equation is typical of the formalism used in the kinetics of phase transitions. This allows us to propose an equivalent Ising model which is described in section 3 below.

Visual input into the system at time t is represented by two light intensity patterns, $I_{L,R}(\mathbf{x}',t)$, on the left and right retinas, where $\mathbf{x}'$ is a two-dimensional retinal coordinate. This input produces an excitation of intensity, $J_{L,R}(\mathbf{x},t)$, at the retinal ganglion cells, which is a convolution of $I_{L,R}$ with a function, $W(\mathbf{x}-\mathbf{x}')$, representing the center–surround receptive field of the ganglion cell,

$$J_{L,R}(\mathbf{x},t) = \int W(\mathbf{x}-\mathbf{x}')I_{L,R}(\mathbf{x}',t)\,d\mathbf{x}'. \tag{1}$$

This induces signals which pass from the ganglion cells via the optic nerve and the lateral geniculate nucleus to cortical cells in layer IV area 17 of the visual cortex. Each cortical cell receives input from both eyes. The input, $K_0(\mathbf{x},t)$, to the cortex from direct stimulation by retinal ganglion cells can be expressed as a weighted sum over the input intensities

$$K_0(\mathbf{x},t) = V_L(\mathbf{x},t)J_L(\mathbf{x},t) + V_R(\mathbf{x},t)J_R(\mathbf{x},t). \tag{2}$$

Here $V_L(\mathbf{x},t)$ and $V_R(\mathbf{x},t)$ are the strengths of the synaptic coupling between retinal and cortical cells for input from the left and right right eyes. The difference between the couplings is the ocular dominance. It varies in strength across the cortex forming a pattern which is usually striped.

A retinotopic map is implicitly assumed in Eq. (2), where spatial coordinates on the retina map conformally onto corresponding coordinates on the cortex. Hence the same coordinate notation can be used for both types of cell, except that the cortical magnification factor [22] must be considered when making quantitative calculations. Because of anisotropy in the cortical magnification factor which varies systematically across the cortex, retinal receptive fields are not radially symmetric when projected onto the visual cortex.

The total response, $K(\mathbf{x},t)$, of a cortical cell depends on the inputs, $K_0(\mathbf{x}',t)$, to neighboring cells. This effect is formally expressed by an intracortical interaction, $U(\mathbf{x}-\mathbf{x}')$, that couples each cortical cell to its neighbors,

$$K(\mathbf{x},t) = \int U(\mathbf{x}-\mathbf{x}')K_0(\mathbf{x}',t)\,d\mathbf{x}'. \tag{3}$$

Equations (2) and (3) can be regarded as a neural network, with $J_{L,R}(\mathbf{x},t)$ as the input and $K(\mathbf{x},t)$ the output. This network evolves by changing the synaptic strengths $V_{L,R}(\mathbf{x},t)$ according to Hebbian learning rules. One formulation of these rules [11] changes the synaptic strength proportionally to the product of the input and the output of the cortical cell,

$$\frac{d}{dt}V_a(\mathbf{x},t) = AK(\mathbf{x},t)J_a(\mathbf{x},t) + \text{non-linear terms}. \tag{4}$$

where $a = L, R$. The "non-linear terms" ensure that the synaptic strength functions saturate at a finite value. Eqs. (1) to (4) give the basic physiological relationships of our theory. In these equations, the inputs, $J(\mathbf{x},t)$, depend on time and the equations cannot be solved unless these inputs are specified.

From Eqs. (1) to (4) we obtain an integral equation for the synaptic strengths

$$\begin{aligned}\frac{d}{dt}V_a(\mathbf{x},t) =& J_a(\mathbf{x},t)\int U(\mathbf{x}-\mathbf{x}')\times \\ & [V_L(\mathbf{x}',t)J_L(\mathbf{x}',t)+V_R(\mathbf{x}',t)J_R(\mathbf{x}',t)]\,d\mathbf{x}' \\ & + \text{non-linear terms.}\end{aligned} \tag{5}$$

This equation can be integrated using specific light intensity patterns, $I_a(\mathbf{x},t)$, but a more general solution is constructed by separating the time dependence into short and long time scales. The long time scale, τ_p, is the characteristic time over which the stripes evolve while the short scale, τ_v, is the time constant characteristic of changes in the visual input. Since τ_p is of the order of weeks and τ_v the order of seconds, $J_a(\mathbf{x},t)J_a(\mathbf{x}',t)$ can be constrained by replacing it by its average over τ_p

$$F_{ab}(\mathbf{x}-\mathbf{x}') = \langle J_a(\mathbf{x},t)J_b(\mathbf{x}',t)\rangle. \tag{6}$$

Then Eq. (5) becomes

$$\begin{aligned}\frac{d}{dt}V_a(\mathbf{x},t) =& \int U(\mathbf{x}-\mathbf{x}')\times \\ & [F_{aL}(\mathbf{x}-\mathbf{x}')V_L(\mathbf{x}',t)+F_{aR}(\mathbf{x}-\mathbf{x}')V_R(\mathbf{x}',t)]\,d\mathbf{x}' \\ & + \text{non-linear terms} + \eta(\mathbf{x},t).\end{aligned} \tag{7}$$

where $\eta(\mathbf{x},t)$ is a Gaussian white noise term representing system noise in the neural network. It is assumed to satisfy the relation $<\eta(\mathbf{x},t)\eta(\mathbf{x}',t')>= \nu\delta(\mathbf{x}-\mathbf{x}')\delta(t-t')$, where ν is a measure of the intensity of the noise. The effect of fluctuations over the short time scale (τ_v) is incorporated in this term.

The ocular dominance is the difference in the strength of the synaptic coupling of the two eyes to the visual cortex: $V_-(\mathbf{x},t) = V_R(\mathbf{x},t) - V_L(\mathbf{x},t)$. It satisfies

$$\begin{aligned}2\frac{d}{dt}V_-(\mathbf{x},t) =& J_-(\mathbf{x},t)\int U(\mathbf{x}-\mathbf{x}')\times \\ & \{V_-(\mathbf{x}',t)J_-(\mathbf{x}',t)+V_+(\mathbf{x}',t)J_+(\mathbf{x}',t)\}\,d\mathbf{x}' \\ & + \text{non-linear terms} + \eta(\mathbf{x},t).\end{aligned} \tag{8}$$

where $V_+(\mathbf{x},t) = V_R(\mathbf{x},t)+V_L(\mathbf{x},t)$ and $J_\pm(\mathbf{x},t) = J_R(\mathbf{x},t) \pm J_L(\mathbf{x},t)$. Thus, the time evolution of ocular dominance depends on two averages:

$$\begin{aligned}\langle J_-(\mathbf{x},t)J_-(\mathbf{x}',t)\rangle =& F_{RR}(\mathbf{x}-\mathbf{x}') + F_{LL}(\mathbf{x}-\mathbf{x}') \\ & - F_{LR}(\mathbf{x}-\mathbf{x}') - F_{RL}(\mathbf{x}-\mathbf{x}')\end{aligned} \tag{9}$$

and

$$\langle J_-(\mathbf{x},t)J_+(\mathbf{x}',t)\rangle = F_{RR}(\mathbf{x}-\mathbf{x}') - F_{LL}(\mathbf{x}-\mathbf{x}') \tag{10}$$

The behavior of these terms depends on the visual environment of the two eyes. Under conditions characteristic of the mature visual system in an ideal viewing environment –all structured information in the same depth plane, both eyes pointed to the same location in that plane– there is right-left symmetry, $J_R(\mathbf{x},t) = J_L(\mathbf{x},t)$. Then both averages, (9) and (10), vanish and the time derivative of the ocular dominance is zero. Under more normal viewing conditions the two eyes see the same visual environment, but imperfect eye-pointing occurs. In immature visual systems the mechanisms that control eye direction are not yet fully developed; even in mature visual systems structure in the visual environment occurs in different depth planes and exact retinal correspondence of all planes is not possible. In this case there is symmetry in the correlation functions:

$$\begin{aligned} F_{RR}(\mathbf{x}-\mathbf{x}') &= F_{LL}(\mathbf{x}-\mathbf{x}') \\ F_{LR}(\mathbf{x}-\mathbf{x}') &= F_{RL}(\mathbf{x}-\mathbf{x}') \end{aligned} \tag{11}$$

but $F_{RR}(\mathbf{x}-\mathbf{x}') \neq F_{RL}(\mathbf{x}-\mathbf{x}')$. The second term in the time evolution equation is zero, but the first is non-zero. When calculated in detail the non-zero term is found to have a three-lobed structure that leads to ocular dominance stripes [9].

Similar conditions apply in utero, after the visual system is well enough established for spontaneous firing in retinal cells to excite the visual cortex. Ocular dominance columns begin forming in the *rhesus* monkey twenty days before birth when no visual input is available. A recent model suggests that the ocular dominance columns form initially via spontaneous firing of retinal cells, since toxic blocking of retinal activity prevents their formation. The visual environment subsequently refines the columns after birth. The formalism presented above can be applied to spontaneous firing of retinal cells in utero. The light intensity patterns, $I_{L,R}(\mathbf{x},t)$, are random, and the correlation functions, $\langle I_{L,R}(\mathbf{x},t)I_{L,R}(\mathbf{x},t)\rangle$, which are averages over the time constant related to the evolution of the stripes, have the same functional form as for random eye pointing.

Formally, these considerations lead to the following equation for ocular dominance in the case of right–left symmetry. From Eqs. (1),(5) and (11) we obtain:

$$\begin{aligned} 2\frac{d}{dt}V_a(\mathbf{x},t) = &\int U(\mathbf{x}-\mathbf{x}')\times \\ &\sum_b V_b(\mathbf{x}',t)\int\int W(\mathbf{x}-\mathbf{y}')\times \\ &W(\mathbf{x}'-\mathbf{y}'')\langle I_a(\mathbf{y}',t)I_b(\mathbf{y}'',t)\rangle\, d\mathbf{x}'\, d\mathbf{y}'\, d\mathbf{y}'' \\ &+ \text{non-linear terms} + \eta(\mathbf{x},t). \end{aligned} \tag{12}$$

The luminance correlation function on the right hand side of Eq. (12) can be written as follows on the basis of a "Born–Oppenheimer (BO) approximation":

$$\langle I_a(\mathbf{y}', t) I_b(\mathbf{y}'', t)\rangle = \sigma^2 \delta(\mathbf{y}' - \mathbf{y}'')\delta_{ab} + L_0 \tag{13}$$

where L_0 is the background luminance. The BO approximation used here is based on the observation that the visual scene has a time constant which varies on the order of seconds while the time constant, τ_{st} related to the progression of the stripes is of the order of days. In consequence, the luminances, $I_{L,R}(\mathbf{y}', t)$, can be treated as random variables over a time τ_{st}, leading to Eq. 13. We also assume that $U(\mathbf{x})$ $(= U_o)$ is a constant over cortical displacements longer than the extent of the convolution of the two centre–surround receptive fields of Eq. (15) below. From Eqs. (12) and (13) the equation for the ocular dominance can be written:

$$\frac{d}{dt} V_-(\mathbf{x}, t) = \int \mathcal{T}(\mathbf{x} - \mathbf{x}') V_-(\mathbf{x}', t)\, d\mathbf{x}' + \text{non-linear terms} + \eta(\mathbf{x}, t). \tag{14}$$

where

$$\mathcal{T}(\mathbf{x} - \mathbf{x}') = \int W(\mathbf{x} - \mathbf{y}) W(\mathbf{y} - \mathbf{x}')\, d\mathbf{y} \tag{15}$$

Another situation occurs when the two eyes experience different visual environments, a condition that can occur naturally as a result of abnormal development of one eye, or that can be produced artificially by eyelid suture or similar treatment. In this case both terms of Eq. (8) are non-zero: the first produces an interaction between ocular dominance at different sites on the cortex, the second an interaction between the ocular dominance at one site and the total synaptic strength at other sites. Unfortunately, suitable models for the development of the total synaptic strength are unavailable. Our earlier work did not require them because it was confined to the case of imperfect eye pointing in normal vision, in which coupling to the total synaptic strength is zero. A reasonable approximation takes the total synaptic strength to be constant; future work will relax this constraint as it leads to the need for large anisotropies in the simulation of monocular deprivation (see section 4). With the total synaptic strength constant the second term is analogous to a uniform bias field, the strength of which depends on the degree of asymmetry between the two eyes. The first term is an interaction term of the same form as that treated in our earlier work. Thus, in this approximation, the field provides an adequate generalization to handle the case where the two eyes are treated asymmetrically.

The full Langevin equation can then be written in terms of the bias field, H, from Eqs. (5) to (10):

$$\frac{d}{dt} V_-(\mathbf{x}, t) = \int \mathcal{T}(\mathbf{x}-\mathbf{x}') V_-(\mathbf{x}', t)\, d\mathbf{x}' + H + \text{non-linear terms} + \eta(\mathbf{x}, t). \tag{16}$$

It is easy to show that $\mathcal{T}(\mathbf{x})$ is a three–lobed function given by (see Fig. 2 and [8])

$$\begin{aligned} 2\mathcal{T}(\mathbf{x}) =& \frac{1}{a_x a_y} \exp - \left\{ \frac{x^2}{a_x^2} + \frac{y^2}{a_y^2} \right\} + \frac{1}{b_x b_y} \exp - \left\{ \frac{x^2}{b_x^2} + \frac{y^2}{b_y^2} \right\} \\ & - \frac{2}{c_x c_y} \exp - \left\{ \frac{x^2}{c_x^2} + \frac{y^2}{c_y^2} \right\} . \end{aligned} \tag{17}$$

where $c_x^2 = (a_x^2 + b_x^2)/2$ and $c_y^2 = (a_y^2 + b_y^2)/2$. x and y are the Cartesian components of the vector, $\mathbf{x}$. $\mathcal{T}(\mathbf{x})$ in Eq. (17) represents an anisotropic interaction which becomes isotropic when $a_x = a_y$ and $b_x = b_y$. In [9] we reported preliminary results of numerical simulations of Eq. (16) for $H = 0$. In the simulations, the "non-linear terms" represent biological constraints of unknown form that confine the function $V_-(\mathbf{x}, t)$ between $+1$ (right eye dominance) and -1 (left eye dominance). Since an abrupt cutoff can lead to anomalous behavior near saturation, a local cubic term, $V_-^3(\mathbf{x}, t)$, is added to the right hand side of Eq. (16). This term ensures a smooth mechanism for saturation. The model is analogous to ϕ^4 field theories that describe phase separation [23]; the resulting non-linear equation for the cortical synaptic strengths, $V(\mathbf{x}, t)$, can be identified with a Langevin equation for a non-conserved order parameter [23].

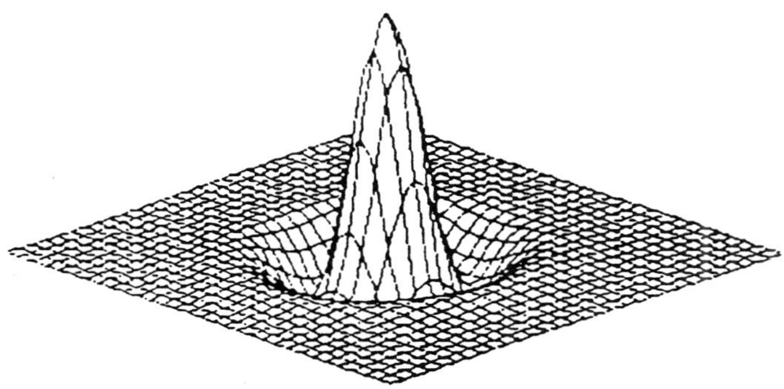

Fig. 2. Three dimensional representation of the intra-cortical interaction, $\mathcal{T}(\mathbf{x})$. This is a three-lobed interaction, whose total integrated weight is zero. The qualitative features only depend on the existence of circular receptive fields with opposed centers and surrounds of equal weight.

3 The Equivalent Ising Model

In order to understand the dynamics of the Langevin equation given in Eq. (16), it is important to study the equilibrium phase behavior of the system it describes. We use an equivalent Ising model for this purpose. The model represented by Eq. (16), including the cubic term, is in the same universality class as an Ising model with a long range interaction given by $\mathcal{T}(\mathbf{x})$ of Eq. (17). The bias field, H, of Eq. (16) is then a uniform magnetic field and the Hamiltonian of the Ising model is written as follows

$$\mathcal{H} = -\sum_{i,j} J_{ij}\sigma_i\sigma_j - H\sum_i \sigma_i. \tag{18}$$

In Eq. (18), i is a site index for a two dimensional lattice, σ_i is a spin variable defined at site i which has values of $+1$ for one eye and -1 for the other eye, and J_{ij} is the interaction between the spins on sites i and j. The numerical values of J_{ij} are obtained by evaluating $\mathcal{T}(\mathbf{x}_i - \mathbf{x}_j)$.

The mean field equation for the thermally averaged site-dependent magnetization, $m_i =< \sigma_i >$, can be written as follows

$$m_i = \tanh\left[\frac{\sum_j J_{ij}m_j + H}{kT}\right], \tag{19}$$

where T is the temperature and k is Boltzmann's constant. The temperature corresponds to the intensity of the white noise in the system. Eq.(19) is identical to the mean field equations used by Moeller [24] and Moeller et al. [25] in their statistical modeling of early visual processing.

The mean field phase diagram for two dimensions is shown in Fig. 3 for an isotropic interaction. It was obtained by calculating the free energy corresponding to Eq. (19) for phases with 2-fold (stripe) symmetry, 3-fold (hexagonal) symmetry and 4-fold (square) symmetries shown in Fig. 4. Fig. 3 shows that, as the field, H, increases, the system exhibits first-order transitions from the striped phase first to a hexagonal "bubble" phase and then to a paramagnetic phase. Fig. 4 shows that a square bubble phase is at best metastable in the mean field approximation for all values of H and T. The first-order phase lines end in a second order critical point at $T = T_c$ and $H = 0$, as shown in Fig. 3. Fig. 3 is basically the same as the phase diagram found in [24,25] by direct iteration of Eq. (19).

The thermal and magnetic field behaviors of the model described by the Hamiltonian of Eq. (18) were also examined by the Metropolis Monte Carlo method with spin-flip dynamics (see [26] and references therein). The simulations were performed on a 2D square lattice. The initial configurations were high temperature states with spatially random distributions. The system was cooled slowly to low temperatures at constant magnetic field (defined with respect to H_c, the critical field for the hexagonal to paramagnetic transition), using a simple annealing schedule with equal temperature decrements.

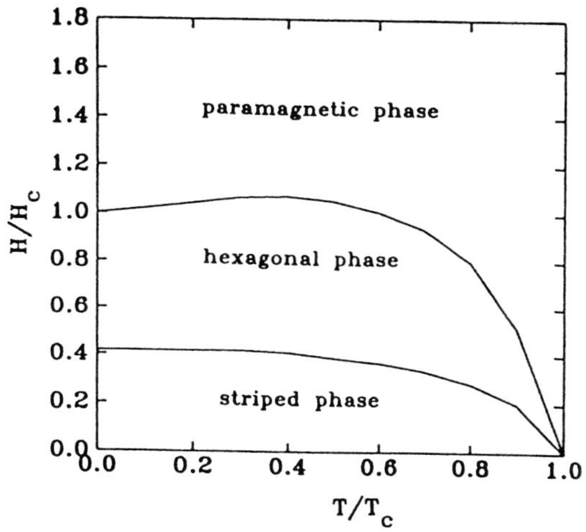

Fig. 3. The mean field phase diagram for the Ising Hamiltonian of Eq.(18) in the $H-T$ plane. The lines represent first order phase transitions and the phases are marked on the diagram.

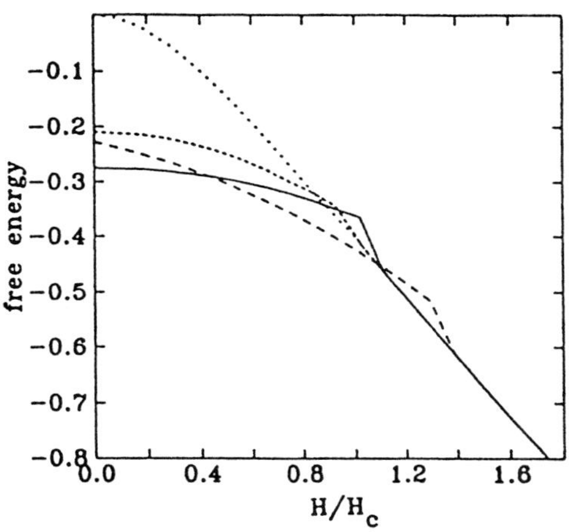

Fig. 4. The free energies of the phases corresponding to Fig. 3 as functions of field for $T/T_c = 0.2$. The continuous line corresponds to the striped phase, the line with long dashes to the hexagonal bubble phase, the line with short dashes to the square bubble phase and the dotted line to the uniform paramagnetic phase.

At each temperature the system was annealed for 4000 Monte Carlo steps per spin (MCS/s). The simulations were performed for isotropic interactions.

Below T_c, the values of the magnetization can be classified in terms of three distinct regimes: negligible values of the magnetization, intermediate values of the magnetization and full magnetization, associated respectively with the striped phase, the hexagonal bubble phase and the paramagnetic phase. Sample configurations illustrating the equilibrium morphology of these phases are shown in Fig. 5. Fig. 5a represents the striped phase at temperatures between $0.1T_c$ and $1.3T_c$, which occurs at field values less than $0.3H_c$. The low temperature phase has smooth interfaces between regions of spin-up and spin-down, and increasing temperature results in an increase in roughness of these interfaces, while the integrity of the stripes disappears altogether above T_c. A similar description applies to the hexagonal bubble phase in Fig. 5b, which occurs at field values $0.3H_c < H < H_c$. The interfacial regions are initially smooth but roughen with increasing temperature, eventually losing all definition. For the paramagnetic phase ($H > H_c$), the magnetization is uniform at low temperatures, while increasing thermal fluctuations induce small clusters of opposite spin, which can be regarded as precursors of the hexagonal bubble phase. This is again destroyed by thermal fluctuations for $T > T_c$.

(a) (b) (c)

Fig. 5. Configurations obtained by simulated annealing with an isotropic interaction showing the three different phases displayed by the model for three different values of the applied field as a function of increasing temperature. The lattice size is 64x64 for Figs. 5 and 6.

Simulations were then performed with anisotropic interactions. Eq. (17) shows that the equi-potentials of the interaction are ellipses, so that the anisotropy can be characterized by the ratio e of the major and minor axes. The anisotropy used in the calculations reported here is uniform whereas that found in the cortical magnification factor varies over the cortex. Monte Carlo simulations were used to calculate thermodynamic quantities as functions of H and T in the case of $e = 1.14$. This value is of the order of the average value of the anisotropy found in the cortical magnification factor. The simulations were performed by quenching the system from high to low temperatures for a large enough value of H to keep the system in the paramagnetic phase. The field, H, was then decreased in small decrements keeping the temperature constant, and 4000 MCS/s were performed at each value of H. For $T = 0.5T_c$, where T_c is the critical temperature at $H = 0$, we found two extremely narrow peaks in the specific heat at points on the phase diagram corresponding to the field value at which the system goes from the striped phase to the hexagonal bubble phase and to the field value at which the system goes from the hexagonal phase to the paramagnetic phase respectively. Within the resolution of our simulations these peaks were essentially delta functions. These results are characteristic of first order phase transitions. Detailed investigations to determine the order of these transitions are in progress.

Fig. 6, which was obtained by simulated annealing for a large value of the anisotropy, $e = 2$, shows that the bubbles no longer exhibit hexagonal symmetry but instead resemble thinned and corroded stripes. This is evidence that the hexagonal phase is eliminated from the phase diagram at sufficiently high anisotropy.

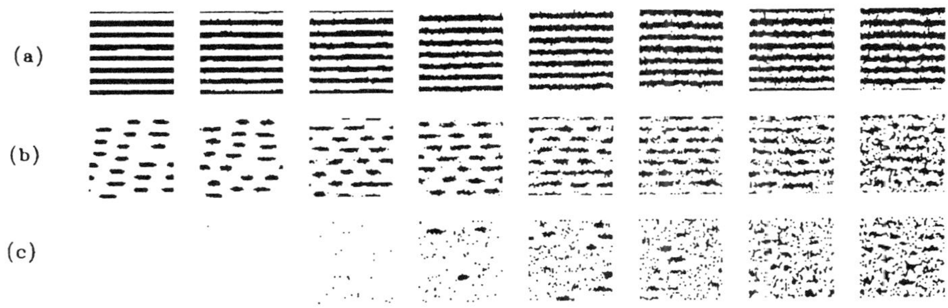

Fig. 6. Configurations obtained by simulated annealing showing the behaviour of the model with a large anisotropy $e = 2$ for three different values of the applied field as a function of increasing temperature.

4 Dynamics of Pattern Formation and Effect of Anisotropy for the Langevin Equation

In this section we examine the dynamics of stripe formation and the effect of anisotropy for both zero and non-zero fields via singular perturbation theory and numerical simulation of the Langevin equation given in Eq. (16). We first present numerical results with $H = 0$ for the case when the non-linear terms are replaced by the condition that $|V_-(\mathbf{x}, t)| \leq 1$. Next we present numerical results for $H = 0$ in the presence of the local cubic term, $V_-^c(\mathbf{x}, t)$, and the approach to equilibrium is studied via the time dependence of the peak in the structure factor. Approximate solutions from singular perturbation theory are then presented for anisotropic interactions when $H = 0$ and for large sample sizes which are not at present amenable to numerical computation. Finally we give results for the case of patterns formed when $H \neq 0$.

The Langevin equation is solved numerically on a square lattice using the Euler method. The discrete version of the linear Langevin equation of (15) is given by the following finite difference equation:

$$V_{ij}^{-}(t+\Delta t) = V_{ij}^{-}(t) + \Delta t \sum_{\ell,\ell'=-N}^{N} \mathcal{T}_{i-\ell,j-\ell'} V_{\ell\ell'}^{-}(t) + \eta_{ij}(t) \tag{20}$$

where (i, j) are the coordinates of a point on the square lattice, Δt is the product of the time interval and the interaction strength and N gives the number of lattice sites over which the discretized interaction, $\mathcal{T}_{i,j}$, extends. The discrete equation was solved on a 128 x 128 lattice for different values of Δt and for both isotropic and anisotropic interactions. With the interaction used here, the lattice is spanned by 16 stripes. It is difficult to obtain the late time growth behavior of such small systems, but the numerical simulations should be adequate for the calculations of equilibrium properties. The linear discrete equation was solved using the constraint $|V_{ij}^{-}| \le V_{max}$ and the initial values of V_{ij}^{-} were obtained from a uniform distribution of random numbers between $\pm\epsilon$, where $|\epsilon| = 0.003 V_{max}$.

The evolution of the stripes was investigated in the presence and absence of system noise. Fig. 7 shows the evolution of the stripes from $10^2 \Delta t$ to $10^3 \Delta t$ for $\Delta t = 7.0$ in the absence of noise and for an isotropic interaction. This value of Δt corresponds to a weak interaction with the system evolving via the nucleation of striped droplets of circular symmetry resembling bull's eyes. Stripes begin to form when the bull's eyes merge, and the system is frozen at $t = 10^3 \Delta t$ with vestiges of the bull's eyes remaining.

For large values of Δt, the system exhibits phenomena which resemble unstable growth at early times. In this case the final pattern exhibits well-defined stripes with a considerable number of branching defects. Fig. 8 gives the effect of an anisotropic interaction for $\Delta t = 7.0$ with $e = 1.14$ on the evolution of the stripes. The figure shows that the system evolves via a nucleation process at early times and that the final stripe pattern is composed of almost parallel stripes with considerably fewer branching defects than in the case of the isotropic interaction of Fig. 7. The configurations of Fig. 8 were generated without system noise ($\nu = 0$).

In order to examine late time dynamic behavior, it is necessary to include in the Langevin equation specific non-linear terms that ensure a smooth passage to equilibrium. The Langevin equation of Eq. (16) was therefore studied with the third order term, $(V_{-}(\mathbf{x}, t))^3$, included in the formalism. The Langevin equation was again solved in the following discrete form on a square lattice using Euler's method in Fourier space:

$$\begin{aligned} V_{ij}^{-}(t+\Delta t) =& V_{ij}^{-}(t) + \Delta t \int \mathcal{T}(\mathbf{k}) V^{-}(\mathbf{k}, t) \exp(-\mathbf{k} \cdot x_{ij})\, d\mathbf{k} - \\ & B(V_{ij}^{-}(t))^3 + \eta_{ij}(t) \end{aligned} \tag{21}$$

where $\mathbf{k}$ is a wave vector and $\mathcal{T}(\mathbf{k})$ is the Fourier transform of $\mathcal{T}(\mathbf{x})$. The equations were solved on a 128 x 128 lattice for large values of Δt with the same initial conditions as above plus a small amount of system noise when $H = 0$. The maximum value of the random numbers used for initialization

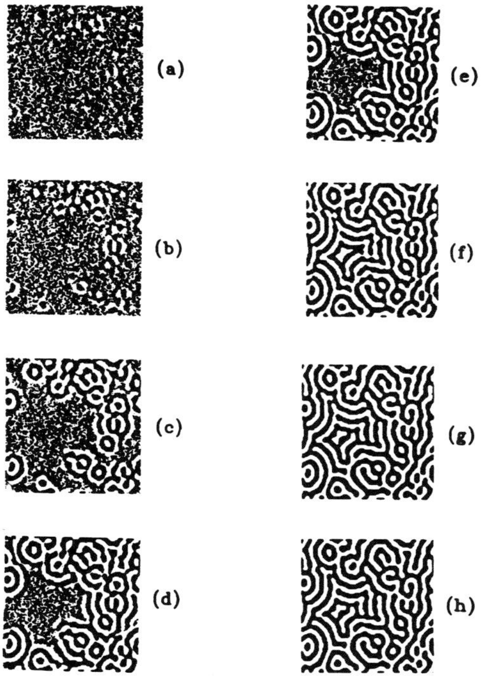

Fig. 7. Time evolution of the striped pattern from $10^2 \Delta t$ to $10^3 \Delta t$ in the absence of noise for an isotropic interaction, with $\Delta t = 7.0$. Figs. 7 through 14 and Figs. 18 - 21 were obtained by numerical simulation of the Langevin equation on a 128 $\times$ 128 lattice.

was $|\epsilon| \leq 0.1$. Fig. 9 gives the final configuration for an isotropic interaction at t $= 2000 \Delta t$ and a noise intensity $\nu = 0.1$. The figure shows that the pattern is isotropic and exhibits many branching defects. Furthermore, the pattern is stuck in this final configuration and does not evolve further. The same is true for the case of a 256 x 256 lattice with the same parameters. Fig. 10 shows the evolution of the one point distribution function for $V_-(\mathbf{x}, t)$ in the isotropic case. The evolution to a bimodal distribution related to stripe formation is evident.

Fig. 11 gives the final configuration for $H = 0$ in the case of an anisotropic interaction with $e = 1.11$ and $\nu = 0.1$ on a 128 $\times$ 128 lattice. The stripes, which align along the axis of anisotropy, form very quickly. The peak of the structure factor in the direction of anisotropy was obtained as a function of time averaged over 160 runs. The corresponding log-log plot does not exhibit a well-defined growth law at late times. This was also found in the case of a 256 $\times$ 256 lattice.

There are several reasons for the difficulty in establishing an asymptotic growth law for the striped phase. As reported in the previous paragraph, small systems lead to saturation effects that can blur the true asymptotic behavior. In addition, there may be long crossover effects due to the competition between the reorientation of stripes and the motion of branching defects. Some analytical calculations by Kawasaki [27] on a similar model

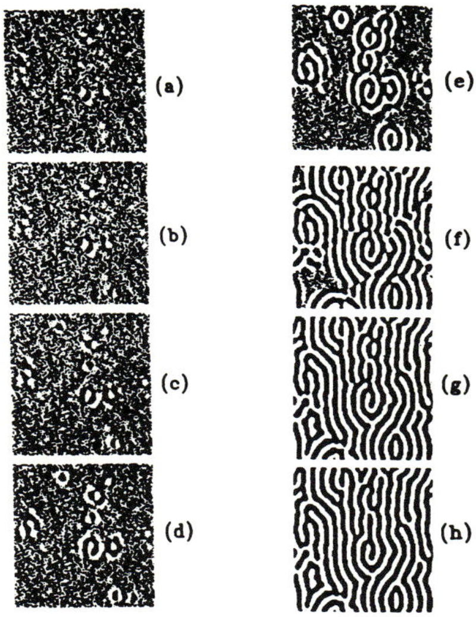

Fig. 8. Time evolution of the striped pattern from $10^2 \Delta t$ to $10^3 \Delta t$ in the absence of noise for an anisotropic interaction, $e = 2$, with $\Delta t = 7.0$.

Fig. 9. The final configuration for an isotropic interaction at $t = 2000\Delta t$ and a noise intensity $\nu = 0.1$.

(the XY model) have shown that the crossover effects can be very strong. To understand the long time behavior of our model a singular perturbative solution of the Langevin equation is presented. The singular perturbation scheme has been applied to a class of models [1, 28] to which the model of section 2 belongs. The first step in applying this method is to expand $V_-(\mathbf{x}, t)$ in terms of the linear solution, $V_o^-(\mathbf{x}, t)$, of the Langevin equation. The Fourier transform of the linear solution is given by

$$V_o^-(\mathbf{k}, t) = \exp(\mathcal{T}(\mathbf{k})t) V_o^-(\mathbf{k}, 0). \tag{22}$$

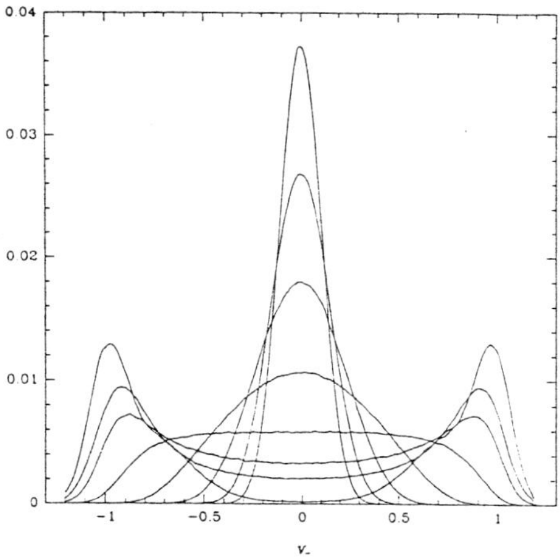

Fig. 10. The evolution of the one point distribution function for $V_-(\mathbf{x}, t)$ in the isotropic case. The distribution develops from the single peak to a bimodal distribution. The curves correspond to $t = 4\Delta t$, $6\Delta t$, $8\Delta t$, $10\Delta t$, $12\Delta t$, $14\Delta t$, $16\Delta t$, $2000\ \Delta t$, as the central peak decreases. The early time data was averaged over 100 independent runs, and the late time data was averaged over 50 independent runs.

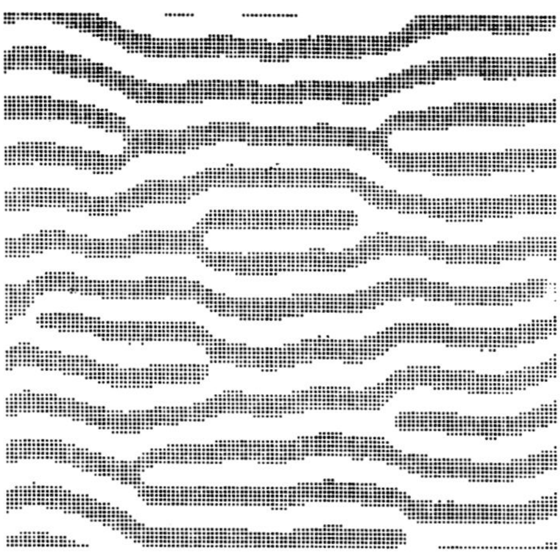

Fig. 11. The final configuration for $H = 0$ in the case of an anisotropic interaction with $e = 1.11$ and $\nu = 0.1$. Stripes align along the direction of the anisotropy.

The series can then be resummed in the late time limit for wavevectors near the wave vector, k_m, at which $\mathcal{T}(\mathbf{k})$ has a maximum. Here k_m is defined by $\partial \mathcal{T}(k)/\partial k = 0$. The reader is referred to reference [1] for details. The main result of this calculation is the solution of the equation for $V_-(\mathbf{x}, t)$ (in the limit $\Delta t = 0$ and $\nu = 0$), which is given by

$$V_-(\mathbf{x}, t) = \frac{V_0^-(\mathbf{x}, t)}{\{1 + \beta V_0^-(\mathbf{x}, t)^2\}^{\frac{1}{2}}}, \tag{23}$$

where $\beta = (B/\mathcal{T}(k_m))^{1/2}$. Eq. (23) provides a complete although approximate solution for the dynamics of $V_-(\mathbf{x}, t)$ in the absence of system noise. In Fig. 12, configurations described by Eq. (23) are plotted at $t = 75$, 250 and 1000. These configurations are similar to those shown in the previous sections. Equation (23) can also be employed to obtain analytic forms for the one point distribution function and the structure factor, $S(\mathbf{k}, t)$.

The prediction for the pair correlation function ($G(\mathbf{x}, t) = \int d\mathbf{k} e^{i\mathbf{k}\cdot\mathbf{x}} S(\mathbf{k}, t)$) from Eq. (23) is,

$$G(\mathbf{x}, t) \approx \frac{2}{B\pi} \arcsin G_o(\mathbf{x}, t)/G_o(0, t) \tag{24}$$

where $G_o(\mathbf{x}, t)$ is the linear solution; i.e., $G_o(\mathbf{x}, t) = e^{2\mathcal{T}(\mathbf{x})t} G(\mathbf{x}, 0)$. At late times, Eq. (22) predicts that $G(x, y = 0, t) \propto f_x(x/t^{1/2})$ and $G(x = 0, y, t) \propto cos(k_m x) f_y(y/t^{1/2})$, where x is the direction perpendicular to the stripes for an anisotropic interaction. Thus the singular perturbation theory predicts a dynamic scaling exponent of 1/2, implying that all lengths grow as a power law in time with exponent 1/2.

Our final results show the development of patterns for monocular deprivation when $H \neq 0$. Analytic solutions are not available and it is at present impossible to examine late time growth laws, since numerical results can be obtained only for small systems. Hence our results are confined to the striped phase $H \neq 0$, since it is pertinent to pattern formation for monocular deprivation (see Fig. 1b). We performed quenches for several values of H and for a considerable range of anisotropies, including the value $e = 1.14$ used in the Monte Carlo simulations of the equivalent Ising model. We found that the stripes for the minority "phase" thin slightly in all cases when the system is quenched from a high temperature disordered configuration for $H < H_c'$ where H_c' is the value of the field for transition from the striped to the hexagonal phase. For H between H_c' and H_c, a hexagonal phase exists for low values of anisotropy. However, as the anisotropy increases, the hexagonal phase exists for a smaller range of fields. At high anisotropy, the system makes a direct transition from the striped phase to the paramagnetic phase. All quenches were performed at low values of the noise intensity and the results agree with those found from simulated annealing which are reported in Section 3.

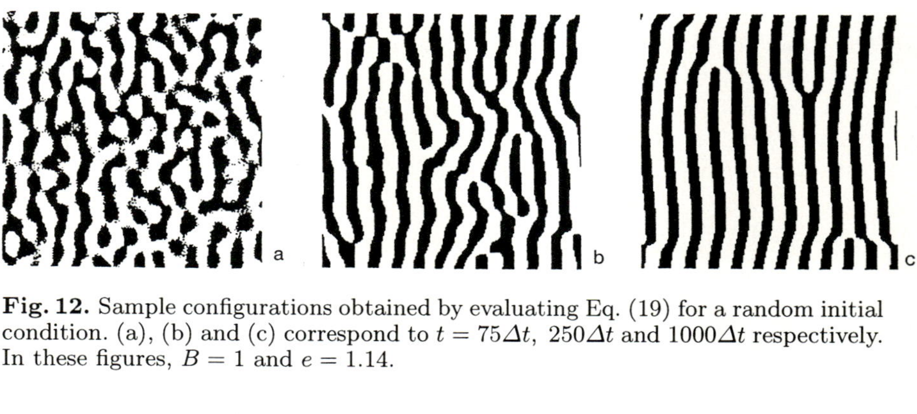

Fig. 12. Sample configurations obtained by evaluating Eq. (19) for a random initial condition. (a), (b) and (c) correspond to $t = 75\Delta t$, $250\Delta t$ and $1000\Delta t$ respectively. In these figures, $B = 1$ and $e = 1.14$.

Monocular deprivation is not, however, equivalent to quenching from a random state. As described in the introduction, stripe development begins in utero and there is an incipient striped pattern in the cortex at birth. At this stage monocular deprivation may be induced by suturing the eyelid of one eye. Furthermore the absence of hexagonal bubble patterns suggests that values of the anisotropy must be large enough that the hexagonal phase is absent. This consideration led us to perform the following quench for a 128×128 system using the Langevin equation. We first chose a value $e = 1.7$ of the anisotropy, high enough so that no hexagonal phase occurs. Then, to simulate random stimulation of the cortex, we initiated the quench at zero field with a high noise intensity $\nu = 1.5$ and continued for a small number of time steps, thereby allowing an incipient striped pattern to form. This represents conditions in utero where patterns are formed on the visual cortex via spontaneous firing of the retinal cells. Next a non-zero field combined with a considerably lower noise intensity $\nu = 0.1$ was imposed on the system and the evolution of the system was continued. Taking H_c'' as the field at which the transition from the striped phase to the paramagnetic phase occurs, we used two fields to simulate monocular deprivation, $H_1 = 4.5H_c''$ and $H_2 = 5.5H_c''$. The equilibrium phase for both fields is the uniform paramagnetic phase. For H_1 we began with a stripe configuration obtained at $t = 20\Delta t$. As shown in Fig. 13, the system "stabilizes" into a long–lived metastable pattern of thinned stripes. The corresponding one–point distribution function shows a considerable degree of imbalance between the thick (white) and thin (black) stripes. Furthermore, the stripes are corroded as expected for monocular deprivation. However, the presence of white patches in the configuration implies that the system is unstable with respect to the paramagnetic phase. For the higher field, H_2, monocular deprivation was simulated with an initial stripe configuration obtained at $t = 50\Delta t$. In this case a more developed stripe configuration was required to prevent the system from passing immediately to the paramagnetic phase. Fig. 14 shows that a pattern of thinned stripes formed but was not stable since the white patches increase in size with increasing time. For comparison we repeated the quench for $H = H_1$, but with an isotropic interaction of the same strength. The system then stabilizes in a hexagonal phase with some residual stripes. The fact that our configurations for the simulation of monocular deprivation are unstable led us to propose an novel version of our model which is an extension of the model of section 2.2 but which is based on a local field rather than a uniform field. This is described in the next section.

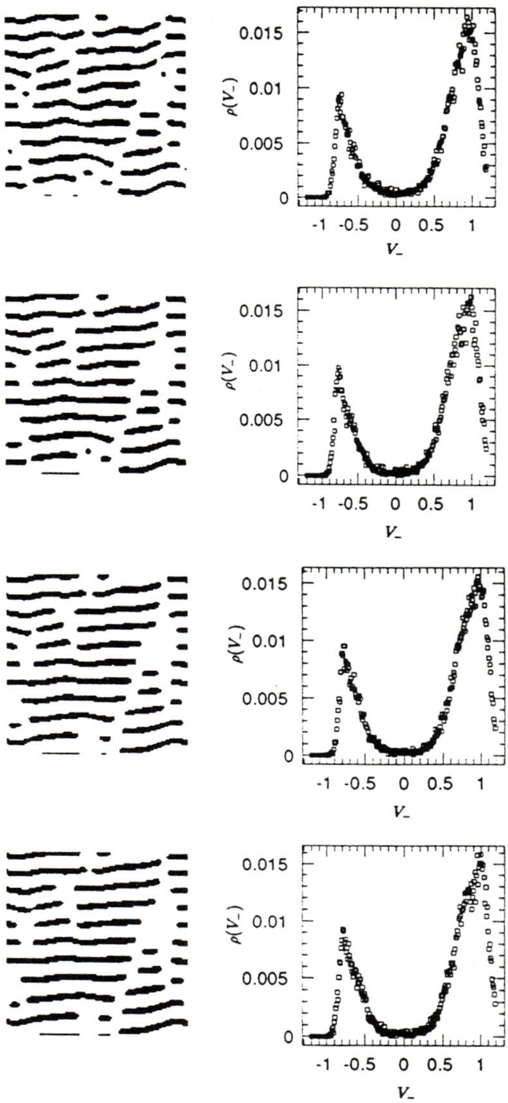

Fig. 13. Configurations and one point distribution functions for $e = 1.7$, in an external field $H = 4.5H_c''$.
Upper configuration: $t = 50\Delta t$.
Second configuration: $t = 100\Delta t$.
Third configuration: $t = 150\Delta t$.
Lowest configuration: $t = 200\Delta t$.

5 A Novel Treatment of Monocular Deprivation

In the previous section we performed simulations of the Langevin equation for $H \neq 0$ since no analytic solution is available for this case. These simulations showed that for weak anisotropy the stripes do not thin with increasing H, but a transition to the hexagonal bubble phase takes place. Significant thinning of the stripes was found only for large anisotropy. The stripe con-

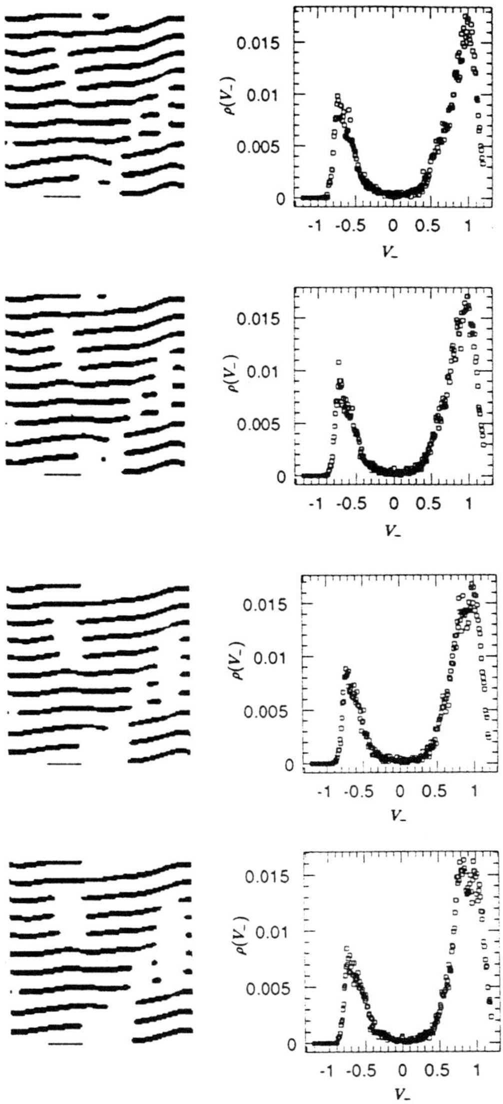

Fig. 14. Configurations and one point distribution functions for $e = 1.7$, in an external field $H = 5.5H_c''$.
Upper configuration: $t = 50\Delta t$.
Second configuration: $t = 100\Delta t$.
Third configuration: $t = 150\Delta t$.
Lowest configuration: $t = 200\Delta t$.

figurations for weak anisotropy in zero-field have a similar appearance to the ocular dominance columns of Fig. 1a. In addition the simulation of monocular deprivation described in section 4 gives thinned and corroded stripes analogous to Fig. 2b. However, a high value of the anisotropy parameter was required to inhibit the formation of the hexagonal bubble phase. When the anisotropy was weaker the thinned stripes were unstable with respect to the hexagonal bubble phase, which at present has no physiological basis. This model therefore would predict that monocular deprivation after the birth of the primate is described in terms of unstable states of the model.

One problem with the $H \neq 0$ simulations is that the thinning of the stripes is related to finite size effects as the simulations were performed on samples which are considerably smaller than the spatial extent of the related region of the visual cortex. Another problem is related to the magnitude of the anisotropy required to required to produce thinning of the stripes, this being of an order of magnitude greater than that observed for the visual cortex. We have therefore developed a novel extension of our model which is based on the supposition that monocular deprivation imposed during the critical period gives rise to the growth of additional afferents to the visual cortex for the undeprived eye as observed for the feline visual cortex. We initially introduced this concept into the model of section 2 and the subsequent simulations via a *phenomenological dynamic* interaction as follows. For the first set of simulations, we initially allowed the stripes to undergo partial developement corresponding to perfect binocular vision. We then imposed a local biasing field corresponding to monocular deprivation as follows. Let the undeprived eye dominance correspond to a positive value of $V_{-}(\mathbf{x}, t)$ at coordinate $\mathbf{x}$ and time t on the visual cortex. Then, during the simulations, if we are at a coordinate where $V_{-}(\mathbf{x}, t) > 0$, we introduce a short range intracortical interaction, $U_{MD}(\mathbf{x} - \mathbf{x}', t) > 0$, which acts on a neighboring site in such a way as to impose the dominance of the undeprived eye. We include a local field strength parameter, H_N, for purposes of symmetry. Formally we write the Langevin equation as follows

$$\begin{aligned}\frac{d}{dt}V_{-}(\mathbf{x},t) = \int \mathcal{T}(\mathbf{x}-\mathbf{x}')V_{-}(\mathbf{x}',t)\,d\mathbf{x}' + \\ H_N\theta(H_N V_{-}(\mathbf{x},t)) \int U_{MD}(\mathbf{x}-\mathbf{x}')V_{-}(\mathbf{x}',t)\,d\mathbf{x}' + \\ H_N\ V_{-}(\mathbf{x},t) \int U_{MD}(\mathbf{x}-\mathbf{x}')\theta(H_N V_{-}(\mathbf{x}',t))\,d\mathbf{x}' + \\ + \text{non-linear terms} + \eta(\mathbf{x},t).\end{aligned} \tag{25}$$

where $\theta(x)$ is the Heavyside theta function. The local field strength, H_N, is taken as positive for the case when the undeprived eye dominance correponds to a positive value of $V_{-}(\mathbf{x}, t)$ and as negative in the opposite situation. In the simulations, H_N is varied but $U_{MD}(\mathbf{x}-\mathbf{x}', t)$ is a fixed positive function. In the second set of simulations, we started from a completely random configuration and then allowed the pattern to develop. The degree of noise was also varied for different simulations. The interaction, $U_{MD}(\mathbf{x} - \mathbf{x}', t)$, ensures that the cells at a given cortical coordinate will be prompted to change to their ocular dominance to the undeprived eye if the local field is large enough.

Formally the local field interaction can be derived if the analog of a shunt function is included in the formalism of section 2 after the manner of Edelman et al. [29]. Eq. 4 would then be written

$$\frac{d}{dt}V_a(\mathbf{x},t) = A\langle K_{MD}(\mathbf{x},t)K(\mathbf{x},t)J_a(\mathbf{x},t)\rangle + \text{non-linear terms}. \tag{26}$$

where

$$K_{MD}(\mathbf{x},t) = \int U_{MD}(\mathbf{x}-\mathbf{x}')\phi(K_0(\mathbf{x}',t))\,d\mathbf{x}'. \tag{27}$$

Here $\phi(y)$ is an analytic fuction which can be expanded in a power series in y. Use of the central limit theorem for the averaging proceedure in Eq. (26) or replacement of $\phi(x)$ by its average with a judicious choice of its analytic form gives Eq. (25) providing all resulting non-linear terms are neglected. The interaction term in Eq. (25) involving U_{MD} can be regarded as a logical developement of the reasoning behind the introduction of the uniform external field, H, of Eq. (16) as it is a field which depends on the nature of the ocular dominance at a given cortical site. From a neural point of view this term represents the stimulation of the growth of afferents related to the undeprived eye which synapse on the cortex and at the same time the atrophy of existing afferents related to the deprived eye.

The simulations for Eq. (25) were performed on a 128×128 grid with a time step $\Delta t = 0.01$ for 8500 time steps. At low values of H_N and for low noise, the stripes thinnned slightly for the deprived eye for both an initial configuration of stripes which were partially formed for $H_N = 0$ and a random initial configuration. Both sets of stripes exhibited defects and an increase in noise level simply caused the stripes to anneal faster.

For high values of H_N and low noise, we first initialized the system by running for 100 time steps at $H_N = 0$. This resulted in a uniform pattern of 'noisy' stripes with many defects. This pattern stabilized very quickly after a value of $H_N = 5.2$ was used and the stripes corresponding to the deprived eye were seen to be thinned and corroded. The thinning of the stripes was confirmed by calculating the corresponding histograms. Electrode penetration scans were simulated for this pattern and showed an oscillating pattern shifted upwards towards the value for the undeprived eye with a few gaps in the wave form. When the noise was significantly increased, a pattern with thinned and corroded stripes was still observed after 500 time steps but both the corroded regions between the thinned stripes and the defects in the stripes had completely annealed out after 6000 time steps yielding a perfect pattern of thinned stripes.

The same result was found when the initial cofiguration was chosen to be random. For example an extremely corroded pattern with very short thinned stripes was observed at early times for sufficiently high noise levels. This pattern then developed into a pattern of thinned uncorroded stipes after approximately 4000 time steps. The related histogram was calculated and confirmed the thinning of the stripes. However, when the noise level was greatly increased, the highly corroded pattern remained and the histograms revealed an almost completely monocular state.

6 Discussion and Conclusion

Section 2.2 presents a new neural network model for the formation of striped ocular dominance patterns on primate visual cortex. The model and its extensions based on the following considerations.

(i) The model is derived from known physiology of the retina and visual cortex. In this regard, the most important aspect of retinal physiology is the center-surround receptive field, $W(\mathbf{x})$, of the retinal ganglion cells.

(ii) The model is simplified by using a Born-Oppenheimer approximation on the retinal outputs, $J(\mathbf{x}, t)$, which then become the inputs to the cortex. The approximation is valid due to the large difference in characteristic times between changes in the visual scene and changes in strengths of cortical synapses.

(iii) When this approximation is incorporated into Hebbian learning dynamics the result is an effective intra–cortical interaction which is responsible for the formation of striped patterns when both eyes receive normal visual stimulation.

(iv) Monocular deprivation is first modeled by including a bias field into the formalism. However since the patterns corresponding to stripe corrosion were clearly unstable, we introduced a novel version of our model for monocular deprivation which extended the original model but was based on a local field.

The dynamical equation derived from the above considerations is a Langevin equation for a one-component non-conserved order parameter with a long range interaction in the presence of an external field.

The model was related to an Ising model with the same interaction, and its equilibrium phase behavior was investigated using mean field theory and numerical simulations. For isotropic interactions, both mean field theory and numerical simulations give three equilibrium phases at sufficiently low temperatures: a stripe phase for low field values, a hexagonal phase composed of a triangular array of bubbles of minority spins, and finally a uniform paramagnetic phase. This phase behavior was also found in mean field calculations and Monte Carlo simulations for weakly anisotropic interactions. Simulated annealing results for strongly anisotropic interactions indicate that the system can be characterized by thinned and corroded stripes for $H > 0$ and that no hexagonal phase is possible.

These conclusions were confirmed by examining the effect of anisotropy on solutions of the Langevin equation. The Langevin equation has the advantage that convergence is generally faster than for the Monte Carlo method and that an approximate but complete analytical solution has been found from singular perturbation theory for zero noise intensity ($\nu = 0$) and zero field ($H = 0$). The numerical solutions were obtained by discretizing and solving the Langevin equation on 128×128 and 256×256 square lattices. The analytic solution provided an approximation to the late stage dynamics not accessible through numerical simulation. This approximation predicts that domains (or regions of stripes with the same orientation) grow at a rate of $t^{1/2}$. This growth rate was implied in the scaling form of the pair correlation function discussed in Section 4. Interestingly, the singular perturbative solution gives similar results for a variety of physical systems, including Rayleigh-Benard convection (i.e., the Swift-Hohenberg equation [30]) and two dimensional magnetic bubbles [3]. The similarity suggests that the visual cortex model developed in this work is in the same dynamic universality class as these other systems. Two generic features of these systems are the selection of a finite wavelength (which leads to a striped equilibrium state) and non-conservation of the order parameter.

The model for monocular deprivation presented in section 4 involves a global symmetry breaking field, H, which was meant to represent the dominance of the undeprived eye. For weak anisotropies, a sufficient increase in H resulted in a change in phase to a hexagonal 'blob' phase even though the histograms resembled those found experimentally. Patterns of thinned, corroded stripes were found when the the anisotropy was made very high and even then the patterns were at best metastable or even unstable relative to a uniform state. This was the concluison of section 4. In section 5 we made an initial attempt to improve this situation by imposing a local field based on the immediate neighbourhood of a cortical cell. This was meant to model, in the simplest way possible, the increase in the number of afferents at the level of the visual cortex from the undeprived eye. This extension of the model produced stable thinned stripes as seen experimentally. However further work needs to be performed to establish the nature of the corroding of the stripes for this model. A further difficulty with monocular deprivation and its effect on primate visual cortex is the lack of reliable data for the spatial change in the patterns related to monocular deprivation.

Several extensions to our model are worth considering. For example, ocular dominance is frozen at the end of the critical period. This may result either from central control of the plasticity of neural connections in the cortex or from an intrinsic termination of the developmental process. Some physiological evidence [31] indicates the former, but further investigation will determine whether the action of non-linear terms can be excluded as an explanation for the intrinsic termination. Finally, several different mathematical expressions of Hebbian learning exist, all possessing the same qualitative features: input activity that results in similar output activity strengthens the input connection; other input activity weakens it. It is important to investigate the robustness of simulation results as the mathematical form of the learning function varies. We are also using a generalised version of this model to study orientational selectivity. Another important direction for research into pattern formation in ocular dominance is the effect of the shape of the relevent parts of the visual cortex and the pre-synaptic sheets on the nature and development of the pattern. A recent paper by Murphy [32] looks at this problem from a connectivity point of view and it is also being examined using the model of section 2.2 in conjuction with an algorithm for placing cortical cells within a shape corresponding very closely to that of layer IV, area 17 of the visual cortex.

The benefit of investigating models of this type is illustrated by two robust predictions. First, combining the size of retinal receptive fields with the cortical magnification factor produces a quantitative prediction of the period of stripes on the cortex with no model-dependent factors. The predicted period is 350 μm, which lies in the range of measured stripe periods. Second, since stripe orientation is controlled by receptive field anisotropy, the anisotropy of the cortical magnification factor can be used to predict the global pattern of orientation stripes throughout area 17, again with no model-dependent factors. Observed patterns are in qualitative agreement with this prediction. The second prediction suggests that it is possible to create a complete model of area 17, including boundary conditions, and observe its global development. The separation of time scales used in this model makes economical simulation of such a system possible. Thus, we believe that models based on the ideas presented here form an important adjunct to the experimental tools of neuroanatomy and neurophysiology.

Acknowledgments

This work was supported by by the NSERC of Canada, and by le Ministère des Affaires Intergouvernementales du Quebec. One of us (MJZ) wishes to thank Angela Brown, Barbara Chapman, Rodney Cotterill, Lionel Harrison, John Hertz, William Hodos, Dale Hogan, Leslie Holden, Michael Lyons and Roger Ward for extremely fruitful discussions and to Professor Wilshaw and Dr Goodhill for sending both relevant articles and a thesis.

References

1 Elder K R and Grant M (1989) J. Phys. (London) **A23** L803
2 Eschenfelder A H (1981) *Magnetic Bubble Technology*, Springer Verlag, Heidelberg
3 Roland C and Desai R C (1990) Phys. Rev. B **42** 6658
4 Murray J D (1989) *Mathematical Biology*, Springer Verlag, Berlin Heidelberg, pp. 372–592
5 LeVay M, Connoly M, Houde J D and van Essen D C (1985) J. Neurosci. **5** 486
6 Wiesel T N (1982) Nobel Lecture, Biosci. Reports **2** 351
7 Details of our theory together with most of the results presented here are contained in three papers [8, 9, 10] the first two of which were published in conference proceedings.

8 Thomson J R (1989) M.Sc. Thesis, McGill University; Thomson J R, Cowan Wm, Zuckermann M J and Grant M (1989) in *Lectures on Thermodynamics and Statistical Mechanics*, Proceedings of the XVIII Winter Meeting on Statistical Physics, Oaxtepec, Mexico (ed. A.E. Gonzalez, M. Medina-Noyola and C. Varea), World Scientific, Singapore, pp. 38–51.
9 Thomson J R, Zhang Z, Cowan Wm, Grant M, Hertz J A and Zuckermann M J (1990) in Proceedings of the Third Nordic Conference on Computer Simulation in Physics, Chemistry, Biology and Mathematics, Lahti, Finland. (ed. K. Kaski and M. Salomaa), Topical Issue of Physica Scripta **T33** pp. 102–109.
10 Thomson J R, Cowan Wm, Elder K R, Daviet Ph, Soga G, Zhang Z, Grant M, and Zuckermann M J (1993) J. Biol. Phys. **18** 217
11 Hebb D O (1949) *Organization of Behavior*, John Wiley and Sons, New York
12 Prestige M C and Wilshaw D J (1975) Proc. Roy. Soc. (London) B **190** 77
13 Malsburg C von der (1973) Kibernetik **14** 85
14 Wilshaw D J and Malsburg C von der (1976) Proc. Roy. Soc. (London) B **194** 431
15 Goodhill G J (1992) Cognitive Science Research Paper Serial No. CSRP 226, School of Cognitive and Computing Sciences, University of Sussex, Great Britain and references therein.
16 Goodhill G J (1988) M.Sc. Thesis, University of Edinburgh.
17 Goodhill G J and Wilshaw D J (1990) Network **1** 41.
18 Swindale N V (1980) Proc. Roy. Soc. Lond. Series B **208** 243: (1989) *ibid.* **245** 605
19 Miller K D, Keller J B and Stryker M P (1980) Science **245** 605
20 Lyons M J and Harrison L G (1990) to be published.
21 Goodhill G J and Barrow H G (1994) Neural Computation **6** 255 and references therein.
22 Dow B M, Vautin R G and Bauer R (1982) J. Neurosci. **324** 221
23 Gunton J D and Droz M (1983) *The Theory of Metastable and Unstable States*, Springer Verlag, Heidelberg
24 Møller P (1987) M.Sc. Thesis, Niels Bohr Inst., University of Copenhagen
25 Møller P, Nylén M and Hertz J A (1988) in *Neural Computers*, (ed. R. Eckmiller and C. von der Malsburg), pp. 139-148, NATO-ASI Series F, Computer Systems Science, Springer Verlag, Heidelberg
26 Mouritsen O G (1984) *Computer Studies of Phase Transitions and Critical Phenomena*, Springer Verlag, Heidelberg
27 Kawasaki K (1985) Phys. Rev. **A31** 3880
28 Suzuki M J (1977) J. Stat. Phys. **16** 11
29 Edelman G M, Gall W E and Cowan W M (1985) Molecular Bases of Neural Development, Wiley, New York.
30 Swift J and Hohenberg P C (1977) Phys. Rev. A **15** 319
31 Singer W, Tretter F and Vinon R (1985) J. Neurosci. **5** 890
32 Murphy, K. private communication: Jones, D.G., van Sluyters, R.C. and Murphy K., (1991) J. Neurosci. **11** 3794

Part III

Dynamics of Biological Macromolecules

Channel Function and Channel-Lipid Bilayer Interactions

Olaf S. Andersen, Jens A. Lundbæk, and Jeffrey Girshman

Abstract

Ion channels are integral membrane proteins that form aqueous pores, which span the lipid bilayer moiety of biological membranes and provide for highly selective transfer of ions across the membrane. Ion transfer through the pore can occur at very high rates, and it is possible to use electrophysiological measuring techniques to record the function of single channels in real time. Ion channels are therefore useful for examining many aspects of macromolecular dynamics. The control of channel function is due to transitions between different channel states (conformations). The distribution between these states is determined by the channel's intrinsic characteristics and by its interactions with the (membrane) environment, neither of which are well understood. We show, using the well-characterized gramicidin A channel as an example, that membrane control of channel function can be rationalized by considering the energetics of channel-bilayer interactions.

1 Introduction

Membrane-spanning channels catalyze solute movement across biological membranes by forming aqueous pores through which small inorganic anions, cations, other polar solutes, and even water (Agre et al., 1993), cross the membrane (Andersen and Koeppe, 1992; Hille, 1992). Ion permeable channels (ion channels) serve as the fundamental elements in the membranes of excitable cells (Hille, 1992), and are important for many other cell functions as well, e.g. the transmembrane movement of fluids across epithelia (Palmer and Sackin, 1992) or cell volume regulation in general (Hoffmann and Simonsen, 1989; Sackin, 1994). Channels are important for normal cell function because of their catalytic power; single channels can mediate large, selective solute fluxes $> 10^6 s^{-1}$ (see below). A given channel type therefore may occur at a low membrane surface density and yet allow for a large solute flux across the membrane. Uncontrolled solute fluxes are deleterious, however, because they may disrupt electrical excitability and compromise a cell's ability to control its volume. A channel's functional state, whether ions or other solutes can permeate or not (whether the channel is conducting or non-conducting), therefore must be under tight control.

The walls of the membrane-spanning pores are formed by integral membrane proteins, as shown schematically in Figure 1.

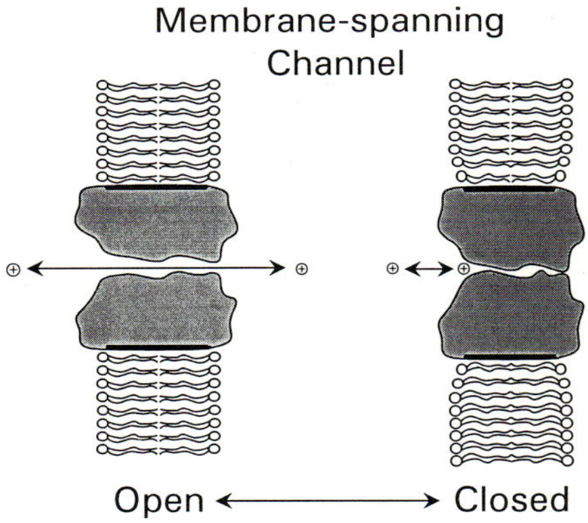

Fig. 1. Schematic representation of a membrane-spanning channel. The channel can exist in two distinct states: open, where solutes (ions) can pass through the pore; and closed, where the pore is occluded, such that solutes cannot pass through. The hydrophobic exterior surface of the channel protein is denoted by the heavy lines. If the open↔closed transitions involve changes in the channel protein's quaternary structure that alter the length of the hydrophobic exterior, the lipid bilayer will adjust to the changing length in order to have hydrophobic matching (Mouritsen and Bloom, 1984).

Transitions between different (non-conducting and conducting) functional states are due to transitions between different quaternary structures. The equilibrium distribution between these states is determined by the difference in their intrinsic conformational energies and by the difference in the membrane deformation energy associated with a quaternary conformational change (cf. Figure 1). The selectivity of a channel (the ability to discriminate among different ions), and the rate at which the selected ion(s) can pass through the pore, are determined by the chemical identity and structural arrangement of the amino acid residues that line the open (conducting) pore.

2 Why Ion Channels?

The current interest in channels arises, in part, because the physiological functions of ion channels depend on features that make this class of integral membrane proteins suitable for examining general questions relating to macromolecular dynamics. First, ion channel can be highly selective. Voltage-dependent cation channels, for example, can discriminate between Na^+ and K^+ with selectivity ratios that vary between $\sim 15:1$ (for Na^+ over K^+ in voltage-dependent sodium channels Hille, 1972) and $\sim 100:1$ (for K^+ over Na^+ in voltage-dependent potassium channels) (Hille, 1973). Second, the catalytic power of ion channels is very high; the turnover number in cation selective channels can exceed $2 \times 10^8 s^{-1}$ (Andersen, 1983; see also Hille, 1992). Third, the functional state of ion channels is tightly regulated - e.g. by the membrane potential difference, chemical transmitter substances, or intracellular second messengers (e.g. Hille, 1992).

The ion selectivity (substrate specificity) means that ion permeable channels are highly specific protein catalysts that mediate transmembrane substrate movement. Ion channels therefore are enzymes (Andersen, 1989; Eisenberg, 1990; Andersen and Koeppe, 1992) - albeit members of a special class of enzymes in which the substrate-to-product turnover does not involve the breaking or making of covalent bonds (Andersen and Koeppe, 1992). The high turnover number means that one can visualize the dynamics of ion channels in real time at the single-molecule level (Figure 2), as a net flux of 10^6 monovalent ion/s corresponds to a single-channel current of 0.16 pA, which is readily measurable. This makes ion channels unique among enzymes - and very attractive tools to examine many aspects of macromolecular dynamics.

The voltage control of channel function means that one can synchronize the transition from a closed to an open state in a large number of channels by a change in membrane potential. This provides for kinetic measurements at high time resolution (e.g. Sigworth, 1993). It also means that many aspects of channel function can be examined only on channels in polarized membranes, which may limit attempts to elucidate a channel's structure.

3 Structure of Ion Channels

As is the case for most other integral membrane proteins, there is only limited information available about the three-dimensional structure of ion channels (Unwin, 1989; Andersen and Koeppe, 1992; Cowan and Rosenbusch, 1994). This ignorance is in striking contrast with the wealth of one-dimensional information that is available about the primary structures of integral membrane proteins (e.g. Numa, 1989; Hille, 1992), and about the kinetic details of channel function obtained from electrophysiological measurements, such as single-channel current recordings (Sine et al., 1990; Auerbach, 1993).

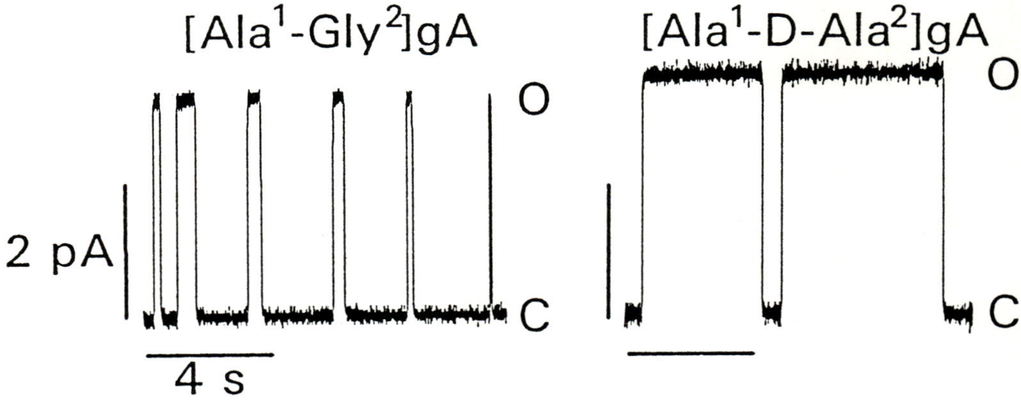

Fig. 2. Single-channel current traces obtained with two different gramicidin analogues [Ala[1] – Gly[2]]gramicidin A and [Ala[1] – D – Ala[2]]gramicidin A. The aqueous electrolyte was 1.0 M NaCl and the bilayers were formed by diphytanoylphosphatidylcholine in n-decane. The applied potential is 200 mV, and the temperature is °25 C. Experimental details as in Andersen (1983).

Atomic resolution structures are known for two channel classes of bacterial origin. One is the porins, a class of proteins that span the outer membrane of Gram-negative organisms and serve to connect the periplasmic space with the extracellular solution proper. The channel structure was obtained from solving the X-ray diffraction pattern (Weiss et al., 1991), and shows that the pore is formed by a β-barrel. The other atomic resolution structure is that of the gramicidin A channel, which was deduced from two-dimensional (Arsen'ev et al., 1985) and solid-state (Ketchem et al., 1993) nuclear magnetic resonance spectra. The common features of these channels are: the membrane-spanning part of the channels are β-sheets, with the outer surface being formed by hydrophobic residues; and there is a very high density of aromatic residues at the membrane-solution interface, which may serve to provide a flexible link between the protein and the surrounding bilayer.

Low-resolution structures are available for two vertebrate ion channel classes, gap junction channels (e.g. Unwin and Ennis, 1984) and nicotinic acetylcholine receptors (nAChR) (Unwin, 1993). In the gap junction channel structure, the resolution is sufficient to resolve the general organization of the subunits, which are organized as a symmetrical hexameric barrel-stave. There is less information about the secondary structure, but the available results are consistent with there being four membrane-spanning α-helices (Milks et al., 1988; Tibbitts et al., 1990). In the nAChR structure the five subunits are organized as a pseudo-symmetrical barrel-stave. The resolution is sufficient (~ 0.9nm) to resolve secondary structural elements, and there is evidence for at least one pore-lining membrane-spanning α-helix[2].

Despite the relatively poor structural resolution of gap junction and nAChR channels, there is good evidence for quaternary structural transitions in the channel structures, which can be related to different functional states (Unwin and Ennis, 1984; Unwin et al., 1988). The conformational changes involve a tilt of the subunits with respect to the pore axis (bilayer normal), which will alter the overall channel length, particularly the length of the channel's hydrophobic exterior surface. It thus appears that changes in the channel's quaternary structure involve the same type of (sliding and rotating) domain movements that control the function of allosteric enzymes (Perutz, 1990). There is not sufficient information to describe the conformational transitions further, but the available information does allow for considerable insights into how the physical state of a lipid bilayer can alter the distribution between different channel states and thus modulate channel function (see below).

4 Ion Permeation

The lipid bilayer moiety of biological membranes serves as a barrier for small ions and polar non-electrolytes because the bilayer's hydrocarbon core is a very poor solvent for polar solutes. In 1.0 M NaCl, for example, the specific conductance (G) of a planar lipid bilayer is $\sim 10^{-9}$S/cm^2 (Hanai et al., 1965a). The membrane/water partition coefficient ($K_{m/w}$) of an ion can be estimated to be $\sim 10^{-14}$ by using the relation (Hodgkin and Katz, 1949):

$$G = N_A \cdot [(ze)^2/kT] P_m C_m \tag{1a}$$

$$\approx N_A \cdot [(ze)^2/kT](D_m/\delta)(K_{m/w} C_{aq}) \tag{1b}$$

where N_A is Avogadro's number, z the ion valence, e the elementary charge, k Boltzmann's constant, and T the absolute temperature. P_m is the (average) ion permeability in the membrane, and C_m the (average) ion concentration in the membrane. $P_m \approx D_m/\delta$, where D_m is the diffusion coefficient in the membrane's hydrocarbon interior, $\sim 10^{-5}$cm^2/s (Schatzberg, 1965; see also Andersen, 1978), and δ the thickness of the membrane's hydrophobic core,

[2] The primary sequences of nAChR receptor subunits contain four strikingly hydrophobic segments (M1-M4), which are long enough to span a lipid bilayer in an α-helical conformation. Unwin (1993) was able to identify only one such element (the pore-lining M2, either because the other hydrophobic segments are not α-helical or because the secondary structure was deduced by rotational averaging the density distribution in the five subunits, in which case small structural differences among the subunits would tend to produce a rather feature-less structure.

$\sim$ 5nm (Hanai et al., 1965b)[3]. $C_m \approx K_{m/w} C_{aq}$, where C_{aq} is the aqueous ion concentration.

$K_{m/w}$ is determined by the electrostatic transfer energy ($\Delta G^o_{\text{trans}}$) associated with moving ions from water, with a dielectric constant ($\epsilon_w \approx 80$), to the membrane interior, with a dielectric constant ($\epsilon_m \approx 2$). To a first approximation, $\Delta G^o_{\text{trans}}$ can be estimated using the charging procedure of Born (cf. Parsegian, 1969):

$$\Delta G^o_{\text{trans}} \approx N_A \cdot [(ze)^2/(8\pi\epsilon_o r)] \left[\frac{1}{\epsilon_m} - \frac{1}{\epsilon_w}\right] \tag{2}$$

where ϵ_o denotes the permittivity of free space, and r the ion radius. Eq. 2 tends to overestimate $\Delta G^o_{\text{trans}}$ because the ion's interaction energy with the aqueous phase will be overestimated (e.g. Bockris and Reddy, 1970); but it serves to illustrate the importance of the ion's electrostatic interactions with its surroundings in determining the magnitude of $\Delta G^o_{\text{trans}}$. Integral membrane proteins catalyze the transmembrane movement of polar solutes by increasing $K_{m/w}$ (decreasing $\Delta G^o_{\text{trans}}$) because the dielectric constant of a protein is higher than that of the bilayer's hydrocarbon core. Experimental estimates of the dielectric constant of proteins vary between 4 and 25 (e.g. Andersen and Koeppe, 1992). In the case of ion channels, water in the pore will serve to further decrease $\Delta G^o_{\text{trans}}$.

The dielectric constant of pore water is an unknown quantity, however, except that it will be less than that of bulk water. The high dielectric constant of bulk water results from the association of water dipoles due to hydrogen-bonded interactions (e.g. Bockris and Reddy, 1970). This hydrogen-bonded association will be less within the pore, which will tend to decrease the effective dielectric constant of water in the pore. This becomes a particular problem for the voltage-dependent cation-selective channels, which have cross sectional areas that are comparable to that of a water molecule (e.g. Hille, 1992), and for channels formed by the linear gramicidins, where the water is organized in a one-dimensional (single-file) array (Rosenberg and Finkelstein, 1978a; Rosenberg and Finkelstein, 1978b). Hydrogen-bonded, or hydrophobic, interactions between the pore wall and water molecules will tend to reduce the importance of water association even when the pore diameter is comparable to a few water diameters; but little appears to be known about that question. Ions in the pore, or charges in the pore wall, will tend to reduce the dielectric constant even further, because the electric field set up by any charge will tend to reorient the water molecules (Mackay et al., 1984; Green and Lewis, 1991) and thereby reduce the dipolar susceptibility. In gramicidin-like channels, for example, the dipolar susceptibility is reduced

[3] Planar lipid bilayers formed from a suspension of lipids in a linear alkane have a thickness that is larger than that of nominally hydrocarbon-free bilayers formed by the same lipids (Benz et al., 1975; White, 1978).

$\sim$ 100-fold, and the effective dielectric constant should be $\sim$ 4 when a monovalent ion is introduced into the pore (Partenskii et al., 1991; Partenskii and Jordan, 1992). In this case, however, the ordered water may actually improve ion solvation in the pore (Mackay et al., 1984) because of favorable ion-dipole interactions between the ion and distant water molecules.

The channel protein's dielectric properties allow for high transfer rates through the pore (across the membrane); but how does a channel select among chemically similar ions, such as the alkali metal ions? The high selectivity arises because the permeating ions interact with the pore walls. Direct evidence for ion-channel interactions can be obtained by examining how the single-channel conductance (g) varies as a function of the aqueous concentration of the permeant ion species. (Such experiments can be done under symmetrical conditions, where the chemical composition of the aqueous phases on the two sides of the membrane are identical, or under asymmetrical conditions, where the permeant ion concentrations in the two solutions differ. We will in the following refer to the symmetrical situation only.) Figure 3 shows results for the cation-selective channels formed by the peptide antibiotic gramicidin A.

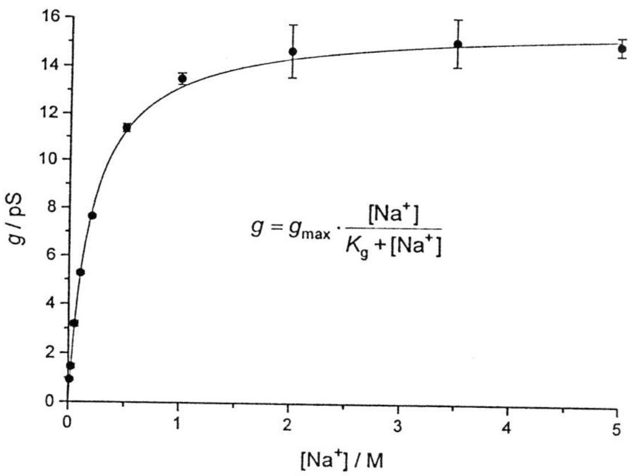

Fig. 3. Conductance-concentration relation for [Val^1]gramicidin A channels using Na^+ as the permeant ion. Diphytanoylphosphatidylcholine/n-decane membranes, 25 °C. The points denote mean standard deviation. The solid line denotes the results of a non-linear least squares fit to Eq. 3: $K_g = 0.202$ M; $g_{\max} = 15.8$ pS.

At low permeant ion concentrations, g increases as an approximately linear function of $[Na^+]$. At high concentrations, g appears to reach a maximal value, and the relation between g and $[Na^+]$ can be approximated as

$$g = g_{\max}[Na^+]/(K_g + [Na^+]), \tag{3}$$

where $g_{\max}$ denotes the maximal conductance, and K_g the concentration for half-maximal conductance. Eq. 3 is the conductance-concentration relation for a channel that can be occupied by at most one ion (Läuger, 1973; Andersen, 1989). Eq. 3 is similar to the Michaelis-Menten relation in enzyme kinetics (e.g. Fersht, 1985). Taking the analogy to enzymes further, the pore corresponds to the active site.

The significance of Eq. 3 is slightly different, however, from that in conventional enzyme kinetics, because channel-mediated ion movement usually occurs at near-equilibrium conditions - meaning that both uni-directional fluxes (left→right and right→left) are finite. In the limit when the membrane potential (V) approaches 0, for example, there will be large uni-directional fluxes through the pore (due to thermal ion movement) but no net flux. The distribution of the permeant between the aqueous phases and pore will be at equilibrium and K_g will denote the dissociation constant for ion binding in the pore.

$g_{\max}/K_g$ is related to the channel's (diffusional) permeability coefficient (p_o) (e.g. Andersen, 1989):

$$g_{\max}/K_g = N_A[(ze)^2/kT]p_o, \tag{4}$$

where z denotes the ion valence, F is Faraday's constant, R the gas constant, and T the temperature in Kelvin. If the pore were a simple aqueous pore, p_o could be approximated by

$$p_o = \pi r_o^2 D/l \tag{5}$$

where r_o denotes the channel's capture radius (Ferry, 1936), the difference between the channel radius (r_c) and the ion radius (r_i):

$$r_o = r_c - r_i, \tag{6}$$

while D is the ion's diffusion coefficient, and l the channel length. (We follow Bean, 1972, and do not consider hydrodynamic wall effects.)

In the case of Na^+ ($r_i \approx 0.1$ nm; $D = 1.3 \times 10^{-5}\text{cm}^2/\text{s}$) moving through gramicidin channels ($r_c \approx 0.2$ nm; $l \approx 2.5$ nm), p_o can be predicted based on Eq. 5 to be $\sim 1.6 \times 10^{-14}\text{cm}^3/\text{s}$. (The units arise because we estimate the permeability coefficient for a single channel.) p_o can also be estimated based on analysis of the g-[Na^+] relation (Fig. 3) using Eqs. 3 and 4. In this case, $g_{\max}/K_g \approx 78$ pS/M and p_o can be estimated to be $\sim 2.1 \times 10^{-14}\text{cm}^3/\text{s}$. The agreement between the two estimates for p_o is remarkable, particularly so because we have neglected hydrodynamic wall effects, which means that Eq. 5 should provide an overestimate of p_o. In fact, the p_o estimate obtained from the g-[Na^+] relation is larger than based on simple geometric considerations, which is particularly surprising because we in the above arguments also neglected the electrostatic barrier for ion movement through the pore[4].

[4] The relation between pore dimensions and the permeability properties is often much more complex than is the case for the gramicidin channels. As summarized

Gramicidin channels are not simple aqueous pores, however, as evidenced by the non-linear g-$[Na^+]$ relation in Figure 3. The saturating behavior indicates that there is an upper limit on the number of Na^+ that can occupy the pore. Assuming that at most one Na^+ can occupy the pore[5], the term $[Na^+]/(K_g + [Na^+])$ denotes the fraction of time the pore is occupied by a Na^+, and $[Na^+]/(K_g + [Na^+]) = 0.5$ when $[Na^+] = 0.2$ M (legend to Figure 3). The effective pore volume (the volume available to the ion's center) is l or $\sim 0.08\ nm^3$, and the time-averaged Na^+ concentration in the pore $[Na^+]_p$ is

$$[Na^+]_p = \left\{[Na^+]/(K_g + [Na^+])\right\}/(\pi r_o^2 l N_A). \tag{7}$$

When $[Na^+] = 0.2$ M, the pore is half-occupied, and $[Na^+]_p \approx 10$ M! That is, the time-averaged Na^+ concentration in the pore is ~50-fold higher than in the bulk aqueous phase. At low permeant concentrations, when $[Na^+] \ll K_g$, the partition coefficient between the pore and the bulk phase ($K_{p/w}$) will be $1/(K_g \cdot \pi r_o^2 l N)$ or ~ 100. The somewhat surprising conclusion is that permeant ions prefer the pore to the bulk phase.

The preference for ions to reside in the pore is due to direct interactions between the permeating ions and polar groups in the pore wall Cations, for example, are solvated better by formamide and dimethylformamide, which mimic peptide groups, than by water (Cox et al., 1974). This is presumably because carbonyl oxygens carry a more negative partial charge than water oxygens. Anions, in contrast, are solvated poorly by such peptide-group mimics (Cox et al., 1974), which means that amino acid side chains are likely to play a crucial role in the solvation of anions. Not surprisingly, therefore, the valence selectivity of cation selective channels is very good, while that of anion selective channels tends to be modest (e.g. Hille, 1992).

The catalytic rate enhancement provided by the channel can be approximated by the ratio $K_{p/w}/K_{m/w}$, which is $\sim 10^{16}$ - similar to the rate enhancement seen for conventional enzymes (Kraut, 1988).

5 Control of Channel Function

The catalytic power (the high turnover number) of conducting channels makes it necessary to have tight control over channel function. The opening (or closing) of only a few channels can have profound effects on the membrane potential, and thus on the propagated action potentials in excitable membranes (Hodgkin and Huxley, 1952; Hille, 1992). Similarly, a sudden change

by Finkelstein (1985), many channels with very wide lumena have surprisingly small single-channel conductances.

5 A more complete analysis (O.S. Andersen, M.D. Becker, and J. Procopio, in preparation) shows that the gramicidin channel can be occupied by two Na^+. This is not, however, of consequence for the following arguments.

in the number of open channels will alter the balance between (passive) electrodiffusive channel-mediated ion fluxes and active ion transport (catalyzed by ATP-driven ion pumps or some other active transport system), which may lead to a net gain (or loss) of solute from the cell and thus to a change in cell volume (Hoffmann and Simonsen, 1989).

As a reflection of the need to minimize uncontrolled ion movement, most membrane-spanning channels possess two different classes of non-conducting states (Figure 4): closed states and inactivated, or desensitized, states[6].

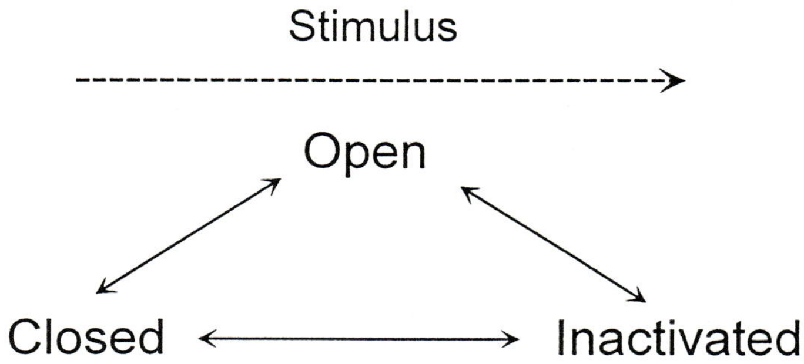

Fig. 4. Interrelation between closed, open and inactivated states and the applied stimulus (membrane potential change or concentration of a chemical transmitter substance).

A physiological stimulus, such as an increase in the membrane potential from the resting value of -60 to -80 mV to some more positive value, will favor transitions from the resting, closed state(s) to an open state. Most channels will after some time re-enter a non-conducting state, which is different from the resting state. The most clear-cut distinction between the two classes of non-conducting states is that transitions to the inactivated states are favored by the same stimulus that causes the channel to open (Horn, 1984; Andersen and Koeppe, 1992). Channels can be deactivated (return from the open to a closed state) by reverting the membrane potential to the resting level and immediately be ready for activation again. If the channel is inactivated, it cannot be activated (opened) by a further positive-going change in the membrane potential and, if the membrane potential is returned to the resting value, it will not be ready for activation until after some time delay.

In practice, channels tend to have many closed states and one, or a few, open states (Sigworth, 1993; Correa and Bezanilla, 1944; Zagotta et al., 1994). The definition of what constitutes a distinct channel state is operational, however, and depends on the frequency response that is available to the

[6] Functionally inactivated and desensitized states are equivalent. Historically, the term inactivated is used when describing voltage-dependent channels, while the term desensitized is used when describing chemically gated channels, such as the nAChR.

experimenter (Andersen and Koeppe, 1992). Disregarding this uncertainty, and focusing only on channel activation, the steady state activation of a channel by a change in membrane potential can be approximated by using a linear kinetic scheme:

$$C_3 \overset{L_2(V)}{\leftrightarrow} C_2 \overset{L_1(V)}{\leftrightarrow} C_1 \overset{L_0(V)}{\leftrightarrow} O \qquad \text{(Scheme I)}$$

where the $L_i(V)$'s denote the voltage-dependent equilibrium constants between the different states, e.g.: $L_2(V) = W_2(V)/W_3(V) = L_2(0)\exp\{z_2eV/kT\}$, where the $W_i(V)$'s denote the probability of being in state i, and z_2 denotes the effective valence associated with the $C_3 \leftrightarrow C_2$ transition. The free energy of channel activation relative to C_3, $\Delta G_a(0)$ is given by $RT\ln\{L_2(0)\cdot L_1(0)\cdot L_0(0)\}$, and the probability that the channel is in the open state, is given by

$$W_0(V) = \frac{L_2(V)\cdot L_1(V)\cdot L_0(V)}{1 + L_2(V) + L_2(V)\cdot L_1(V) + L_2(V)\cdot L_1(V)\cdot L_0(V)}. \tag{8}$$

Figure 5 shows examples of $W_0(V)$ for different choices of $L_2(0)$, $L_1(0)$, and $L_0(0)$, with $z_2 = z_1 = z_0 = 2$ and the product $L_2(0)L_1(0)L_0(0)$ kept fixed at 10^9.

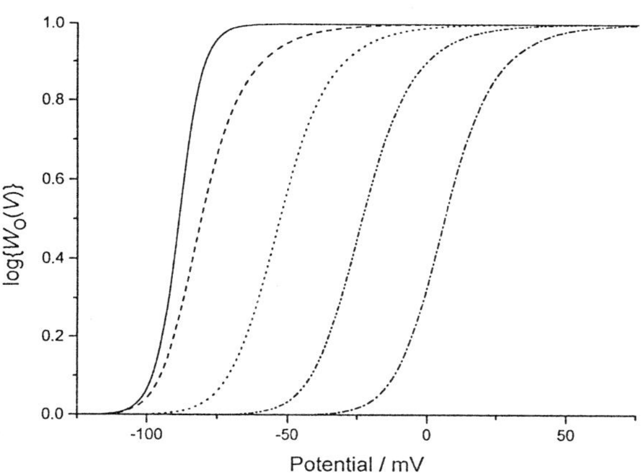

Fig. 5. Activation for a voltage-dependent channel with three closed and one open state (Scheme I). See text for details. The five different simulations were done as follows (from left to right): $L_2 = 1$, $L_1 = 1000$, $L_0 = 1000000$; $L_2 = 1000$, $L_1 = 1000$, $L_0 = 1000$; $L_2 = 100000$, $L_1 = 100$, $L_0 = 100$; $L_2 = 10000000$, $L_1 = 10$, $L_0 = 10$; $L_2 = 1000000000$, $L_1 = 1$, $L_0 = 1$.

Not surprisingly, the shape (the steepness) of the activation curve varies as a function of how $\Delta G_a(0)$ is partitioned among the three steps. What is surprising is that different combinations of $L_i(0)$ values can give rise to almost parallel shifts in the $W_0(V)$ relation! Such shifts in the activation curve has traditionally been ascribed to alterations in the electrostatic potential at the channel/solution interface (e.g. Hille et al., 1975), but that is only one of several possible interpretations!

The free energy difference between any two states, e.g. C_1 and C_2, is given by

$$\Delta G_{1,2}(V) = RT \ln\{L_1(0) \exp\{z_1 eV/kT\}\} = \Delta G_{1,2}(0) + z_1 FV, \tag{9a}$$

$$= \Delta G^p_{1,2}(0) + \Delta G^m_{1,2}(0) + z_1 FV, \tag{9b}$$

where $\Delta G_{1,2}(0)$ has been further divided into a contribution due to the intrinsic change in the protein's structure ($\Delta G^p_{1,2}(0)$), and a contribution due to interactions between the protein and the surrounding lipid bilayer ($\Delta G^m_{1,2}(0)$), which could arise because fixed charges at the membrane-solution interface will tend to modify the potential profile across the membrane (Gilbert and Ehrenstein, 1984; Green and Andersen, 1991) or because changes in the channel's structure may be associated with a deformation of the adjacent membrane (e.g. Figure 1).

6 Membrane Deformation and Ion Channel Function

The lipids of cell membranes regulate the function of embedded membrane proteins in a seemingly nonspecific manner (Sandermann, 1978; Devaux and Seigneuret, 1985; McMurchie, 1988; Bienvenue and Marie, 1994). Such membrane-dependent changes in protein function have been ascribed to changes in the fluidity of the surrounding lipid bilayer (e.g. (Stubbs and Smith, 1984; Szöllösi 1994). But the precise role of the lipid bilayer, and of membrane fluidity, in determining membrane protein function is unclear, except for the fact that the lipid composition of living cells is controlled such that the bilayer is in a liquid-crystalline phase (e.g. Singer and Nicolson, 1972; McElhaney, 1985).

Once the lipid bilayer is in a liquid crystalline state the causal relationship between membrane fluidity and membrane protein function is obscure. It is, in fact, not clear that membrane fluidity *per se* is a useful term for describing the interactions between a lipid bilayer and its embedded proteins (McElhaney, 1985; Lee, 1991; Zakim et al., 1992; Lundbæk & Andersen, 1994). Probably the strongest argument in this respect is that the *equilibrium distribution* of a membrane protein between different states can be affected by the lipid composition of the host bilayer (Lee, 1991; Gibson and Brown, 1993; Keller et al., 1993; Lundbæk and Andersen, 1994); but the equilibrium distribution between conformational states of a protein cannot be influenced by the membrane fluidity, which would affect the forward and backward rate constants to the same extent (e.g. Schurr, 1970a; Schurr, 1970b; see also Lee, 1991). If a transition involves the protein's membrane-spanning domain the associated rate constants could be affected by changes in membrane fluidity, but that would affect function only to the extent that the kinetic step is significantly rate determining. Most methods for determining membrane fluidity (e.g. ESR, fluorescence depolarization) are sensitive not only to the rate, but also to the extent of motion of lipid molecules in the membrane (strictly only the rate of motion is a measure of membrane fluidity).

If not fluidity, what then? As alterations in the membrane lipid composition can alter the equilibrium distribution between different protein states, there has been increasing emphasis on the importance of the mechanical properties of the host bilayer for the function of embedded proteins (Gruner, 1991; Andersen et al., 1992; Gibson and Brown, 1993; Keller et al., 1993; Lundbæk and Andersen, 1994). This development is based on the recognition that different functional states of a protein correspond to different structural states, and that quaternary conformational changes in the membrane-spanning part of an integral membrane protein will alter the protein-lipid interface. Such changes in quaternary structure have been observed (see section on Structure of Ion Channels). The lipid bilayer will tend to adjust to this change in the protein's exterior surface in order to minimize the hydrophobic mismatch between bilayer and protein (e.g. Mouritsen and Bloom, 1984; see also Figure 1). A conformational change in an integral membrane protein that involves the membrane-spanning part of the protein therefore is associated with a membrane deformation, and the energetic cost of membrane deformation will be reflected in the equilibrium distribution among different conformational states, which in turn will affect the protein's function.

Insights into the physico-chemical basis for the control of membrane protein function is best obtained by direct experimental manipulations of the membrane environment, because dietary manipulations may be associated with homeostatic control of membrane properties (Hazel, 1988; McMurchie, 1988; Vigh et al., 1993). Lipid metabolites usually are sufficiently water-soluble that one can incorporate them into membranes by adding these compounds to the aqueous solution, and amphipathic lipid metabolites (such as free fatty acids and lysophospholipids) alter the function of widely diverse membrane proteins including ion channels (Kiyosue and Arita, 1986; Ordway et al., 1989; Hwang et al., 1990; Oishi et al., 1990; Swarts et al., 1990; Burnashev et al., 1991; Eddlestone and Ciani, 1991; Wallert et al., 1991; Shimada and Somlyo, 1992). The underlying mechanisms have not been determined, but the functional changes are induced at fairly high aqueous metabolite concentrations, where they are likely to directly modify the physical properties of the lipid bilayer. Indeed, the available data suggests that membrane proteins can be classified into two groups: those whose function are affected by lipid metabolites and those that have not yet been examined. As a specific example, we illustrate this control of channel function by the membrane environment using the channels formed by the linear gramicidins, a system where receptor-mediated biochemical events can be excluded.

The linear gramicidins, as exemplified by valine gramicidin A (gA), have the sequence (Sarges and Witkop, 1965):

`Formyl – L – Val – – – Gly – L – Ala – L – Leu – L – Ala–`
`–D – Val – L – Val – L – Val – D – Trp – D – Leu – L – Trp–`
`–D – Leu – L – Trp – D – Leu – L – Trp – ethanolamine.`

Gramicidin channels are formyl-NH-to-formyl-NH-terminal dimers of two right handed $\beta^{6.3}$ helical monomers that are stabilized by six intermolecular hydrogen bonds (for recent reviews see Andersen and Koeppe, 1992; Killian, 1992; Busath, 1993). The conducting gramicidin channels form by the transmembrane dimerization (O'Connell et al., 1990) of two non-conducting monomers that seem to diffuse freely in the monolayers (Tank et al., 1982). All dimers appear to be conducting (Veatch et al., 1975). The length of the channel's hydrophobic exterior surface is $\sim$2.2 nm (Elliott et al., 1983), less than the thickness of the membrane's hydrophobic core, and dimer formation is associated with a deformation (local compression) of the host bilayer (Figure 6 top).

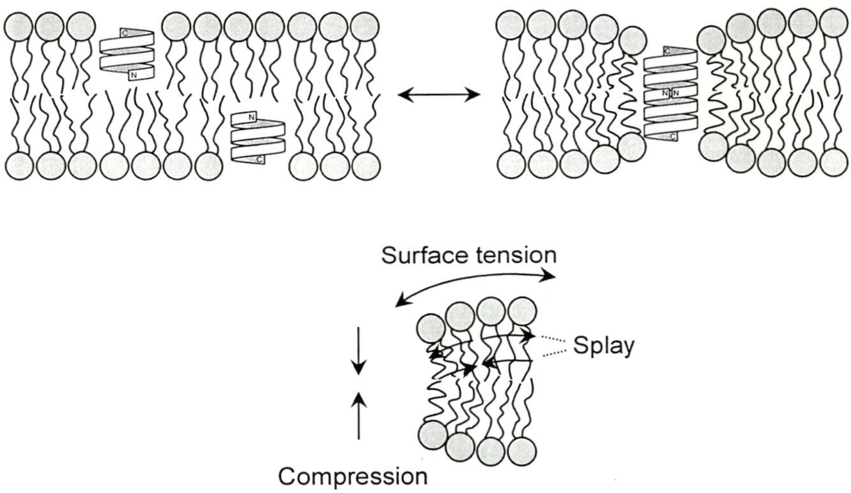

Fig. 6. Formation of gramicidin channels: Top: gramicidin channels form by the dimerization of non-conducting monomers from opposite monolayers. The channel length is less than that of the unmodified bilayer, and channel formation will be associated with a membrane deformation. Bottom: different contributions to the bilayer deformation energy.

The energetics of this membrane deformation has been examined using continuum liquid-crystal theories (Huang, 1986; Helfrich and Jakobsson, 1990), in which the overall deformation energy is broken down into contributions from three different terms (Figure 6 bottom): the compression energy -

due to the change in membrane thickness (compression of the lipid molecules in a direction perpendicular to the membrane-solution interface); the surface energy - due to the increase in the area of the membrane/solution interface, which has a finite surface tension; and the monolayer splay energy - due to the changes in the cross-sectional area available for the acyl chains along their length (from the interface to the terminal methyl group). The strict separation of these contributions to the membrane deformation energy is problematic in cases where the radii of curvature are comparable to the linear dimensions of the lipid molecules (cf. de Gennes and Prost, 1993), as is the case here. Nevertheless, the continuum analysis provides good insights into the possible mechanisms that underlie the control of membrane function by the membrane environment.

One can use gramicidin channels as molecular force-transducers to measure the energetic cost of deforming a lipid bilayer around a membrane protein, because the elastic energy stored in the membrane deformation will tend to pull the monomers apart (break the hydrogen bonds connecting the monomers). When examined using single-channel methods, the formation of a gramicidin channel is seen as a distinct increase in the magnitude of the current through the membrane; dimer dissociation is seen as a corresponding decrease in the current magnitude (e.g. Figure 2). Measurements of single-channel currents thus provide information about how the channel and the membrane interacts, because the membrane deformation energy will be a determinant of the dimerization constant as well as the average channel (dimer) duration. A change in membrane deformation energy will alter channel activity, the equilibrium constant between conducting dimers and non-conducting monomers. (This information is best obtained in ensemble-averaged measurements using membrane with many channels (Lundbæk and Andersen, 1994).) A change in the average channel duration could be due to an altered deformation energy, which will alter the "pull" on the dimer and thus the activation energy for dimer dissociation, or to a change in membrane fluidity, in the lateral diffusion of the molecules.

Following Helfrich (1973), Huang (1986), and Helfrich and Jakobsson (1990), the deformation free energy of each monolayer (ΔG_{def}) can be expressed as:

$$\Delta G_{\mathrm{def}} = 2\pi \int \left[B \cdot u^2/a + a \cdot K\{(du/dr)/r + d^2u/dr^2 - C_0\} + \gamma (du/dr)^2\right] r dr, \tag{10}$$

where a is the thickness of the unperturbed monolayer, r the radial distance from the channel/lipid interface, $u(r)$ the depth of the deformation relative to the relaxed membrane/solution interface, C_0 is the intrinsic curvature of the relaxed monolayer[7], and the three terms under the integral correspond

[7] The curvature of the planar bilayer will be zero; but the intrinsic curvature of a monolayer may well deviate from zero.

to the contributions from compression (deformation coefficient B), splay (deformation coefficient K), and surface energy (surface tension), respectively. The surface energy term is small (Huang, 1986; Helfrich and Jakobsson 1990) and will be ignored. When C_0 is zero, the compression and splay contributions combine to give the approximate expression (Huang, 1986; Durkin et al., 1993):

$$\Delta G_{\mathrm{def}} = A u_c^2, \tag{11}$$

where u_c denotes the deformation depth at the channel/lipid interface. When the intrinsic curvature of the monolayer is not zero, the deformation energy can be approximated as:

$$\Delta G_{\mathrm{def}} = A (u_c - u_0)^2, \tag{12}$$

where u_0 denotes the deformation that would be induced by the monolayer's intrinsic curvature.

The intrinsic curvature will, among other factors, be related to the shape of the molecules in the membranes (Gruner, 1985; Hui and Sen, 1989; Gruner, 1989; Keller et al., 1993; Lundbæk and Andersen, 1994). The mechanical properties of a bilayer will correspondingly reflect the different shapes of its lipid moieties.

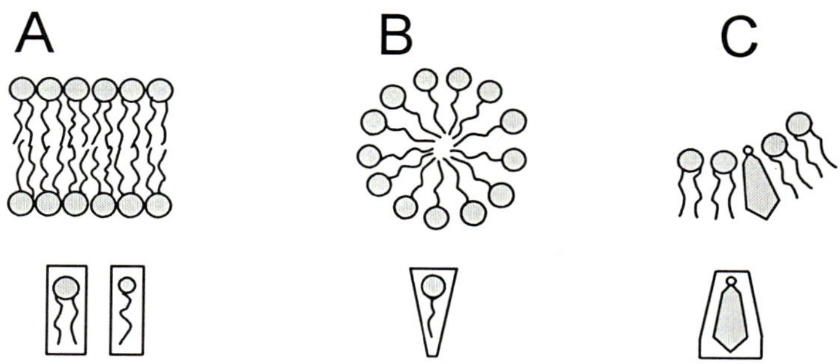

Fig. 7. Shapes of amphipathic molecules and the structures they tend to form. According to their molecular "shapes" lipid and detergent molecules will tend to form either: (A) bilayers, e.g. phospholipid and glycerolmonooleate; (B), micelles, e.g. lysophospholipids and many detergents; or (C) HII-like structures, e.g. cholesterol and long-chain alcohols.

Phospholipids constitute the majority of lipids in a cell membrane. Phospholipids have roughly cylindrical "shapes" (Cullis and de Kruijff, 1979) and form stable planar bilayers, as the polar head group occupies a cross-sectional area in the bilayer that is comparable to that of the hydrocarbon part of the molecule (Figure 7A). In micelle forming lysophospholipids (Figure 7B), the polar head group occupies a larger cross-sectional area than the single acyl chain. Such molecules can therefore be described as cone "shaped". The presence of such cone "shaped" molecules in a membrane will tend to alter the spontaneous curvature of each monolayer and decrease the energetic cost of a convex monolayer deformation around a membrane protein, which should promote channel conformations associated with such deformations. Consequently, the presence of lysophosphatidylcholine (LPC) in a bilayer will tend to increase the appearance rate and average duration of gramicidin channels (Figures 8 and 9).

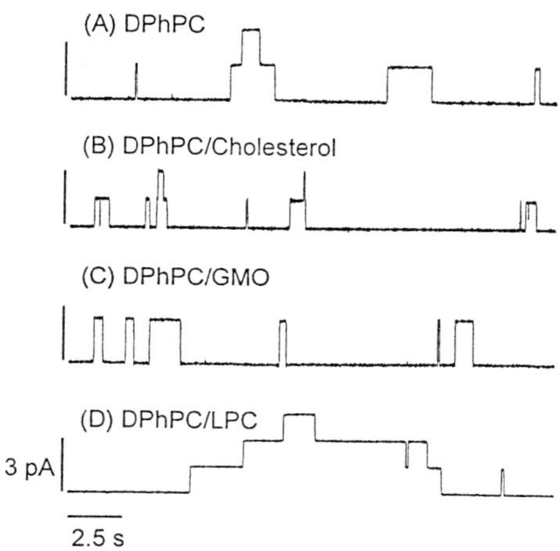

Fig. 8. Gramicidin A single-channel current traces obtained in bilayers formed from: (A) diphytanoylphosphatidylcholine (DPhPC); (B) DPhPC + cholesterol at a molar ratio of 1:2 in the membrane-forming solution; (C) DPhPC + glycerolmonooleate (GMO) at a molar ratio of 3:1; and (D) DPhPC – lysophosphatidylcholine (LPC) at a molar ratio of 4:1. 1.0 M NaCl, 200 mV, 25°C.

This effect is general: the ability of different lysophospholipids (LPLs) to modify gramicidin channels demonstrate that there is a general relation between the degree of cone shape of an LPL an its ability to stabilize the gramicidin channel (i.e. the larger the head group, the larger the effect) (Lundbæk and Andersen, 1994). Moreover, the bilayer-forming glycerolmonooleate (GMO) that also has a single acyl chain, but a very small polar head group, does not induce a corresponding increase in the average channel duration (Figures 8 and 9). In contrast to LPC, the cholesterol molecule has its smallest cross-sectional area at the membrane-solution interface (Figure 7C). Cholesterol in a membrane would therefore be expected to *increase* the energetic cost associated with a convex membrane deformation. In accordance with this notion, cholesterol decreases the duration and appearance rate of gramicidin channels in a DPhPC membrane (Figures 8 and 9). The shape-related inverse effect of cholesterol and LPC on the cost of

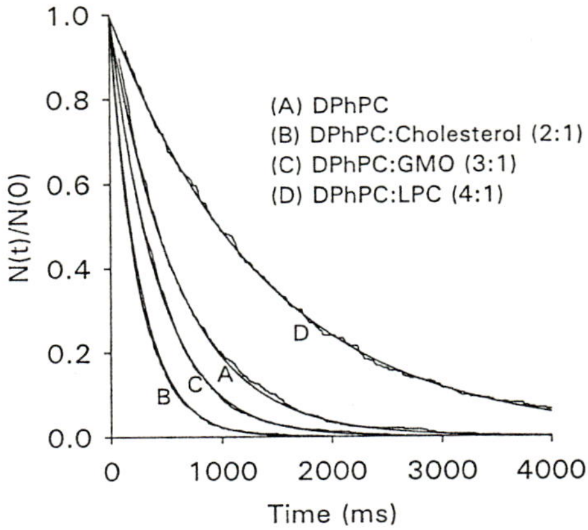

Fig. 9. Duration distributions for gramicidin channels in bilayers formed from: (A) diphytanoylphosphatidylcholine (DPhPC); (B) DPhPC + cholesterol at a molar ratio of 1:2 in the membrane-forming solution; (C) DPhPC + glycerol-monooleate (GMO) at a molar ratio of 3:1; and (D) DPhPC + lysophosphatidylcholine (LPC) at a molar ratio of 4:1. The results are displayed as normalized survivor plots. The smooth curves depict the best fits of a single exponential distribution: $N(t) = N(0)\exp\{-t/\tau\}$ to the results, where $N(t)$ is the number of channels with a duration longer than t, and τ is the average duration. τ was: (A) 600 ms; (B) 230 ms; (C) 430 ms; and (D) 1400 ms. 1.0 M NaCl, 200 mV, 25°C.

a convex monolayer deformation may explain why cholesterol in small lipid vesicles prefers the (concave) inner bilayer (Huang et al., 1974; Thompson et al., 1974) whereas LPC tends to prefer the outer (convex) monolayer (de Kruyff et al., 1977; Kumar et al., 1989)).

The relative effects of cholesterol and LPC can be rationalized by considering their general "shapes." But their different shapes may exert several different effects on a bilayer. For example, when LPC molecules are introduced into a bilayer, the surface area/hydrophobic volume ratio of the membrane will be altered, which will tend to decrease the membrane thickness (both the equilibrium thickness and thickness fluctuations). The thickness of DPhPC/n-decane bilayers is not measurably affected by the addition of LPC (Lundbæk and Andersen, 1994), however, and cholesterol actually decreases the thickness of decane-containing bilayers (Hanai et al., 1965b), which in its own right would tend to *increase* channel lifetime.

In the above we have emphasized the importance of the mechanical (energetic) interactions between gramicidin channels and their membrane environment. Traditionally, the effects of LPLs and cholesterol on membrane proteins are ascribed to the fact that LPLs decrease the order parameter of membrane phospholipids (DaTorre et al., 1991), whereas cholesterol has the opposite effect (Stockton and Smith, 1976; Yeagle, 1988). Changes in order parameter are often interpreted to signify that LPLs increase the membrane fluidity whereas cholesterol decreases it. But an increased membrane fluidity would be expected to decrease the average gramicidin channel duration. Conversely, a decrease in membrane fluidity caused by cholesterol would be expected to *increase* the average channel duration. Given the results presented in Figures 8 and 9 we conclude that fluidity changes *per se* cannot account for the changes in gramicidin channel function.

7 Membrane Proteine-Membrane Interactions

The reciprocal effects of LPC and cholesterol on gramicidin channel function suggest that these compounds exert their actions by altering bilayer deformation energy, which in turn can alter the distribution between an imbedded membrane protein conformational states.

Is there a general relationship between the cost of deforming the membrane and the function of membrane proteins? Studies on well-characterized channels suggest so. In the nAChR, the transition from the closed to the desensitized state is associated with a tilt of the of the subunits relative to the plane of membrane (Unwin et al., 1988), which decreases the length of the membrane-spanning part of the protein. It is thus of more than passing interest that the micelle-forming detergent Triton X-100, which stabilizes gramicidin channels, (Sawyer et al., 1989), promotes the desensitized state of the nAChR (Kasai and Podleski, 1970; McCarthy and Moore, 1992). Correspondingly, cholesterol promotes channel activity (McNamee and Fong, 1988). In gap junction channels, where the closed channel state is longer than the open state (Unwin and Ennis, 1984), addition of octanol promotes the closed state (e.g. Spray and Burt, C195). Octanol has a very small polar group, and would be expected to promote HII phases and is known to decrease the average duration of gramicidin channels (Pope et al., 1982). The simple notions of lipid shape and membrane deformation energy seem to provide insights into how the membrane environment can alter the function of ion channels, which is likely to pertain to integral membrane proteins in general.

Acknowledgments

We wish to thank R. E. Koeppe II, L. L. Providence, G. Saberwal, and D. B. Sawyer for helpful discussion.

This work was supported by NIH and by the Danish Medical Research Council.

References

Agre, P., G. M. Preston, B. L. Smith, J. S. Jung, S. Raina, C. Moon, W. B. Gugggino, and S. Nielsen. 1993. Aqyaporin CHIP: the archetypical molecular water channel. Am. J. Physiol. 265:F463-F476.

Andersen, O. S. 1978. Permeability properties of unmodified lipid bilayer membranes. In Membrane Transport in Biology, G. Giebisch, D. C. Tosteson, and H. H. Ussing, ed. Springer-Verlag, Berlin. 369-446.

Andersen, O. S. 1983. Ion movement through gramicidin A channels. Single-channel measurements at very high potentials. Biophys. J. 41:119-133.

Andersen, O. S. 1989. Kinetics of ion movement mediated by carriers and channels. Meth. Enzymol. 171:62-112.

Andersen, O. S., and R. E. Koeppe, II. 1992. Molecular determinants of channel function. Physiol. Rev. 72:S89-S158.

Andersen, O. S., D. B. Sawyer, and R. E. Koeppe, II. 1992. Modulation of channel function by the host bilayer. In Biomembrane Structure and Function, B. P. Gaber, and K. R. K. Easwaran, ed. Adenine Press, Schenectady, NY. 227-244.

Arsen'ev, A. S., I. L. Barsukov, V. F. Bystrov, A. L. Lomize, and Y. A. Ovchinnikov. 1985. Proton NMR study of gramicidin A transmembrane ion channel. Head-to-head right- handed, single-stranded helixes. FEBS Lett. 186:168-174.

Auerbach, A. 1993. A statistical analysis of acetylcholine receptor activation in Xenopus myocytes: stepwise versus concerted models of gating. J. Physiol. 461:339-378.

Bean, C. P. 1972. The physics of porous membranes–neutral pores. In Membranes, A Series of Advances, 1-54.

Benz, R., O. Fröhlich, P. Läuger, and M. Montal. 1975. Electrical capacity of black lipid films and of lipid bilayers made from monolayers. Biochim. Biophys. Acta. 394:323-334.

Bienvenue, A., and J. S. Marie. 1994. Modulation of protein function by lipids. Curr. Top. Membr. 40:319-354.

Bockris, J. O. '. M., and A. K. N. Reddy. 1970. Modern Electrochemistry. Volume 1. Plenum Press, New York.

Burnashev, N. A., A. I. Undrovinas, I. A. Fleidervish, J. C. Makielski, and L. V. Rosenshtraukh. 1991. Modulation of cardiac sodium channel gating by lysophosphatidylcholine. J. Mol. Cell. Cardiol. 23 (Suppl. I):23-30.

Busath, D. D. 1993. The use of physical methods in determining gramicidin channel structure and function. Ann. Rev. Physiol. 55:473-501.

Correa, A. M., and F. Bezanilla. 1944. Gating of the squid sodium channel at positive potentials. II. Single channels reveal two open states. Biophys. J. 66:1864-1878.

Cowan, S. W., and J. P. Rosenbusch. 1994. Folding pattern diversity of integral membrane proteins. Science. **264**:914-916.

Cox, B. G., G. R. Hedwig, A. J. Parker, and D. W. Watts. 1974. Solvation of ions. XIX Thermodynamic properties for transfer of single ions between protic and dipolar aprotic solvents. Aust. J. Chem. 27:477-501.

Cullis, P. R., and B. de Kruijff. 1979. Lipid polymorphism and the functional roles of lipids in biological membranes. Biochim. Biophys. Acta. 559:399-420.

DaTorre, S. D., M. H. Creer, S. M. Pogwizd, and P. B. Corr. 1991. Amphipathic lipid metabolites and their relation to arrythmogenesis in the ischemic heart. J. Mol. Cell. Cardiol. 23 (Suppl. I):11-22.

de Gennes, P. G., and J. Prost. 1993. The Physics of Liquid Crystals, 2nd Ed. Clarendon Press, Oxford.

de Kruyff, B., A. M. H. P. van den Besselaar, and L. L. M. van Deenen. 1977. Outside- inside distribution and translocation of lysophosphatidylcholine in phosphatidylcholine vesicles as determined by 13C-NMR using (N-13CH3)-enriched lipids. Biochim. Biophys. Acta. 465:443-453.

Devaux, P. F., and M. Seigneuret. 1985. Sepcificity of lipid-protein interactions as determined by spectroscopic techniques. Biochim. Biophys. Acta. 822:63-125.

Durkin, J. T., L. L. Providence, R. E. Koeppe, II, and O. S. Andersen. 1993. Energetics of heterodimer formation among gramicidin analogues with an NH2-terminal addition or deletion. Consequences of a missing residue at the join in channel. J. Mol. Biol. 231:1102- 1121.

Eddlestone, G. T., and S. Ciani. 1991. Lysophospholipids modulate the K(ATP) channel. Biophys. J. 59:16a (Abstract).

Eisenberg, R. S. 1990. Channels as enzymes. J. Membrane Biol. 115:1-12.

Elliott, J. R., D. Needham, J. P. Dilger, and D. A. Haydon. 1983. The effects of bilayer thickness and tension on gramicidin single-channel lifetime. Biochim. Biophys. Acta. 735:95-103.

Ferry, J. D. 1936. Statistical evaluation of sieve constants in ultrafiltration. J. Gen Physiol. 20:95-104.

Fersht, A. 1985. Enzyme structure and mechanism, 2nd Ed. W. H. Freeman and Company, New York.

Finkelstein, A. 1985. The ubiquitous presence of channels with wide lumens and their gating by voltage. Ann. N. Y. Acad. Sci. 456:26-32.

Gibson, N. J., and M. F. Brown. 1993. Lipid headgroup and acyl chain composition modulate the MI-MII equilibrium of rhodopsin in recombinant membranes. Biochemistry. 32:2438-2455.

Gilbert, D. L., and G. Ehrenstein. 1984. Membrane surface charge. Curr. Top. Membr. Transp. 22:407-421.

Green, M. E., and J. Lewis. 1991. Monte carlo simulation of the water in a channel with charges. Biophys. J. 59:419-426.

Green, W. N., and O. S. Andersen. 1991. Surface charges and ion channel function. Ann. Rev. Physiol. 53:341-359.

Gruner, S. M. 1985. Intrinsic curvature hypothesis for biomembrane lipid composition: arole for nonbilayer lipids. Proc. Natl. Acad. Sci. USA. 82:3665-3669.

Gruner, S. M. 1989. Stability of lyotropic phases with curved interfaces. J. Phys. Chem. 93:7562-7570.

Gruner, S. M. 1991. Lipid membrane curvature elasticity and protein function. In Biologically Inspired Physics, L. Peliti, ed. Plenum Press, New York. 127-135.

Hanai, T., D. A. Haydom, and J. Tayulor. 1965b. The influence of lipid composition and some adsorbed proteins on the capacitance of black hydrocarbon membranes. J. Theoret. Biol. 9:422-423.

Hanai, T., D. A. Haydon, and J. Taylor. 1965a. The variation of capacitance and conductance of bimolecular lipid membranes with area. J. Theoret. Biol. 9:433-443.

Hazel, J. R. 1988. Homeoviscous adaptation in animal cell membranes. In Physiological Regulation of Membrane Fluidity, R. C. Aloia, C. C. Curtain, and L. M. Gordon, ed. Alan R. Liss, Inc., New York. 149-188.

Helfrich, P., and E. Jakobsson. 1990. Calculation of deformation energies and conformations in lipid membranes containing gramicidin channels. Biophys. J. 57:1075-1084.

Helfrich, W. 1973. Elastic properties of lipid bilayers: theory and possible experiments. Z. Naturforsch. 28C:693-703.

Hille, B. 1972. The permeability of the sodium channel to metal cations in myelinated nerve. J. Gen. Physiol. 59:637-658.

Hille, B. 1973. Potassium channels in myelinated nerve: Selective permeability to small cations. J. Gen. Physiol. 61:669-686.

Hille, B. 1992. Ionic Channels of Excitable Membranes, 2nd Ed. Sinauer Associates, Inc., Sunderland, Massachusetts.

Hille, B., A. M. Woodhull, and B. I. Shapiro. 1975. Negative surface charge near sodium channels of nerve: divalent ions, monovalent ions, and pH. Phil. Trans. R. Soc. Lond. B. 270:301-318.

Hodgkin, A. L., and A. F. Huxley. 1952. A quantitative description of membrane current and its application to conduction and excitation in nerve. J. Physiol. 117:500-544.

Hodgkin, A. L., and B. Katz. 1949. The effect of sodium ions on the electrical activity of the giant axon of the squid. J. Physiol. 108:37-77.

Hoffmann, E. K., and L. O. Simonsen. 1989. Membrane mechanisms in volume and pH regulation in vertebrate cells. Physiol. Rev. 69:315-382.

Horn, R. 1984. Gating of channels in nerve and muscle: A stochastic approach. Curr. Top. Membr. Transp. 21:53-97.

Huang, C.-H., J. P. Sipe, S. T. Chow, and R. B. Martin. 1974. Differential interaction of cholesterol with phosphatidylcholine on the innner and outer surface of lipid bilayer vesicles. Proc. Natl. Acad. Sci. USA. 71:359-362.

Huang, H. W. 1986. Deformation free energy of bilayer membrane and its effect on gramicidin channel lifetime. Biophys. J. 50:1061-1070.

Hui, S.-W., and A. Sen. 1989. Effects of lipid packing on polymorphic phase nehavior and membrane properties. Proc. Natl. Acad. Sci. USA. 86:5825-5829.

Hwang, T.-C., S. E. Guggino, and W. B. Guggino. 1990. Direct modulation of secretory chloride channels by arachidonic and other cis unsaturated fatty acids. Proc. Natl. Acad. Sci. USA. 87:5706-5709.

Kasai, M., and T. R. J.-P. C. Podleski. 1970. Some structural properties of excitable membranes labelled by fluorescent probes. FEBS Lett. 7:13-19.

Keller, S. L., S. M. Bezrukov, S. M. Gruner, M. W. Tate, I. Vodyanoy, and V. A. Parsegian. 1993. Probability of alamethicin conductance states varies with nonlamellar tendency of bilayer phospholipids. Biophys. J. 65:23-27.

Ketchem, R. R., W. Hu, and T. A. Cross. 1993. High-resolution conformation of gramicidin A in a lipid bilayer by solid state-state NMR. Science. 261:1457-1460.

Killian, J. A. 1992. Gramicidin and gramicidin-lipid interactions. Biochim. Biophys. Acta. 1113:391-425.

Kiyosue, T., and M. Arita. 1986. Effects of lysophosphatidylcholine on resting potassium conductance of isolated guinea pig ventricular cells. Pflügers Arch. 406:296-302.

Kraut, J. 1988. How do enzymes work? Science. 242:533-540.

Kumar, V. V., B. Malewicz, and W. J. Baumann. 1989. Lysophosphatidylcholine stabilizes small unilamellar phosphatidylcholine vesicles. Phosphorus-31 NMR evidence for the "wedge" effect. Biophys. J. 55:789-792.

Lee, A. G. 1991. Lipids and their effects on membrane proteins: Evidence against a role for fluidity. Prog. Lipid Res. 30:323-348.

Lundbæk, J. A., and O. S. Andersen. 1994. Lysophospholipids modulate channel function bu altering the mechanical properties of lipid bilayers. J. Gen Physiol. 104:in press.

Läger, P. 1973. Ion transport through pores. Rate-theory analysis. Biochim. Biophys. Acta. 311:423-441.

Mackay, D. H. J., P. H. Berens, K. R. Wilson, and A. T. Hagler. 1984. Structure and dynamics of ion transport through gramicidin A. Biophys. J. 46:229-248.

McCarthy, M. P., and M. A. Moore. 1992. Effects of lipids and detergents on the conformation of the nicotinic acetylcholine receptor from Torpedo californica. J. Biol. Chem. 267:7655-7663.

McElhaney, R. N. 1985. Membrane lipid fluidity, phase state, and membrane function in prokaryotic microorganisms. In Membrane Fluidity in Biology. Vol. 4, Cellular Aspects, R. C. Aloia, and J. M. Boggs, ed. Academic Press, New York. 147-208.

McMurchie, E. J. 1988. Dietary lipids and the regulation of membrane fluidity and function. In Physiological Regulation of Membrane Fluidity, R. C. Aloia, C. C. Curtain, and L. M. Gordon, ed. Alan R. Liss, Inc., New York. 189-237.

McNamee, M. G., and T. M. Fong. 1988. The effects of membrane lipids and fluidity on acetylcholine receptor function. In Lipid domains and the Relationship to Membrane Function, R. C. Aloia, C. C. Curtain, and L. M. Gordon, ed. Alan R. Liss, New York. 43- 62.

Milks, L. C., N. M. Kumar, R. Houghtopn, N. Unwin, and N. B. Gilula. 1988. Topology of the 32-kd liver gap junction protein determined by site-directed antibody localizations. EMBO J. 7:2967-2975.

Mouritsen, O. G., and M. Bloom. 1984. Mattress model of lipid-protein interactions in membranes. Biophys. J. 46:141-153.

Numa, S. 1989. A molecular review of neurotransmitter receptors and ionic channels. Harvey Lect. 83:121-165.

O'Connell, A. M., R. E. Koeppe, II, and O. S. Andersen. 1990. Kinetics of gramicidin channel formation in lipid bilayers: transmembrane monomer association. Science. 250:1256-1259.

Oishi, K., B. Zheng, and J. F. Kuo. 1990. Inhibition of Na,K-ATPase and sodium pump by protein kinase C regulators sphingosine, lysophosphatidylcholine, and oleic acid. J. Biol. Chem. 265:70-75.

Ordway, R. W., J. V. Walsh, Jr., and J. J. Singer. 1989. Arachidonic acid and other fatty acids directly activate potassium channels in smooth cells. Science. 244:1176-1179.

Palmer, L. G., and H. Sackin. 1992. Electrophysiological analysis of transepithelial transport. In The Kidney: Physiology and Pathophysiology, 2nd Ed., D. W. Seldin, and G. Giebisch, ed. Raven Press, New York. 361-405.

Parsegian, A. 1969. Energy of an ion crossing a low dielectric membrane: Solutions to four relevant electrostatic problems. Nature. 221:844-846.

Partenskii, M. B., M. Cai, and P. C. Jordan. 1991. Influence of pore-former charge distribution on the electrostatic properties of dipolar water chains in transmembrane ion channels. Electrochim. Acta. 36:1753-1756.

Partenskii, M. B., and P. C. Jordan. 1992. Nonlinear dielectric behavior of water in transmembrane ion channels: Ion energy barriers and the channel dielectric constant. J. Phys. Chem. 96:3906-3910.

Perutz, M. F. 1990. Mechanisms of cooperativity and allosteric regulation in proteins. Cambridge University Press, Cambridge.

Pope, C. G., B. W. Urban, and D. A. Haydon. 1982. The influence of n-alkanols and cholesterol on the duration and conductance of gramicidin single channels in monoolein bilayers. Biochim. Biophys. Acta. 688:279-283.

Rosenberg, P. A., and A. Finkelstein. 1978a. Interaction of ions and water in gramicidin A channels. Streaming potentials across lipid bilayer membranes. J. Gen. Physiol. 72:327- 340.

Rosenberg, P. A., and A. Finkelstein. 1978b. Water permeability of gramicidin A-treated lipid bilayer membranes. J. Gen. Physiol. 72:341-350.

Sackin, H. 1994. Stretch-activated ion channels. In Cellular and Molecular Physiology of Cell Volume Regulation, K. Strange, ed. CRC Press, Boca Raton. 215-240.

Sandermann, H., Jr. 1978. Regulation of membrane enzymes by lipids. Biochim. Biophys. Acta. 515:209-237.

Sarges, R., and B. Witkop. 1965. Gramicidin A. V. The structure of valine- and isoleucine- gramicidin A. J. Am. Chem. Soc. 87:2011-2019.

Sawyer, D. B., R. E. Koeppe, II, and O. S. Andersen. 1989. Induction of conductance heterogeneity in gramicidin channels. Biochemistry. 28:6571-6583.

Schatzberg, P. 1965. Diffusion of water through hydrocarbon liquids. J. Polymer Sci., Part C. 10:87-92.

Schurr, J. M. 1970a. The role of diffusion in bimolecular solution kinetics. Biophys. J. 10:701-716.

Schurr, J. M. 1970b. The role of diffusion in enzyme kinetics. Biophys. J. 10:717-727.

Shimada, T., and A. P. Somlyo. 1992. Modulation of voltage-dependent Ca channel current by arachidonic acid and other long-chain fatty acids in rabbit intestinal smooth muscle. J. Gen. Physiol. 100:27-44.

Sigworth, F. J. 1993. Voltage gating of ion channels. Q. Rev. Biophys. 27:1-40.

Sine, S. M., T. Claudio, and F. J. Sigworth. 1990. Activation of torpedo acetylcholine receptors expressed in mouse fibroblasts: Single channel current kinetics reveal distinct agonist binding affinities. J. Gen. Physiol. 96:395-437.

Singer, S. J., and G. L. Nicolson. 1972. The fluid mosaic model of the structure of cell membranes. Science. 175:720-731.

Spray, D. C., and J. M. Burt. 1990. Structure-activity relations of the cardiac gap junction channel. Am. J. Physiol. 258:C205.

Stockton, G. W., and I. C. P. Smith. 1976. A deuterium nuclear magnetic resonance study of the condensing effect of cholesterol on egg phsophatidyl;choline bilayer membranes. I. Perdeuterated fatty acid probes. Chem. Phys. Lipids. 17:251-263.

Stubbs, C. D., and A. D. Smith. 1984. The modification of mammalian membrane polyunsaturated fatty acid composition in relation to membrane fluidity and function. Biochim. Biophys. Acta. 779:89-137.

Swarts, H. G. P., F. M. A. H. S. Stekhoven, and J. J. H. H. M. de Pont. 1990. Binding of unsaturated fatty acids to Na+,K+-ATPase leading to inhibition and inactivation. Biochim. Biophys. Acta. 1024:32-40.

Szöllösi, J. 1994. Fluidity/viscosity of biological membranes. In Mobility and Proximity in Biological Membranes, S. Damjanovich, M. Edidin, J. Szöllö, and L. Trón, ed. CRC Press, Boca Raton. 137-208.

Tank, D. W., E. S. Wu, P. R. Meers, and W. W. Webb. 1982. Lateral diffusion of gramicidin C in phospholipid multibilayers. Effects of cholesterol and high gramicidin concentration. Biophys. J. 40:129-135.

Thompson, T. E., C. Huang, and B. J. Litman. 1974. Bilayers and biomembranes: compositional asymmetries induced by surface curvature. In The Cell Surface in Development, A. A. Moscona, ed. John Wileay & Sons, New York. 1-16.

Tibbitts, T. T., D. L. D. Caspar, W. C. Philips, and D. A. Goodenough. 1990. Diffraction diagnosis of protein folding in gap junction connexons. Biophys. J. 57:1025-1036.

Unwin, N. 1989. The structure of ion channels in membranes of excitable cells. Neuron. 3:665-676.

Unwin, N. 1993. Nicotinic acetylcholine receptor at 9 Å resolution. J. Mol. Biol. 229:1101- 1124.

Unwin, N., C. Toyoshima, and E. Kubalek. 1988. Arrangement of the acetylcholine receptor subunits in the resting and desentisitized states, determined by cryoelectron microscopy of crystalllized Torpedo postsynaptic membranes. J. Cell. Biol. 107:1123-1138.

Unwin, P. N. T., and P. D. Ennis. 1984. Two configurations of a channel-forming membrane protein. Nature. 307:609-613.

Veatch, W. R., R. Mathies, M. Eisenberg, and L. Stryer. 1975. Simultaneous fluorescence and conductance studies of planar bilayer membranes containing a highly active and fluorescent analog of gramicidin A. J. Mol. Biol. 99:75-92.

Vigh, L., D. A. Los, M. Horváth, and N. . 1993. The primary signal in the biological perception of temperature: Pd-catalyzed hydrogenation of membrane lipids stimulated the expression of the desA gene in Synechocystis PCC6803. Proc. Natl. Acad. Sci. USA. 90:9090-9094.

Wallert, M. A., M. J. Ackerman, D. Kim, and D. E. Clapham. 1991. Two novel cardiac atrial K+ channels, IK,AA and IK,PC. J. Gen. Physiol. 98:921-939.

Weiss, M. S., U. Abele, J. Weckesser, W. Welte, E. Schiltz, and G. E. Schulz. 1991. Molecular architecture and electrostatic properties of a bacterial porin. Science. 254:1627- 1630.

White, S. H. 1978. Formation of "solvent-free" black lipid bilayer membranes from glyceryl monooleate dispersed in squalene. Biophys J. 23:337-347.

Yeagle, P. L. (ed.). 1988. Biology of Cholesterol. CRC Press, Boca Raton.

Zagotta, W. N., T. Hoshi, and R. W. Aldrich. 1994. Shaker potassium channel gating III: evaluation of kinetic models for activation. J. Gen. Physiol. 103:321-362.

Zakim, D., J. Kavecansky, and S. Scarlata. 1992. Are membrane enzymes regulated by the viscosity of the membrane environment. Biochemistry. 31:11589-11594.

Dynamics of Nucleic Acids and Nucleic Acid:Protein Complexes

Lennart Nilsson

Abstract

Molecular dynamics simulations can shed light on subnanosecond phenomena in nucleic acids interacting with proteins. Results from studies of Ribonuclease T1, the glucocortoid receptor DNA binding domain, and short model nucleotides are reported, and compared to experimental NMR and fluorescence data. Events occurring on different timescales will be discussed, together with an analysis of the the role of water, especially in the interface between protein and DNA.

1 Introduction

One of the key elements in understanding the function of biomolecules is their dynamic behavior. All the structural information, at atomic resolution, gathered over the last 30 years on a number of proteins and nucleic acids is only a starting point for an explanation of how these molecules work. The specific aspect of biomolecular action that will be the focus of this report is molecular recognition, an essential component in almost all biomolecular processes, specifically in processes relating to transcription and translation of the genetic material. The long term aims of our effort are to shed light on the nature of interactions stabilizing the structural elements as well as the complexes, and to understand how small sequence changes can lead to differences in structure or affinity. Since most biomolecular reactions take place in an aqueous environment it is also very important to consider the effects of the solvent, which may be of several kinds: competition for hydrogen bonds, screening of electrostatic interactions or favoring structural arrangements with minimal exposure of hydrophobic groups, to name a few possibilities. There are also cases in which water molecules have been seen to supply crucial and specific contacts between protein and DNA.

The enormous amounts of sequence data soon to emerge from the Human Genome Project also call for methods of analysis, capable of providing 3D structures and stability estimates, not only for proteins (corresponding to a small fraction of the genome), but also for the RNA products. Secondary structure prediction methods are of limited reliability on their own, and need to be complemented by experimental structure related information, as well as by theoretical and simulation methods which can give a very detailed description of biomolecular systems, not only to fill in the gaps between the experiments, but also to provide predictions and suggestions for further studies. It is very clear from a study of enzyme:substrate interactions [1], where interaction energies (enthalpies) between different substrates and proteins were calculated and compared with both simulated free energy values and with experimental binding data, that straightforward, intuitive guesses of the outcome of mutation experiments in such complicated systems are unreliable - slight structural changes, interplay with solvent and ions, and entropic effects are very difficult to guess. Therefore more precise methods, like free energy perturbation or potential of mean force calculations are necessary.

Motions in biomolecules occur on a wide range of timescales, from (sub)-picosecond librational motions, and isomerization reactions to microseconds or longer for folding of the native 3D structure of a protein. Present day computer technologies allow simulations in the nanosecond range for systems consisting of several thousand atoms, which means that we can usually only investigate the properties of the system in a small neighborhood of the native state.

Simulation studies of nucleic acids are not as common as of protein systems, mainly for the two reasons that 1) there are fewer structures available, and 2) nucleic acids are more difficult to handle due to their highly charged and non-globular nature. With the discovery of RNAs with catalytic activity, "ribozymes", it has become clear that there is much more to nucleic acid structure than the canonical DNA duplexes, but still there are very few experimentally determined structures other than some tRNAs and short oligomers of DNA and RNA. We have worked on several aspects of simulating nucleic acid systems, and here some results from molecular dynamics simulations on small model systems (oligonucleotides) and on protein:nucleic acid complexes (ribonuclease T1, glucocorticoid receptor DNA binding domain) will be presented and compared with experimental observations of dynamical properties.

2 Methodology

In molecular dynamics simulations Newton's equations of motions are integrated for a system of N interacting particles, usually the atoms in the system. This requires knowledge of the structure of the molecular system and of the potential energy as a function of the structure. Structures can be obtained from experimental studies, by X-ray crystallography or NMR, or from model building; the potential energy, or rather its gradient which gives the force, has to be evaluated many times and thus has to be as simple as possible. Commonly used empirical energy functions have terms corresponding to chemically intuitive forces:

$$V = \frac{1}{2}\sum_{\text{bonds}} K_b(b - b_0)^2 + \frac{1}{2}\sum_{\text{bond-angles}} K_\theta(\theta - \theta_0)^2 +$$

$$\sum_{\text{dithetral angles}} K_\phi[1 + \cos(n\phi - \delta)] + \sum_{\text{nonbonded pairs}} (\frac{A}{r^{12}} - \frac{B}{r^6} + \frac{q_i q_j}{4\pi\epsilon r}) \quad (1)$$

The parameterization of such functions is made using data for model compounds, representative of the macromolecular systems that will eventually be studied. This is a very important step, providing the connection between subsequent simulations and reality; the simulations themselves only probe the properties of the given energy function, and the biological relevance of the results depends on how well this energy function represents reality.

All simulations presented here have been performed with the macromolecular simulation program CHARMM [2], using two generations of parameter sets, PARAM19 [2,3] and PARAM22 (MacKerell & Karplus, in preparation). The Verlet algorithm [4] is used for the integration, with a timestep of 2fs and all bonds involving hydrogen kept fixed with the SHAKE algorithm [5]. Nonbonded interactions are smoothly shifted to zero at a cutoff distance of $9-10$ Å[2]. Simulations in water use the TIP3P water model [6], either with periodic boundary conditions, or with a spheric stochastic deformable boundary [7].

Fluorescence time-dependent anisotropies and $^1\mathrm{H} - ^{15}\mathrm{N} -$ NMR relaxations both measure molecular motions obeying correlation functions of the form:

$$C(t) = \frac{2}{5}C_I(t)e^{-t/\tau_R} \quad (2)$$

where the overall rotational motion characterized by τ_R is assumed to separate from the internal motion

$$C_I(t) = < P_2(\mathbf{h_i}(0)\mathbf{h_j}(t)) > \quad (3)$$

with $P_2(x) = (3x^2 - 1)/2$, and $\mathbf{h_i}$, $\mathbf{h_j}$ are unit vectors along an amide bond in the NMR case or along absorption and emission dipoles, respectively, in the fluorescence case. This internal correlation function can often be approximated by an exponential decaying to a plateau value, and can be described by an effective correlation time τ_e and a generalized order parameter S^2 [8]:

$$C_I(t) = S^2 + (1 - S^2)e^{-t/\tau_e} \quad (4)$$

We have analyzed chromophore dynamics and backbone amide motions by calculating the same quantities from the simulated trajectories.

3 Results and Discussion

Results from a series of simulations of increasing complexity, ranging from a short DNA duplex in vacuo, via solvated oligonucleotides and a small protein, to a large protein:DNA complex in water (30977 atoms) will be discussed. For the smaller systems simulations can be run long enough to thoroughly sample even overall diffusive motions, and indeed give quantitative agreement with experimental data.

3.1 EcoRI Recognition Sequence

Restriction endonucleases have been carefully studied genetically, biochemically and structurally, because they are interesting in themselves, and also because of their sequence specific recognition of DNA. EcoRI, the first one for which a 3D structure was obtained [9,10], is a dimeric type II restriction endonuclease (2 × 276 amino acids) and requires Mg^{2+} as a cofactor for hydrolysis of the DNA, even though it can bind specifically also in the absence of Mg^{2+}. It specifically recognizes the double stranded sequence $d(GAATTC)_2$, and hydrolyzes the DNA backbone between the G and A nucleotides. Binding to the DNA is very tight and discrimination against single basepair changes, or against sequences with methylated adenines, is extremely good. We have simulated the self complementary DNA decamer duplex $d(CTGAATTCAG)_2$, and a modified variant where the innermost adenine has been replaced with 2-aminopurine, a highly fluorescent analogue to adenine [11]. The simulations were performed for 120ps under "pseudo-vacuum" conditions, with a friction term and a stochastic term added to the force in Newton's equations of motion. Consistent with NMR experiments, the modified sequence was stable in a B-type conformation during the simulation, and showed mainly a rapid sliding type of motion in the plane of the basepairs, without breaking any hydrogen bonds, even though the basepairs move several Å laterally.

3.2 Mono- and Dinucleotides in Aqueous Solution

Quantitative simulations of small (mono- and dinucleotides) biomolecular systems in aqueous solution have been made to study the influence of the water model on dynamic and structural properties of the biomolecules, as well as to monitor the conformational dynamics of dinucleotide models in stacked and unstacked conformations. For this combined goal to be attainable a close comparison with experimental results concerning one aspect of the problem, the rotational motion of the solute as measured in fluorescence depolarization studies using 2-aminopurine as one of the nucleotides (Kulinski & Rigler, personal communication), is crucial. Figure 1 shows the correlation function corresponding to a time-dependent fluorescence anisotropy measurement obtained from a 0.5 ns simulation of the mononucleotide 3'-guanosine monophosphate in a sphere of TIP3P [6] water. This curve gives the correlation time for rotational diffusion as 25 ps, which is somewhat faster than experiments on 2-aminopurine seem to indicate [12]; these experiments are, however, of limited resolution and one cannot tell whether the difference is significant.

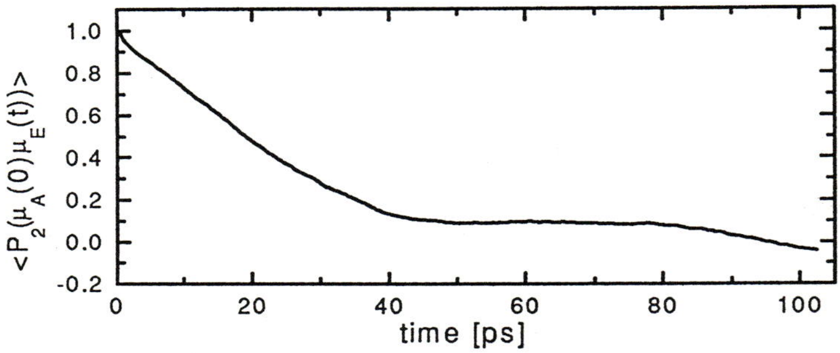

Fig. 1. Rotational correlation of 3'GMP in water.

From the first series of simulations using the dinucleotide GpU and the water model TIP3P in a periodic box, the molecule seems to exist predominantly in a stacked conformation (1ns simulation), yielding an overall rotation tumbling time of 140 ps (experimental 150 ps), whereas the extended conformation (2ns simulation) results in rotational motion times around 250 ps [13]. Simulations in a canonical (N,P,T)-ensemble of GpU at pressures of 1, 5000 and 12000 atm, indicate significantly lowered atomic fluctuations around the average structure at 12000 atm, but not quite as much as the self-diffusion coefficient of water is reduced [14].

3.3 Ribonuclease T_1

Ribonuclease T_1 is a small enzyme which selectively cleaves single-stranded RNA on the 3'side of guanines, and for which the 3D structure is known with and without bound substrate. This makes the enzyme ideal for the study of protein-nucleic acid interactions. Our work on this system has so far focused on changes in the structure of the wild-type enzyme, particularly in its active site, upon binding of the inhibitor 2'GMP. Initial experimental studies using time-resolved fluorescence showed the presence of a picosecond motion of Trp59 (the only tryptophan in the enzyme) which was altered by the binding of the inhibitor 2'GMP [15]. This was also observed in a series of molecular dynamics simulations of the same systems, in water as well as *in vacuo* [16-18].

The main influence of the water in the 2'GMP:Ribonuclease T1 complex was to help maintain the structural integrity of the protein around the active site cleft, which collapsed in the vacuum simulation, whereas the atomic fluctuations were less affected by the solvent. The guanine base of the inhibitor/substrate is held rigidly in place in the active site, sandwiched between two tyrosines; several hydrogen bonds are also found to the phosphate group. In the solvated simulation an equilibrium fluctuation in the position of His92, whereby it moves away for a short time from its hydrogen bonding contact with the phosphate of the inhibitor, was found and suggested to represent a situation where a small shift in the conformation of the enzyme would offer an opportunity for the inhibitor to leave the enzyme.

3.4 DNA Binding Domain of the Glucocorticoid Receptor

The glucocortiocoid receptor (GR) belongs to the family of ligand-inducible transcription factors, which also includes the steroid hormones, thyroid hormones, vitamin D_3, and retinoic acid receptors. These proteins have a common amino acid sequence organization composed of discrete functional domains for ligand binding, a highly conserved domain for DNA binding (DBD), consisting of about 70 residues [19], and a domain for transcriptional regulation. The binding of a ligand to the GR induces a transformation such that it translocates from the cytoplasm into the nucleus and associates with DNA to regulate transcriptional activity of certain proteins. The DNA binding sites for the GR are termed glucocorticoid response elements (GREs) and the naturally occurring consists of a 15 base pair, partially palindromic sequence comprising two hexameric half sites with a 3 base pair intervening gap [20]. The DBD binds as a dimer to the GRE sequence, with initial binding preferentially occurring to one of the two half sites and a previous binding to one half site has been shown to facilitate binding to the other half site [21,22]. This is consistent with a two- site cooperative model, where DNA binding is dependent on DBD-DNA interactions as well as on interactions between the monomers at the dimerisation interface.

The recently solved 3D structure of the DNA binding domain (GR DBD) of the glucocortioid receptor [23] (Figure 2)and of the GR DBD-DNA complex [24] provides an excellent opportunity to study the complex between DNA and DBD.

We have performed simulations of a monomeric DBD and of the complex, in aqueous solution, to look into the nature of the specific interactions between the molecules (DNA-protein and protein-protein). A first series of simulations on one monomer of GR DBD *in vacuo* and in aqueous solution has just been completed [26] as well as of the dimeric complex with DNA in aqueous solution[27]; results on backbone dynamics correlate well with NMR relaxation data from ^{15}N-labelled proteins. Figure 3 shows the generalized order parameter S^2, as defined in equation 4, from $\{^1H\} -^{15} N - NMR$ relaxations and calculated from the simulation of the monomer in solution

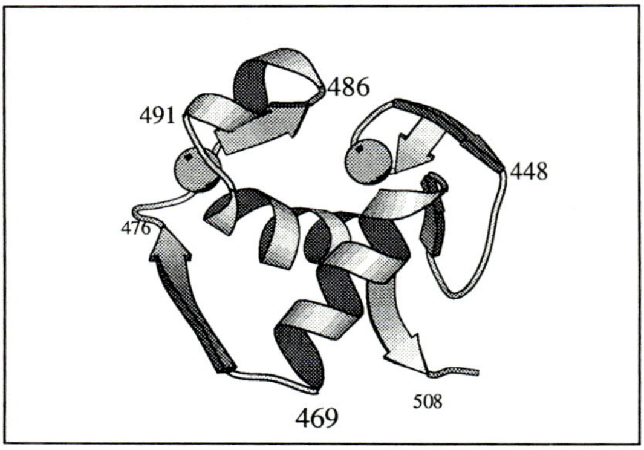

Fig. 2. Schematic drawing [25] of the 3D structure of GR DBD; α-helices I,II and III, and the two Zn^{++}-ions are shown.

[26]. Overall a fairly uniform rigidity, with $2 \approx 0.8$, is seen for the protein backbone experimentally, as well as in the simulations, indicating a stable and well defined protein structure. Discrepancies between the simulated and experimental S^2 values, as seen for a handful residues, could be rationalized in terms of under-sampled motions where the amide jumps between two hydrogen bonded states with different acceptor groups; such events, when occurring two-three times during a 200ps simulation are too infrequent to give reliable statistics.

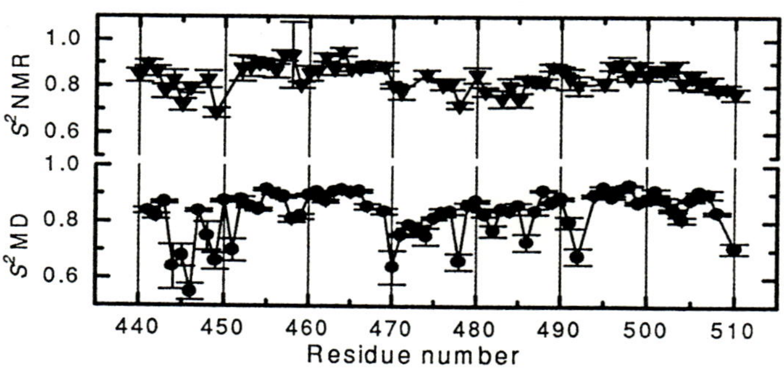

Fig. 3. Generalized order parameters for backbone amides in GR DBD. Upper curve from NMR relaxation data, lower curve from simulation of the monomer in water.

Structural analyses of the simulated GR DBD:DNA complex [27] and comparisons with the crystallographic data, show increased mobilities in the free monomer mainly at sites where there are contacts either with the DNA or with the other protein monomer in the crystal. A large body of experimental genetic and thermodynamic data is available for this system, relating both to DNA-protein interactions (specific and unspecific binding modes) and to protein-protein interactions (allosteric affects). It has, for instance, been shown [28] that in the glucocorticoid receptor single amino acids in the recognition region can form both positive and negative contacts with specific basepairs with the cognate DNA, but only negative contacts with non-cognate DNA; there are also synergistic effects of combinations of amino acids. These studies are of a statistical nature, and for a deeper, mechanistic, understanding of these findings a detailed study is necessary.

References

1. A. Elofsson, L. Nilsson "Free Energy Perturbations in Ribonuclease T_1 Substrate Binding. A Study of the Influence of Simulation Length, Internal Degrees of Freedom and Structure in Free Energy Perturbations" Mol. Sim. **10**, 255-276 (1993).
2. B.R. Brooks, R.E. Bruccoleri, B.D. Olafson, D.J. States, S. Swaminathan, M. Karplus "CHARMM: A Program for Macromolecular Energy, Minimization, and Dynamics Calculations" J. Comp. Chem. **4**, 187-217 (1983).
3. L. Nilsson, M. Karplus "Empirical Energy Functions for Energy Minimizations and Dynamics of Nucleic Acids" J. Comp. Chem. **7**, 591-616 (1986).
4. L. Verlet "Computer 'Experiments' on Classical Fluids. I. Thermodynamical Properties of Lennard-Jones Molecules" Phys. Rev. **159**, 98-103 (1967).
5. J.P. Ryckaert, G. Ciccotti, H.J.C. Berendsen "Numerical Integration of the Cartesian Equations of Motion of a System with Constraints: Molecular Dynamics of n-alkanes" Journal of Computational Physics **23**, 327-341 (1977).
6. W.L. Jorgensen, J. Chandrasekhar, J. Madura, R.W. Impey, M.L. Klein "Comparison of Simple Potential Functions for Simulating Liquid Water" J. Chem. Phys. **79**, 926-935 (1983).
7. C.L. Brooks, A.T. Brünger, M. Karplus "Active Site Dynamics in Protein Molecules: A Stochastic Boundary Molecular Dynamics Approach" Biopolymers **24**, 843-865 (1985).
8. G. Lipari, A. Szabo "A Model-free Approach to the Interpretation of Nuclear Magnetic Resonance Relaxation in Macromolecules. I. Theory and Range of Validity" J. Am. Chem. Soc. **104**, 4546-4559 (1982).
9. C.A. Frederick, J. Grable, M. Melia, C. Samudzi, L. Jen-Jacobson, B.C. Wang, P. Greene, H.W. Boyer, J.M. Rosenberg "Kinked DNA in Crystalline Complex with EcoRI Endonuclease" Nature **309**, 327-331 (1984).
10. Y. Kim, J.C. Grable, R. Love, P.J. Greene, J.M. Rosenberg "Refinement of EcoRI Endonuclease Crystal Structure: A Revised Chain Tracing" Science **249**, 1307-1309 (1990).
11. T. Nordlund, S. Andersson, L. Nilsson, R. Rigler, A. Gräslund, L.W. McLaughlin "Structure and Dynamics of a Fluorescent DNA Oligomer Containing the EcoRI Recognition Sequence: Fluorescence, Molecular Dynamics, and NMR Studies" Biochemistry **28**, 9095-9103 (1989).
12. R. Rigler, F. Claesens. "Picosecond Time Domain Spectroscopy of Structure and Dynamics in Nucleic Acids", .in Structure and Dynamics of RNA, eds.P.H. van Knippenberg and C.W. Hilbers (Plenum Press, New York 1986) pp. 45-54

13. J. Norberg, L. Nilsson. "Stacking-unstacking of the Dinucleoside Monophosphate Guanylyl-3'-5'-Uridine Studied with Molecular Dynamics" Biophysical Journal in press, 1994
14. J. Norberg, L. Nilsson "High Pressure Molecular Dynamics of a Nucleic Acid Fragment", Chem. Phys. Lett. in press
15. A.D. MacKerell, R. Rigler, L. Nilsson, U. Hahn, W. Saenger "Protein Dynamics: A Time-Resolved Fluorescence, Energetic and Molecular Dynamics Study of Ribonuclease T1" Biophys. Chem. **26**, 247-261 (1987).
16. A.D. MacKerell, R. Rigler, L. Nilsson, U. Heinemann, W. Saenger "Molecular Dynamics of Ribonuclease T1: Effect of Solvent on the Interaction with 2'GMP" Eur. Biophys. J. **16**, 287-297 (1988).
17. A.D. MacKerell, L. Nilsson, R. Rigler, U. Heinemann, W. Saenger "Molecular Dynamics of Ribonuclease T1: Comparison of the Free Enzyme and the 2'GMP-Enzyme Complex" Proteins **6**, 20-31 (1989).
18. A.D. MacKerell, L. Nilsson, R. Rigler, W. Saenger "Molecular Dynamics Simulations of Ribonuclease T1: Analysis of the Effect of Solvent on the Structure, Fluctuations and Active Site of the Free Enzyme" Biochemistry **27**, 4547-4556 (1988).
19. T. Härd, J.-Å. Gustafsson "Structure and Function of the DNA-Binding Domain of the Glucocorticoid Receptor and other Members of the Nuclear Receptor Supergene Family" Acc. Chem. Res. **26**, 644-650 (1993).
20. M. Beato Cell **56**, 335-344 (1989).
21. T. Härd, E. Kellenbach, R. Boelens, R. Kaptein, K. Dahlman, J. Carlstedt-Duke, L.P. Freedman, B.A. Maler, E. Hyde, J.-Å. Gustafsson, K.R. Yamamoto "1H NMR Studies of the Glucocorticoid Receptor DNA-binding Domain: Sequential Assignments and Identification of Secondary Structure Elements" Biochemistry **29**, 9015-9023 (1990).
22. S.Y. Tsai, J. Carlstedt-Duke, N.L. Weigel, K. Dahlman, J.-Å. Gustafsson, M.-J. Tsai, B.W. O'Malley "Molecular Interactions of Steroid Hormone Receptor with its Enhancer Element: Evidence for Receptor Dimer Formation" Cell **55**, 361-369 (1988).
23. T. Härd, E. Kellenbach, R. Boelens, B.A. Maler, K. Dahlman, L.P. Freedman, J. Carlstedt-Duke, K.R. Yamamoto, J.-Å. Gustafsson, R. Kaptein "Solution Structure of the Glucocorticoid Receptor DNA-binding Domain" Science **249**, 157-160 (1990).
24. B.F. Luisi, W.X. Xu, Z. Otwinowski, L.P. Freedman, K.R. Yamamoto, P.B. Sigler "Crystallographic Analysis of the Interaction of the Glucocorticoid Receptor with DNA" Nature **352**, 497-505 (1991).
25. P.J. Kraulis "MOLSCRIPT: A Program to Produce both Detailed and Schematic Plots of Protein Structures" J. Appl. Cryst. **24**, 946-950 (1991).
26. M.A.L. Eriksson, H. Berglund, T. Härd, L. Nilsson "A Comparison of 15N NMR Relaxation Measurements With a Molecular Dynamics Simulation: Backbone Dynamics of the Glucocorticoid Receptor DNA-Binding Domain" Proteins **17**, 375-390 (1993).
27. M. Eriksson, T. Härd, L. Nilsson. "Molecular Dynamics Simulation of a DNA Binding Protein Free and in Complex With DNA", .in Computational Approaches to Supramolecular Chemistry, ed.G. Wipff (Kluwer, 1994)
28. J. Zilliacus, A.P.H. Wright, U. Norinder, J.-Å. Gustafsson, J. Carlstedt-Duke "Determinants for DNA-binding Site Recognition by the Glucocorticoid Receptor" J. Biol. Chem. **267**, 24941-24947 (1992).

Part IV

Physiological Control Systems

Models of Renal Blood Flow Autoregulation

N.-H. Holstein-Rathlou, K.H. Chon, D.J. Marsh, and V.Z. Marmarelis

Abstract

A major drawback in the analysis of experimental data on the dynamics of renal autoregulation has been the use of linear methods for calculating transfer functions. This chapter gives a brief introduction to our ongoing work on making the Volterra-Wiener method more efficient for obtaining information on the higher order nonlinearities of input–output systems. To assess the dynamic characteristics of renal blood flow autoregulation, we have applied broad–band ('white' noise) perturbations of the arterial pressure, using renal blood flow as the response variable. To interpret the experimental data we have constructed a mathematical model of the nephron. To approximate the situation *in vivo* where a large number of individual nephrons act in parallel, each run of the model was performed with 125 parallel versions of the model. The key parameters of the 125 versions of the model were chosen randomly within the physiological range. The model reproduced most of the features of the linear transfer functions obtained experimentally.

1 Introduction

The functional unit of the kidney is the nephron [11]. The human kidney consists of approximately 1 million nephrons, while the rat kidney has approximately 30,000 nephrons. A nephron has 2 functionally distinct parts, the glomerulus and the tubular system. The glomerulus consists of a network of capillaries supplied with blood through the afferent arteriole. Because of a high pressure within the glomerular capillaries, an ultrafiltrate of the plasma is formed. As the ultrafiltrate moves down the tubular system, its volume and composition are changed due to active and passive secretory and reabsorptive processes before it finally emerges as urine.

The kidney is exposed to a constantly fluctuating arterial blood pressure, see Fig. 1.

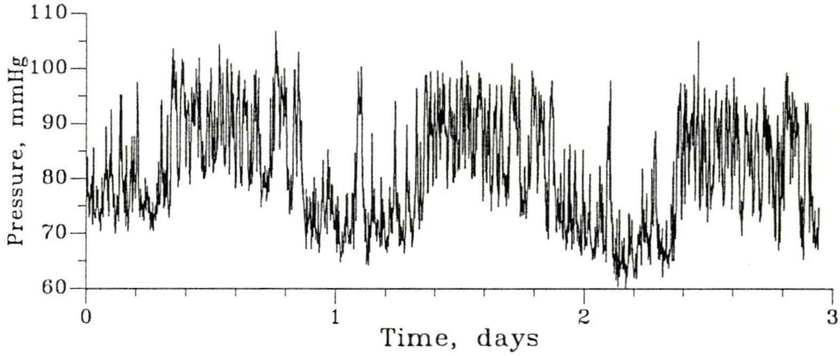

Fig. 1. Arterial blood pressure fluctuations. The recording shows the spontaneous blood pressure variation in a normal Sprague-Dawley rat. The pressure was recorded in a conscious freely moving rat over a 72 hour period using an implantable pressure transducer.

These fluctuations would, if unopposed, result in large perturbations of renal function. This is due to the pressure dependence of the ultrafiltration process. A change in the rate of ultrafiltration would result in a perturbation of the flow in the tubular system, and this could lead to serious, life threatening changes in the salt and water balance of the organism.

Autoregulation is the process that minimizes changes in renal function caused by fluctuations in the arterial pressure. Autoregulation involves an active adjustment of the resistance of the renal vessels. An increase in the arterial pressure causes an increase in the hemodynamic resistance of the preglomerular vessels preventing, or at least reducing, pressure mediated increases in renal blood flow, and glomerular filtration rate (GFR). Experimental studies have demonstrated that each nephron is equipped with at least two control mechanisms that could contribute to autoregulation: the tubuloglomerular feedback mechanism (TGF), and the myogenic mechanism.

TGF is an intranephronal feedback system that stabilizes nephronal blood flow, single nephron GFR, and the tubular flow rate [20]. The anatomical basis for TGF is the return of the tubule (the ascending limb of the loop of Henle (ALH)) to its own glomerulus. The macula densa, which is the sensor mechanism for the TGF, is a plaque of specialized epithelial cells in the wall of the ALH. It is localized at the site where the tubule establishes contact with the glomerulus. Because of a flow dependence of NaCl reabsorption in the ALH, a change in tubular flow rate, elicited for example by a change in the arterial pressure, will lead to a change in the NaCl concentration of the tubular fluid. This is sensed by the macula densa, and through unknown mechanisms results in a change in the hemodynamic resistance of the afferent arteriole. Fig. 2 shows a causal loop diagram for the TGF mechanism.

The myogenic mechanism is an intrinsic response of the smooth muscle cells of the vascular wall. It is widely accepted that the myogenic response

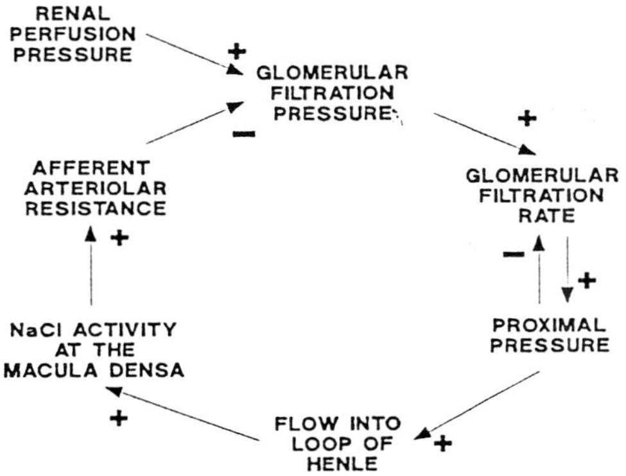

Fig. 2. Causal loop diagram of the TGF system. A + indicates a direct variation, and a − an opposite variation between variables.

acts to minimize changes in vessel wall tension. An increase in pressure elicits a vasoconstriction. The reduction in vessel diameter reduces the wall tension (Laplace's law) towards the control value. The regulated variable is therefore probably not flow itself but rather wall tension, flow being stabilized because of the adjustments needed to minimize changes in wall tension.

The dynamic properties of these control systems have been characterized in experimental studies [3, 22], and both the TGF and the myogenic mechanism have been shown to have complicated dynamics, including autonomous self-sustained oscillations [13, 22]. Clearly, it is the dynamics at the level of the single nephron that determines the dynamics observed at the level of the whole organ. However, because of the huge number of nephrons, and because of possible interactions between nephrons [5], it becomes an important physiological task to elucidate how the dynamics at one level of organization, that is at the level of the single nephron, affects the dynamics at the higher level of organization, that of the whole organ. Understanding this relationship would contribute significantly to our understanding of the dynamic interactions between the kidney and the cardiovascular system.

To gain insight into this problem we have constructed a multinephron model of the rat kidney. The model is based on our current understanding of the physical and biological characteristics of the nephron. By exposing the mathematical model to the same procedures as used in experimental studies on the dynamics of renal blood flow autoregulation we can begin to investigate the role that the various subsystems play in autoregulation, and how their intrinsic dynamic properties translate into the dynamic properties of the whole organ.

Most studies on renal blood flow autoregulatory dynamics have used classical linear methods for systems identification [3, 8, 18]. Because of the nonlinearities of the underlying control mechanisms these methods are not well suited for this purpose. The presence of self-sustained oscillations in the TGF and the myogenic responses make it likely that significant nonlinear behavior will be present when the kidney is exposed to pressure fluctuations with frequencies close to those of the self-sustained oscillations. It would therefore be desirable to have a general method for nonlinear systems identification that is useful in an experimental setting. The techniques for nonlinear systems identification developed by Volterra and Wiener [14] could be such an alternative to the classical linear methods. We therefore present a brief introduction to these methods, including some recent computational improvements that have enhanced their practical usefulness. We also demonstrate that these methods are able to capture essential nonlinearities of the renal autoregulatory processes.

For didactic reasons we refer to the first type of models as parametric models, and to the second type, the Volterra-Wiener models, as nonparametric models.

2 Parametric Model of the Nephron

This section presents an outline of the rat nephron model. For additional details see [7, 9].

2.1 Afferent Arteriolar Model

The description of the dynamics of the afferent arteriole is based on a model proposed by Ursino and Fabbri [21]. The preglomerular vascular bed is modeled as three segments arranged in series. Segment 1 represents the larger renal vessels, and its resistance is assumed to be constant, R_{A1}. The two more distal segments have a variable resistance under the control of both the tubuloglomerular feedback (TGF), and the myogenic mechanism. The resistance in each segment is given as $R_{Aj} = \Lambda_j / r_v^4$ where Λ_j is a constant and r_v the inner radius of the vascular segment. The inner radius is calculated from the relationship $P_v r_v = T$, where P_v is the intravascular pressure in the segment, and T is vessel wall tension. The latter is a sum of three components: an elastic term, T_e, stemming from stretch of the elastic elements in the vessel wall; an active tension, T_m, generated by the contraction of the vascular smooth muscle, and a viscous tension, T_v; thus $T = T_e + T_m + T_v$. The elastic and the muscular terms are given by:

$$T_e = \sigma_e h(r_v)[\exp(k_e[r_v - r_e]) - 1] \tag{1}$$

and

$$T_m = \sigma_m h(r_v) f(x_t) \exp(-k_m[r_v - r_m]^2) \tag{2}$$

where $h(r_v)$ is thickness of the vessel wall, and σ_e, σ_m, k_e, k_m, r_e, and r_m are empirical constants. Wall thickness, $h(r_v)$, varied with the radius so that the cross sectional area of the vessel wall remained constant [21]. $f(x_t)$ is an empirical function, expressing the active contraction of the vascular smooth muscle:

$$f(x_t) = \frac{3\exp(x_t)}{\exp(x_t) + 2\exp(-0.5x_t)} \tag{3}$$

where x_t is the degree of activation of the vascular smooth muscle. It is the sum of contributions from the TGF system, and the myogenic mechanism:

$$x_t = x_{tgf} + x_{myo} \tag{4}$$

The viscous tension is proportional to the rate of change of the vessel radius:

$$T_v = \frac{\sigma_v h(r_v)}{r_n}\frac{dr_v}{dt} \tag{5}$$

where σ_v and r_n are constants.

2.2 Myogenic Mechanism

The degree of activation of the vascular smooth muscle arising through the myogenic mechanism, x_{myo}, is given by a linear first order differential equation

$$\frac{dx_{myo}}{dt} = -\frac{1}{\tau}[x_{myo} - \lambda G(T - T_o)] \tag{6}$$

where τ is the time constant, and G the gain of the response, λ a constant that sets the ratio of the gain between segments 2 and 3, and T_o a reference tension.

2.3 Tubuloglomerular Feedback

The action of tubuloglomerular feedback was described by a logistic equation relating the activation of the vascular smooth muscle, x_{tgf}, to the NaCl concentration at the macula densa, $C_S(md)$:

$$x_{tgf} = \frac{1}{\lambda}(\xi_{\max} - \frac{\psi}{1 + \exp(k[C_S(md) - C_{1/2}])}) \tag{7}$$

where $C_{1/2}$, is the concentration giving a half maximal response, $\xi_{\max}$ is the maximal response, and ψ is the dynamic range. λ is identical to the constant in equation 6. The parameter k determines the steepness or strength of the TGF response.

Because there is a delay between a change in the NaCl concentration at the macula densa, and a change in the vascular resistance of the afferent arteriole, the action of the TGF, x_{tgf}, was passed through a third order delay function,

$$\frac{dy_{i+1}}{dt} = \frac{T_d}{n}(y_i - y_{i+1}), \quad n = 3, \quad i = 0, .., 2 \tag{8}$$

where $y_0 = x_{tgf}$, T_d is the delay in the transmission of the TGF signal from the macula densa to the afferent arteriole. The actual value added to x_{myo} to give x_t (equation 4) was therefore y_3 from equation 8 above.

2.4 Glomerular Model

Time dependent processes in the glomerular capillaries are very rapid compared to tubular processes, and we therefore assume that the glomerular system is in a quasi steady state with respect to the rest of the system.

As blood flows through the glomerular capillaries, fluid is filtered through membranes that are not permeable to proteins in the plasma. The filtration rate, $Q_T(0)$, can be calculated from the fractional change in protein concentration, C, and the plasma flow, Q_A, entering the capillaries as $Q_T(0) = (1 - C_A/C_E)Q_A$.

The filtration process is proportional to the local pressure difference, which is given by the sum of the capillary hydrostatic pressure, P_{GC}, the tubular pressure, P_T, and the osmotic pressure due to proteins, $\pi(C)$:

$$\frac{dC}{dx} = \frac{K_f C^2}{LQ_A C_A}[P_{GC} - P_T(0) - \pi(C)] \tag{9}$$

where x is fractional position along the glomerular capillary, K_f is the filtration coefficient, and L is the length of the capillary.

The plasma flow to the glomerular capillaries is given by

$$Q_A R_A = (P_{Art} - P_{GC})(1 - Ht_A) \tag{10}$$

where Ht_A is the hematocrit, P_{Art} the arterial pressure, and $R_A = R_{A1} + R_{A2} + R_{A3}$, the preglomerular resistance, see afferent arteriolar model. A similar relation holds for the blood flow through the efferent arteriole [7]. The resistance of the efferent arteriole is assumed constant.

2.5 Tubular Model

The equations governing the temporal and spatial behavior of the tubular model are based on local conservation of mass and momentum. Because of the low Reynolds' number, one dimensional approximations of the Navier-Stokes equations are suitable.

$$\frac{\partial P_T}{\partial z} = -\frac{\rho}{\pi r^2}\frac{\partial Q_T}{\partial t} - \frac{8\eta}{\pi r^4}Q_T \tag{11}$$

and

$$\frac{\partial Q_T}{\partial z} = -2\pi r \frac{\partial r}{\partial P_T}\frac{\partial P_T}{\partial t} - J_V \tag{12}$$

r is tubular radius, ρ and η are fluid density and viscosity, z is position along the tubule, and Q_T is tubular flow rate. One boundary condition, the glomerular filtration rate, $Q_T(0)$, is calculated from the equations of the glomerular model, and the second boundary condition, the tubular pressure at the end of the loop of Henle, $P_T(Z)$, is calculated from an empirical relationship between $Q_T(Z)$ and $P_T(Z)$ [17].

Tubular fluid reabsorption, J_V, in the proximal tubule was assumed to be an exponential function of distance (z): $J_V(z) = \kappa exp(-\theta z)$, where κ and θ are empirical constants [17]. In the descending limb of Henle's loop, $J_V(z) = L_V(C_I(z) - C_S(z))$, where L_V is the water permeability of the tubule, and C_I and C_S the concentration of NaCl in the interstitium and the tubule, respectively. In the ascending limb of Henle's loop fluid reabsorption was set to zero.

Tubular radius was a function of transtubular pressure, $r(z) = \gamma P_T(z) + r_o$, where γ, and r_o are empirical constants [7].

The tubules reabsorb NaCl and water. Because of the isosmotic transport in the proximal tubule this process causes no change in the NaCl concentration in this segment. In the loop of Henle, where the NaCl concentration ($C_S(z)$) does change:

$$\frac{\partial A C_S}{\partial t} = -\frac{\partial Q_T C_S}{\partial z} - J_S \tag{13}$$

where A is the cross sectional area of the tubular lumen. The boundary condition needed to solve this equation is $C_S = 150$ mmol/l at the end of the proximal tubule. The rate of tubular NaCl reabsorption, $J_S(z)$, was specified as the sum of a diffusive term and an active term that follows Michaelis-Menten kinetics:

$$J_S(z) = L_S(C_S(z) - C_I(z)) + \frac{V_{\max} C_S(z)}{K_m + C_S(z)} \tag{14}$$

The interstitial NaCl concentration, $C_I(z)$, varied linearly from 150 mmol/l at the cortico-medullary boundary to 300 mmol/l at the bend of the loop of Henle. For the cortical thick ALH, C_I was set to 150 mmol/l.

2.6 Model simulations

The model equations were solved numerically using a centered difference scheme in both space and time [17]. The parameters of the model were obtained, where possible, from experimental measurements. The values of the parameters are given in [9].

Experimental studies in rats have found oscillations with a frequency around 20–40 mHz in both tubular pressure and flow, distal tubular NaCl concentration, and in peritubular capillary blood flow [5], see Figures 3 and 6.

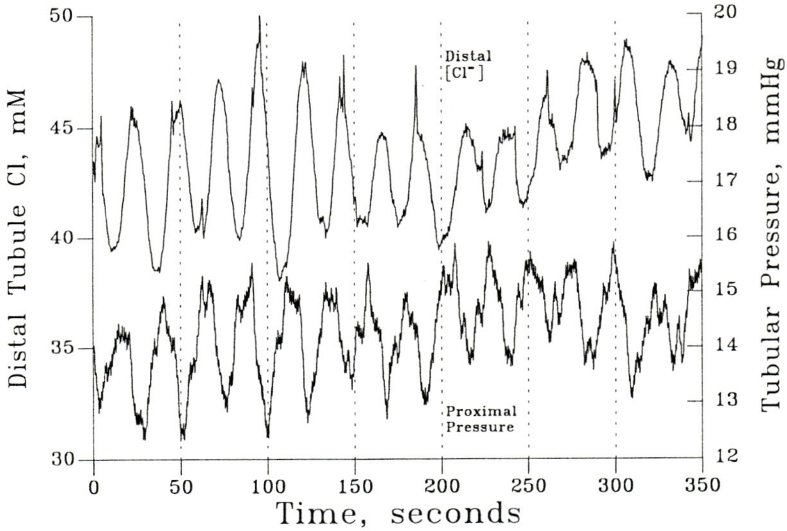

Fig. 3. Oscillations of proximal tubular pressure (lower trace), and early distal tubular Cl^- concentration (upper trace). The measurements were performed in halothane anesthetized Sprague-Dawley rats.

Substantial experimental evidence indicates that the oscillations are caused by the operation of the TGF system [5]. The role of TGF in generating the tubular oscillations have also been investigated in several model studies [6, 7, 10, 12]. The studies have shown that with typical parameter values the TGF system will be in an oscillatory state, but that the system is near enough to the boundary of the nonoscillatory region that small changes in parameter values could result in nonoscillatory behavior [6, 7, 10, 12]. This is in good agreement with the experimental observations that only approximately 80% of the investigated nephrons oscillate, and that the same nephron may shift between an oscillatory and a stable state [5].

Like the previous dynamic models of the nephron, the present model also shows self–sustained oscillations in its various variables. The primary determinants of whether the model oscillates or whether it reaches a stable state are the strength of the TGF response (k), and the size of the transmission delay (T_d) across the macula densa to the afferent arteriole. These results confirm the results of previous model studies, and the reader is referred to those for further details [6, 7, 12].

Previous experimental studies have characterized the dynamic properties of whole kidney blood flow regulation [3, 8, 16, 18]. The experimental approach has been to induce broad–band fluctuations in the arterial pressure while simultaneously measuring the resulting fluctuations in the renal blood flow [3, 8, 16, 18]. Using the arterial pressure as the input, and the renal blood flow as the output, the transfer function can be estimated [14].

A major problem has been the identification of the mechanisms responsible for the various characteristics of the obtained transfer functions. In earlier studies, this problem has been approached using models of single nephrons [8, 17]. This approach is unsatisfactory because of the self-sustained oscillations in the single-nephron models. A single-nephron model predicts strong oscillations in the blood flow, a feature not normally found in whole kidney blood flow. In the intact kidney, the individual nephrons oscillate with different frequencies [5], and they tend to average out each other, resulting in a stable renal blood flow. To investigate how the dynamics at the single nephron level influences the dynamic behavior of the whole kidney, it is therefore necessary to use more elaborate models involving more than one nephron.

The rat kidney consists of approximately 30,000 nephrons. To simulate this situation we used 125 versions of the above single nephron model. The nephrons were arranged in parallel, and renal blood flow was computed as the sum of the contributions from the individual nephrons The 125 nephrons falls short of the number of nephrons in the rat kidney, but because of computational limitations this number was considered an acceptable compromise that ensures a large enough population so that significant averaging effects will be present. To secure that each nephron was unique, we chose the delay across the macula densa, (T_d; equation 8), the strength of the TGF response, (k; equation 7), the time constant of the myogenic response, (τ; equation 6), and the gain of the myogenic response, (G; equation 6), at random.

The model was tested by comparing the linear transfer function for the renal blood flow as predicted by the model to that obtained from experiments performed in halothane anesthetized rats (see [8] for details regarding the experimental data). The model was perturbed by a broad–band ('white' noise) forcing of the arterial pressure. The forcing was done using a Gaussian random number generator, with a mean and standard deviation of the arterial pressure of 106 $\pm$ 4 mmHg. This input closely simulates the broad–band forcing of the arterial pressure used in the experimental studies [8].

Figure 4 shows the transfer functions for the renal blood flow calculated from the model simulations and the experimental data. The model captured many of the characteristic features of the transfer function obtained experimentally. It had a broad maximum in the admittance gain in the frequency range above 0.1 Hz. In the experimental series the gain maximum reached 5–6 dB, whereas the value predicted by the model was somewhat lower, 2–4 dB. In good agreement with the experimental observations, the model showed a gradually increasing attenuation at frequencies below 0.1 Hz, reaching a value between -10 – -15 dB at the lowest frequencies investigated in this study. The peak in the gain in the frequency range between .025 – 0.06 Hz is due to the self–sustained oscillations of the individual nephron models [7]. This peak is more pronounced in the model than in the experimental series. However, simulating several parallel versions of the single nephron model diminishes

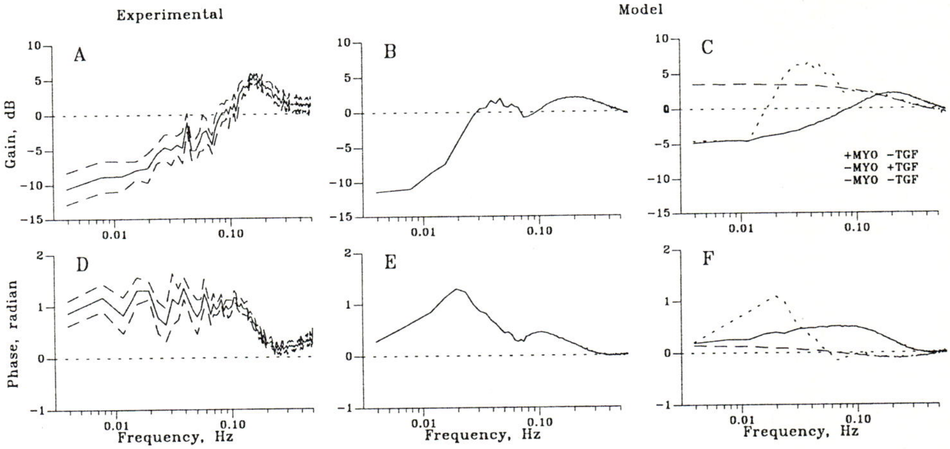

Fig. 4. Response of renal blood flow to broad-band forcing of arterial pressure expressed as admittance gain and phase. Panels A, and D experimental results from [8]. Solid line is mean value, dashed lines are SE. Panels B, and E simulation results obtained by averaging over 125 independent versions of the single nephron model. Panels C and F simulation results with only the myogenic mechanism present (solid line), with only TGF present (short dashed line), and with no regulatory mechanism present (long dashed line).

this peak. In the case where there is only one nephron in the model, the peak typically reaches a value of 5 – 15 dB [8].

The phase curve also showed reasonable good agreement between experiment and model. The major discrepancy between the phase curve predicted by the model and the one obtained from the experimental data was that the former had a pronounced bimodal shape whereas the latter appeared unimodal (Figure 4 panels D and E).

The curves in Figure 4 panels C and F show the predicted transfer functions when only TGF is included, when only the myogenic mechanism is included, and when no regulatory mechanism is included. It is clear that none of the two regulatory mechanisms (TGF and myogenic) alone can explain the observed transfer function.

3 Nonparametric Models

For the nonparametric modeling of the system, we have chosen the Volterra-Wiener approach. Here system identification is based solely on input and output data. It is a generalization of the well known linear technique where the output of the system is expressed as a convolution integral of the input with the impulse response of the system [14]. In the linear case, system identification consists of determining the impulse response of the system from input-output data. The Volterra-Wiener approach generalizes this method to nonlinear systems.

For a nonlinear, time invariant system with finite memory, the relationship between input, $x(t)$ (blood pressure), and output, $y(t)$ (renal blood flow), can be written as

$$y(t) = \sum_{n=0}^{\infty} \int_0^{\infty} k_n(\tau_1, \ldots, \tau_n) x(t - \tau_1) \ldots x(t - \tau_n) d\tau_1, \ldots, d\tau_n \tag{15}$$

where the Volterra kernels $\{k_n\}$ characterize the dynamic properties of the system. The zeroth-order kernel is simply a constant. The first, second, and third order kernels represents the linear, the quadratic, and the cubic components of system dynamics, respectively. The Volterra series is, however, not useful for practical applications because the individual terms in the series are not orthogonal with respect to each other. The expansion in equation 15 may however be reformulated into a series where the individual terms are orthogonal when the input to the system is Gaussian White Noise (GWN). A GWN signal is an attractive stimulus because it contains every possible frequency. The advantage of orthogonalizing the Volterra series is that the higher-order kernels have no effect on the lower-order kernels; each kernel can therefore be estimated independently of the other. The orthogonalized version is called the Wiener series. As a result of orthogonalization, the kernels span the space of all possible input/output maps most efficiently, and the Wiener series therefore has stronger convergence than the Volterra series, and provides the best system representation (in the least squares sense). The Wiener kernels will in general be different from the Volterra kernels, but a specific analytical relation exists between the two [14].

Because of its orthogonality with respect to GWN the Wiener series is well suited for the study of nonlinear input–output systems that can be subjected to a broad–band input approximating GWN. The most commonly used method for kernel estimation has been the cross-correlation technique developed by Lee and Schetzen [14]. It is however far from being optimal. It requires long data records, and strict whiteness of the input, both of which may be difficult to fulfill in the experimental setting. In addition, a heavy computational burden is associated with the estimation of higher order kernels.

Marmarelis recently suggested an alternative approach that appears to solve the above problems [15]. It employs an expansion of the kernels using the orthonormal set of Laguerre functions $\{L_j(\tau)\}$

$$k_n(\tau_1, \ldots, \tau_n) = \sum_{j_1} \cdots \sum_{j_n} c_n(j_1, \ldots, j_n) L_{j_1}(\tau_1) \ldots L_{j_n}(\tau_n) \tag{16}$$

where $\{c_n(j_1, \ldots, j_n)\}$ are the unknown expansion coefficients that are to be estimated from the input–output data, and $L_j(\tau)$ is the j–th order Laguerre function [15].

The Volterra series of equation 15 then becomes

$$y(t) = \sum_{n=0}^{\infty} \sum_{j_1} \cdots \sum_{j_1,\ldots,j_n} c_n(j_1,\ldots,j_n) v_{j_1}(t) \ldots v_{j_n}(t) \tag{17}$$

where,

$$v_j(t) = \int_0^{\mu} L_j(\tau) x(t-\tau) d\tau \tag{18}$$

and μ is the extent of the system memory (i.e., the kernels attain negligible values for $\tau > \mu$). A similar expression may be derived for the Wiener series. In this case equation 17 will be in the form of a multi-dimensional Hermite polynomium in the variables $\{v_j\}$ [15]. The unknown coefficients, $\{c_n\}$, are estimated by a least–squares method using the discretized signals $y(t)$ and $\{v_j(t)\}$ (equation 18).

We tested the ability of first, second, and third order Wiener models to accurately predict renal blood flow during broad–band forcing of the arterial pressure [16]. Three different levels (as measured by the variance of the pressure fluctuations) of arterial pressure forcing were used in each of 4 experiments. The experimental data were obtained in halothane anesthetized Sprague-Dawley rats over 256 seconds with a sampling rate of 2 Hz. The results summarized in Table 1 clearly demonstrate the significance of the 3rd–order kernel in representing the the nonlinear pressure–flow dynamics. The contribution of the 3rd–order kernel was significant at all three power levels.

Table 1. Average normalized mean-square errors of first, second, and third order Wiener model predictions for 4 experiments. In each experiment the forcings were performed at three different power levels.

	Model order		
Power level	1st order	2nd order	3rd order
High	6.86 %	6.67 %	1.51 %
Medium	15.47 %	15.47 %	4.46 %
Low	19.89 %	24.80 %	6.62 %

These results show that estimation of 3rd–order Volterra-Wiener models from short experimental records are now possible with the present kernel estimation technique.

3.1 Frequency Domain Representation of the Kernels

For further insight into the meaning of kernels, it is advantageous to shift from the *time domain* representation in equation 15 to a *frequency domain* representation of the nonlinear system. The latter is obtained by considering the response of the Volterra model described by equation 15 to an oscillatory input. Because all real periodic signals can be expressed as a sum of periodic analytic functions of the form $e^{j\omega t}$ (Fourier analysis), where $j = \sqrt{-1}$, ω is the frequency, and t is time, the most general expression is obtained by considering the response to such an input. The output of the system, $y(t)$, can viewed as a sum of the outputs of the individual terms, i.e., $y(t) = y_0 + y_1(t) + y_2(t) + y_3(t) + \ldots$, where y_0 is a constant, $y_1(t)$ the contribution to the output from the 1st–order kernel, and so on. For the 1st-order kernel it is easily seen that

$$\begin{aligned} y_1(t) &= \int_0^{\infty} k_1(\tau)e^{(j\omega(t-\tau))}d\tau = e^{(j\omega t)} \int_0^{\infty} k_1(\tau)e^{(-j\omega\tau)}d\tau \\ &= e^{(j\omega t)}\mathcal{F}[k_1(t)] = K_1(\omega)e^{(j\omega t)} \end{aligned} \tag{19}$$

where $\mathcal{F}$ is the Fourier transform. $K_1(\omega)$ is thus the Fourier transform of the 1st–order kernel, and it represents the linear frequency response function of the system [15]. Thus, when the system is exposed to a single oscillatory input with frequency ω the output from the 1st–order kernel will oscillate at the same frequency. The magnitude, $|K_1(\omega)|$, of the linear frequency response function expresses the amplitude of the output oscillation as a function of the frequency ω, i.e., it represents the linear system gain [15].

Likewise, if we consider the response of the 2nd–order kernel to the same oscillatory input, $e^{(j\omega t)}$, we obtain

$$\begin{aligned} y_2(t) &= \int_0^{\infty}\int_0^{\infty} k_2(\tau_1,\tau_2)e^{(j\omega(t-\tau_1))}e^{(j\omega(t-\tau_2))}d\tau_1 d\tau_2 \\ &= e^{(j2\omega t)}\int_0^{\infty}\int_0^{\infty} k_2(\tau_1,\tau_2)e^{(-j\omega\tau_1)}e^{(-j\omega\tau_2)}d\tau_1 d\tau_2 \\ &= e^{(j2\omega t)}\mathcal{F}[k_2(\tau_1,\tau_2)] = e^{(j2\omega t)}K_2(\omega,\omega) \end{aligned} \tag{20}$$

where $K_2(\omega,\omega)$ is the 2-dimensional (2D) Fourier transform of the 2nd–order kernel. Eq. 20 shows that in the case of a single oscillatory input, the effect of the 2nd–order kernel is to cause the appearance of an oscillation at twice the frequency in the output. The amplitude of this second harmonic is given by the absolute value of $K_2(\omega,\omega)$. The diagonal elements of the 2D Fourier transform of the 2nd–order kernel therefore provides a quantitative measure of the amplitudes of the second harmonics in the response of the system. In a similar manner the diagonal elements of the 3D Fourier transform of the 3rd–order kernel provides a quantitative measure of the amplitudes of the third harmonics, and so on. The diagonal elements of the Fourier transform of the higher order kernels therefore expresses the nonlinear response of the system at the corresponding frequency. If there are large diagonal elements at some frequency, a sinusoidal input at this frequency will result in a nonsinusoidal output due to the addition of higher harmonics.

If the input signal contains more than one frequency, there will be an interaction between the various frequencies depending on the values of the off–diagonal elements of the higher order kernels. To see this consider for simplicity a signal consisting of two frequencies, ω_1 and ω_2. In this case the output from the 1st–order kernel will simply be the scaled sum of the 2 inputs:

$$y_1(t) = K_1(\omega_1)e^{(j\omega_1 t)} + K_1(\omega_2)e^{(j\omega_2 t)} \tag{21}$$

This is to be expected, since the 1st–order kernel represents the linear part of the response. By insertion and rearranging as in Eq. 20 above it is easily seen that the corresponding output from the 2nd–order kernel will be

$$\begin{aligned} y_2(t) =& K_2(\omega_1, \omega_1)e^{(j2\omega_1 t)} + K_2(\omega_2, \omega_2)e^{(j2\omega_2 t)} + \\ & 2K_2(\omega_1, \omega_2)e^{(j(\omega_1+\omega_2)t)} \end{aligned} \tag{22}$$

As before, we see that the diagonal elements expresses the nonlinear response at the individual frequencies. The last term is due to the quadratic nonlinear interaction between the 2 inputs, and it will add a component to the output at the sum of the two input frequencies $\omega_1+\omega_2$. The amplitude of this component will be determined by the absolute value of the off-diagonal value of the 2D Fourier transform of the second order kernel, $|K_2(\omega_1, \omega_2)|$. Because of the symmetry of the 2nd-order kernel [15] (i.e., $K_2(\omega_1, \omega_2) = K_2(\omega_2, \omega_1)$) only one such term will be present. A similar approach will show that the off-diagonal elements of the 3D Fourier transform of the 3rd-order kernel will express the strength of all possible combinations of cubic nonlinear interactions between the frequencies present in the input signal. An examination of the off-diagonal elements of the Fourier transformed kernels can therefore be used to probe for possible nonlinear interactions in the system under study.

Because of the above considerations Volterra-Wiener analysis is well suited for detecting interactions between the myogenic and the TGF mechanisms. If they did not interact with each other one would not expect to find off-diagonal elements in the Fourier transformed kernels at the intersection of the frequency ranges of the two mechanisms. The only off-diagonal elements would be within the frequency ranges of the individual mechanisms. On the other hand, if the myogenic and the TGF mechanisms interact with each other, i.e., if the response of one affects the response of the other, then by necessity a nonlinear interaction in the form of significant off-diagonal elements should be present at the intersection of the frequency ranges of the two mechanisms.

3.2 Experimental Results

The above approach was applied to experimental measurements of renal blood flow in 4 normotensive rats. Figure 5 shows contour and surface plots of the absolute value of the 2-dimensional Fourier transform (2D-FFT) of the 2nd-order kernel.

The figure was obtained by averaging 4 independently estimated 2nd-order kernels, and then computing the 2D-FFT of the averaged kernels. The

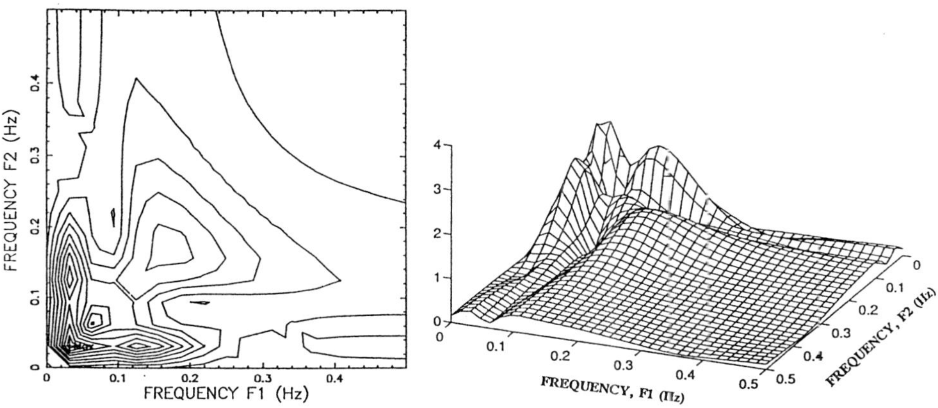

Fig. 5. Contour plot (left panel) and surface plot (right panel) of the absolute value of the 2D-FFT of the averaged 2nd-order kernel obtained in halothane anesthetized Sprague-Dawley rats. The axes of the plots represent the two frequency components f_1 and f_2, respectively. Notice the strong nonlinear interaction peak present at the frequency components $f_1 = 130$ mHz and $f_2 = 30$ mHz.

kernels showed a strong peak at $f_1 = f_2 = 30$–50 mHz, the frequency of the slow oscillations generated by the TGF mechanism. A nearly identical picture was seen when slices of the 3rd-order kernel were examined. It is not surprising to find strong quadratic and cubic nonlinearities at this frequency, since one would expect the system to behave nonlinearly in a frequency range where the subsystems (the nephrons) exhibits a stable self-sustained oscillation. Indeed, in a previous study we showed that forcing of the arterial pressure with a sinusoidal oscillation with a frequency of 33 mHz leads to a strong resonance (low impedance) in the blood flow [8]. In addition, at this frequency the response was highly nonlinear. As the forcing level was increased, higher harmonics were detected in the renal blood flow [8], as expected from the present finding.

In addition to the strong diagonal peak at $f_1 = f_2 = 30$–50 mHz, the FFT of the 2nd-order kernel (Figure 5) also showed a strong off-diagonal peak at $f_1 = 130$ mHz, and $f_2 = 30$ mHz. A similar peak was present in slices of the 3rd-order kernel (not shown). This observation is interesting because a recent study [22] has identified a second, faster oscillation in single nephron blood flow, see Figure 6.

This faster oscillation has a frequency of around 120–140 mHz, and it has been suggested that it represents an intrinsic vascular oscillation caused by the myogenic mechanism [22]. The oscillation is too fast to be caused by TGF, since both experimental and model studies have shown that the upper limit for TGF is 50-60 mHz [4, 7, 8]. The limited frequency response of TGF is due to the damping caused by the transmission of the signal through the tubular system [4, 7, 8].

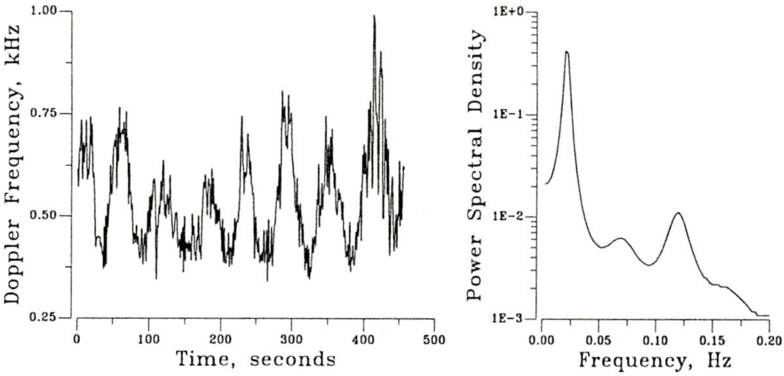

Fig. 6. Left panel: Laser–Doppler measurement of the spontaneous variation in single nephron blood flow in a halothane anesthetized Sprague-Dawley rat. Right panel: Corresponding power spectrum. The peak at 30 mHz represents the slow TGF mediated oscillation. The peak at 120 mHz is probably due to the myogenic mechanism.

The strong off-diagonal peak located at $f_1 = 130$ mHz, $f_2 = 30$ mHz is a typical example of a nonlinear cross–talk between spectral components of the input signal. The fact that the nonlinear interaction occurs at exactly the frequencies where independent studies have identified the presence of self-sustained oscillatory modes at the level of the single nephron, leads us to suggest that this interaction is a reflection, at the level of the whole kidney, of an interaction between TGF and the myogenic response at the single nephron level. This is supported by experimental studies that have shown such an interaction at the single nephron level [2, 19]. Thus, by using a nonlinear systems identification procedure, we are able to detect at the level of the whole kidney an interaction between the two regulatory mechanisms involved in renal autoregulation of blood flow. Such an interaction had previously been observed only at the single nephron level.

4 Conclusion

The multinephron model presented in this paper is able to reproduce many of the dynamic features of renal blood flow regulation. Thus, it captures the TGF-mediated regular oscillations seen at the level of the single nephron, and when forced with a broad–band input reproduces the linear frequency response function obtained experimentally. Whereas the regular oscillations in tubular pressure and flow can be reproduced by a single nephron model that only includes the TGF mechanism [5], reproduction of the linear frequency response function requires a multinephron model. This is not surprising since the regulation of renal blood flow is the result of the concerted action of all the nephrons within the kidney, and experimental studies have demonstrated a rather large variation in the dynamical properties of the individual nephrons [5]. Within a kidney the frequency of the spontaneous oscillations can vary between 25 and 40 mHz [5]. This is probably the reason why oscillations cannot be detected in total renal blood flow. Because of the frequency variation, the oscillations at the level of the single nephrons will tend to average out each other. To capture this a model with a large number of nephrons is needed.

One of the principal questions under debate has been the relative roles of TGF and the myogenic response in autoregulation. The spectrum of published opinions ranges from little or no role for TGF in autoregulation [1] to complete reliance on it [18]. The results of the present model show that both contributes significantly to autoregulation of renal blood flow. If either TGF or the myogenic response are the only mechanism present, the calculated transfer functions deviates from those seen experimentally.

Experimental studies have recently demonstrated a second oscillation at the level of the single nephron. This oscillation has been demonstrated both in the local capillary blood flow using Laser-Doppler flowmetry [22], and in the tubular pressure [5]. It has a frequency around 120–140 mHz, which is somewhat faster than that of the TGF-mediated oscillation. This faster oscillation is currently believed to be due to the myogenic mechanism [22].The present model of the myogenic response of the afferent arteriole does not appear to have autonomous oscillations within a range of physiologically reasonable parameter values. Although it would be highly desirable to have the model reproduce this higher frequency oscillation, this is difficult because of our lack of understanding of the underlying oscillatory mechanism. Progress in this direction will have to await experimental studies on the specific mechanism behind the oscillation.

The fact that it is now possible to calculate reliable estimates of 3rd-order Volterra-Wiener models, together with the demonstration that such models capture most of the dynamics of renal blood flow autoregulation makes this approach highly attractive for use in future experimental studies on the dynamics of renal blood flow autoregulation. The Volterra-Wiener model showed that the major nonlinearity of the renal autoregulatory response was located in the frequency range of the TGF-mediated oscillations. This demonstrates directly how the dynamics at the level of the individual nephrons affect the dynamics of the response of the whole kidney. In addition, the Volterra-Wiener model revealed an interaction between the two oscillations detected in the renal microvasculature, the slow TGF-mediated oscillation at 30–50 mHz, and the faster, presumably myogenically mediated oscillation at 120–140 mHz. This indicates an interaction between the TGF and the myogenic mechanisms. That the two vascular control mechanisms interact is not surprising, since they both act to regulate the diameter of a single arteriole, and so the action of one mechanism will have an effect on the local vascular pressure sensed by the other, providing one basis for the interaction. Whether such a purely physical, pressure mediated effect is sufficient for explaining the interaction is not clear. Expanding the multinephron model to also reproduce the faster oscillation would make it possible to test this hypothesis in model studies.

In summary, the present work has shown that a multinephron model reproduces many dynamical features of renal blood flow regulation. However, the model is still incomplete since it fails to reproduce an experimentally observed oscillation in single nephron blood flow, and the interaction of this oscillation with the slower TGF-mediated oscillation. Future models should incorporate this faster oscillation to provide a means for studying the mechanism whereby the two oscillations interact.

5 Acknowledgments

The present work was supported by grants HL-45623, DK-15968, and RR-01861 from the National Institutes of Health, and by grants from the Danish Medical Research Council (12-3164-1), the Novo-Nordisk Foundation, and the Whitaker Foundation.

References

1 K. Aukland, and A.H. Øien: Renal autoregulation: models combining tubuloglomerular feedback and the myogenic response. Am. J. Physiol. **252**, F768-F783 (1987)

2 D. Casellas, and L.C. Moore: Autoregulation and tubuloglomerular feedback in juxtamedullary glomerular arterioles. Am. J. Physiol. **258**, F660-F669 (1990)

3 F.H. Daniels, W.J. Arendshorst, and R.G. Roberds: Tubuloglomerular feedback and autoregulation in spontaneously hypertensive rats. Am. J. Physiol. **258**, F1479-F1489 (1990)

4 F.H. Daniels, and W.J. Arendshorst: Tubuloglomerular feedback kinetics in spontaneously hypertensive and Wistar-Kyoto rats. Am. J. Physiol. **259**, F529-F534 (1990)

5 N.-H. Holstein-Rathlou: Dynamic aspects of the tubuloglomerular feedback mechanism. Dan. Med. Bull. **39**, 134-154 (1992)

6 N.-H. Holstein-Rathlou, and P.P. Leyssac: Oscillations in proximal intratubular pressure: a mathematical model. Am. J. Physiol. **252**, F560-F572 (1987)

7 N.-H. Holstein-Rathlou, and D.J. Marsh: A dynamic model of the tubuloglomerular feedback mechanism. Am. J. Physiol. **258**, F1448-F1459 (1990)

8 N.-H. Holstein-Rathlou, A.W. Wagner, and D.J. Marsh: Tubuloglomerular feedback dynamics and renal blood flow autoregulation in rats. Am. J. Physiol. **260**, F53-F68 (1991)

9 N.-H. Holstein-Rathlou, and D.J. Marsh: A dynamic model of renal blood flow autoregulation. Bull. Math. Biology, in press (1994)

10 K.S. Jensen, E. Mosekilde, and N.-H. Holstein-Rathlou: Self-sustained oscillations and chaotic behavior in kidney pressure regulation. Monde en Developpement **54-55**, 91-109 (1986).

11 W. Kritz, and B. Kaissling: Structural organization of the mammalian kidney. In: *The Kidney: Physiology and Pathophysiology, 2nd Edition.* (D. Seldin and G. Giebisch, editors), Raven Press, New York 1992, pp. 707-778

12 H.E. Layton, E.B. Pitman, and L.C. Moore: Bifurcation analysis of TGF-mediated oscillations in SNGFR. Am. J. Physiol. **261**, F904-F919 (1991)

13 P.P. Leyssac, and L. Baumbach: An oscillating intratubular pressure response to alterations in Henle flow in the rat kidney. Acta Physiol. Scand. **117**, 415-419 (1983)

14 P.Z. Marmarelis, and V.Z. Marmarelis: Analysis of Physiological systems. Plenum Press, New York 1978

15 V.Z. Marmarelis: Identification of nonlinear biological systems using Laguerre expansions of kernels. Ann. Biomed. Eng. **21**, 573-589 (1993)

16 V.Z. Marmarelis, K.H. Chon, Y.-M. Chen, N.-H. Holstein-Rathlou, and D.J. Marsh: Nonlinear analysis of renal autoregulation under broadband forcing conditions. Ann. Biomed. Eng. **21**, 591-603 (1993)

17 T. Sakai, D.A. Craig, A.S. Wexler, and D.J. Marsh: Fluid waves in renal tubules. Biophys. J. **50**, 805-813 (1986)

18 T. Sakai, E. Hallman, and D.J. Marsh: Frequency domain analysis of renal autoregulation in the rat. Am. J. Physiol. **250**, F364-F373 (1986)

19 J. Schnermann, and J.P. Briggs: Interaction between loop of Henle flow and arterial pressure as determinants of glomerular pressure. Am. J. Physiol. **256**, F421-F429 (1989)

20 J. Schnermann, and J.P. Briggs: Function of the juxtaglomerular apparatus: local control of glomerular hemodynamics. In: *The Kidney: Physiology and Pathophysiology, 2nd Edition.* (D. Seldin and G. Giebisch, editors), Raven Press, New York 1992, pp. 1249-1290

21 M. Ursino, and G. Fabbri: Role of the myogenic mechanism in the genesis of microvascular oscillations (vasomotion): analysis with a mathematical model. Microvasc. Res. **43**, 156-177 (1992)

22 K.P. Yip, N.-H. Holstein-Rathlou, and D.J. Marsh: Mechanisms of temporal variations in single-nephron blood flow in rats. Am. J. Physiol. **264**, F427-F434 (1993)

Dynamics of Bone Remodelling

Li. Mosekilde, J.S. Thomsen, and E. Mosekilde

Abstract

The skeleton is formed during childhood through the constant influence of daily mechanical loading. Later, the skeleton is remodelled in order to renew bone tissue and reorganise bone structure. Bone remodelling is a dynamic process, with several types of cells working close together in time and space. It occurs in anatomically discrete sites, which are active for a few months and then rest for several years. During each remodelling process, some bone mass is lost, causing the normal age-related bone loss. The bone remodelling process mediates at any time the effect of both hormonal and mechanical agents that act on the skeleton. Different naturally-occurring events or therapeutic regimens can influence the activation frequency, the balance, and all phases of the remodelling process. This dynamic remodelling process can be simulated during normal aging, the menopause, and also during different therapeutic regimens. A simulation model thereby provides "non-invasive" information concerning the influence of the remodelling process on bone mass, architecture, and thereby bone strength – and it also provides a tool for evaluating existing and new regimens for the treatment of osteoporosis.

1 Introduction

The human skeleton is formed and modelled during childhood and youth through a constant influence of daily mechanical loading. In young adulthood the skeleton achieves its maximum mass and strength (Frost 1973; Parfitt 1988). The skeleton is designed in such a way that this maximum bone mass is as small as possible consistent with providing sufficient strength for its mechanical support of the body. From the age of 25–30 years and throughout adult life, bone mass is lost at an almost constant rate as a consequence of normal aging. Therefore, to maintain a sufficient mechanical support, despite declining bone mass, bone structure has to be redesigned and redistributed continuously (Frost 1983B, 1987). This process is called bone remodelling and, like growth and bone modelling, is controlled by daily mechanical loading and through a number of hormonal factors. If an optimal redesign fails, bone strength declines critically and the fracture threshold is exceeded – osteoporosis has occurred. Osteoporosis, which is the most common bone disease in the western world, is defined as a systemic skeletal disease characterized by low bone mass and microarchitectural deterioration of bone tissue, with a consequent increase in bone fragility and susceptibility to fracture.

A detailed description of the changes in bone mass and structure with normal aging in both sexes is crucial to the understanding of how extreme loss of bone competence can be prevented during aging. Such a detailed description of the trabecular bone characteristics could take form of a lattice model for the simulation of changes in bone strength during aging (Jensen et al. 1990). Additionally, a simulation model of the bone remodelling process based on human biological data would add information concerning microarchitectural changes during normal aging, during the menopause, and also during different treatment regimens (Thomsen et al. in press).

In this review, we have focussed on the following subjects: 1. bone biology; 2. bone remodelling; 3. changes in the vertebral trabecular network during aging; 4. computer simulation of the vertebral trabecular lattice. Finally, we have combined these subjects, all based on human bone, in a simulation model for vertebral trabecular bone remodelling.

2 Bone Biology

2.1 Gross Anatomy

Bone is a tissue formed of two separate components – compact and trabecular bone. The compact bone provides a dense shell around all bones and thereby gives support to the internal trabecular network and at the same time encloses the bone marrow. Cortical bone constitutes approximately 80% of the skeleton. In the diaphyses of long bones, the cortical or compact bone is very thick and gives these bones the extreme strength needed for their support of the body during daily dynamic loading. At other places (e.g. the vertebral bodies and the femoral neck) the cortical bone provides a shell only a few hundred μm thick. It is at these places that osteoporotic fractures most commonly occur. It is also at these places that trabecular bone dominates and provides the main strength (Frost 1973). In sum, trabecular bone constitutes only 20% of the total bone mass.

Trabecular bone consists of a network of bone plates, columns, and struts with diameters varying from 40–400 μm. In loadbearing trabecular bone (such as the vertebral body and the femoral neck) the trabecular lattice is obviously designed to withstand dominating vertical loads and strains (Atkinson 1967; Jensen et al. 1990; Radin 1983; Townsend et al. 1975B). In the vertebral body, the trabecular network has all its thick trabecular columns orientated vertically so as to withstand the dominating compressive forces. Thinner, horizontal struts provide support for the vertical columns to prevent their bending when under load (Townsend et al. 1975A). The trabecular network in the vertebral bodies (as in other loadbearing sites) is anisotropic in all directions apart from the horizontal plane. In this plane the network is isotropic (Fig. 1).

2.2 Ultrastructural and Microscopic Anatomy

Bone tissue consists of 80% mineral and only 20% organic material. At both microscopic and ultrastructural levels, the relationship between the inorganic and organic components of bone tissue is extremely close and well-organised – again, so as to provide maximum mechanical support with minimum mass. The inorganic components are mainly calcium-hydroxyapatite crystals and amorph calcium phosphate. The size of the crystals is 2nm × 30nm and they are situated in close relationship to the major component of the organic material (the collagen fibres).

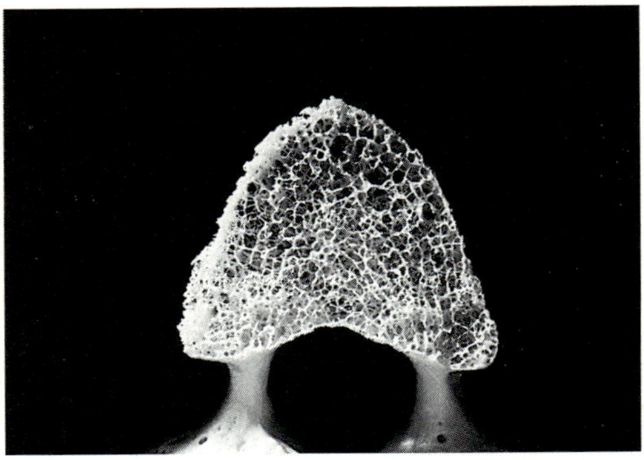

Fig. 1. Horizontal section through a normal human vertebral body. The trabecular network is isotropic in this plane. The cortical shell is very thin.

The perfect design of the ultrastructure, with the strict alignment of crystals and collagen fibres giving maximum mechanical competence, is further confirmed at the microscopic level. In normal lamellar bone, the collagen fibres are arranged in a parallel pattern in a layer of thickness of approximately 3μm (one lamella). In the adjacent lamella, the parallel pattern of the collagen fibres has a different angle (45°) to the first – and so on throughout the bone structural unit – the osteon. This pattern again provides optimum biomechanical properties for the bone tissue.

In cortical bone, the osteons, identical with the Haversian systems, are cylindrical structures with a diameter of 100–200 μm and a length of 1500–4000 μm. The cortical osteons are often aligned parallel to the long axis of the bones and consist of 10–15 lamellar rings arranged around a central vascular channel. The cortical osteons are demarcated by a cement line, which is ground substance with few collagen fibres. In between the perfect cortical osteons are fragments of previous osteons.

The same units – trabecular osteons or packets – are found in trabecular bone, and the lamellar thickness and arrangement are the same as in cortical bone. However, whereas the cortical osteons are cylindrical structures, the trabecular osteons are flat, disc shaped or branching structures covering the trabecular network (Kragstrup and Melsen 1983). The thickness of the trabecular osteons is fairly constant (approximately $50\mu m$). The number of lamellae is, therefore, constant too, and numbers 15–20. The trabecular osteons are large and cover surface areas of more than $1000\mu m \times 1000\mu m$, and in contrast to the cortical osteons do not contain any vascular channels.

2.3 Bone Cells

The four types of bone cells that control the dynamic processes in the bone are: 1. lining cells; 2. osteoblasts; 3. osteocytes; 4. osteoclasts.

The resting bone surface, where no dynamic process takes place, is covered by a layer of flat, inactive *lining cells.* These cells lie on a thin layer of non-mineralised matrix (collagen type I). The thickness of this layer is normally only 100–200 nm. Despite its thinness, this layer seems very important, since bone resorption is never initiated on osteoid covered surfaces or on surfaces covered with any other form of non-mineralised matrix, therefore this layer has to be removed before any dynamic process can occur; (Chambers and Fuller 1985). Lining cells (osteoblastic lineage cells) are also involved in the bone-resorption process in that they are believed to secrete the enzyme (collagenase) necessary for the removal of the thin non-mineralised surface layer (Chambers et al. 1985).

The bone matrix is secreted by *osteoblasts* (activated preosteoblasts) that lie on the surface of existing matrix and deposit new bone onto it. During active bone formation, the cuboidal osteoblasts lie closely together on the bone matrix and produce osteoid onto it at a generally constant rate. Some of the osteoblasts remain free on the surface, while others (some 10%) gradually become embedded in their own secretion. The newly formed material (osteoid) is rapidly converted into hard bone matrix by the deposition of hydroxyapatite crystals between the closely packed collagen fibres.

Once captured in the hard bone matrix, the original bone-forming cells, now called *osteocytes*, have no opportunity to secrete further matrix, but the osteocytes, each lying in a small lacuna, are not isolated. Tiny channels in the bone tissue enable the cells to form gap-junctions with adjacent osteocytes. This network probably plays a major role in controlling both the activity of the osteocytes themselves and the activity of the osteoblasts and lining cells on the surfaces. The osteocyte network seems to be able to sense piezoelectric fields and thereby mechanical stresses and strains applied to the bone. The signals from these cells are transmitted not only to the osteoblasts and lining cells but also to the fourth bone cell type – the osteoclasts.

Osteoclasts are derived from circulating mononuclear precursors and seem able to erode bone as soon as the thin protective layer of non-mineralised matrix has been removed. Osteoclasts are large (20–100 μm), multi-nucleated and mobile cells with a refined enzymatic system (Baron 1989; Chambers 1985). During bone resorption they are capable of engulfing both collagen, mineral, and osteocytes (Elmardi et al. 1990). The osteoclasts move quickly over the surface as they excavate the cavities (Boyde and Jones 1987) – the process normally stops at a depth of 45–50 μm (Eriksen et al. 1984). The cells leave the surface raw, with collagen fibres visible in the bottom of the cavities. If coupling takes place, this surface is soon covered by ground substance (cement line) and the formative process begins (Baron et al. 1981). Like osteoblasts, osteoclasts are also controlled from the network of underlying osteocytes – continuously sensing and signalling the mechanical stresses and strains applied to the bone.

3 Bone Remodelling

The design and redesign of bone involve three different processes:

- Growth
- Modelling
- Remodelling

3.1 Growth

During childhood and the early years of adulthood, the skeleton grows in length. The bones also expand in diameter and achieve their external shape (are modeled)(Frost 1988).

3.2 Modelling

During bone modelling, osteoblasts and osteoclasts work independently of each other in time and space – often over large surface areas (Parfitt 1988). The net balance is positive (i.e. there is increased bone mass) and bones reach their final external form and high bone density during this period.

Around the age 20–25 years, peak bone mass is achieved as a result of these processes (Gilsanz et al. 1988). Individual peak bone mass is dependent on genetic, racial and hormonal factors and also on external factors: physical activity and nutrition. Peak bone mass is normally 20–25% higher in men than in women (Parfitt 1988).

From the age of 20–25 years, bone mass begins to decline (Gilsanz et al. 1988; Marcus et al. 1983; Riggs et al. 1986). This early, age-related loss is especially detectable at places where trabecular bone dominates e.g. the vertebral bodies (Eastell et al. 1990). Loss of bone mass with age is unavoidable and is caused by the third process – bone remodelling.

3.3 Remodelling

During the remodelling process, osteoclasts and osteoblasts work closely together in time and space (coupling), and they work in units called BMU (Basic Multicellulary Units)(Frost 1964). When a remodelling process has just been completed, a small amount of bone (trabecular or cortical) has been removed and has been replaced by new bone. The process seems designed to renew old bone – possibly having microfractures or with dead osteocytes (Frost 1963) – and to reorganize trabecular structures to achieve the maximum strength obtainable with the remaining bone mass (Frost 1983A). The remodelling process has another role: it ensures the maintenance of calcium homeostasis.

Each remodelling process (Fig. 2) is initiated by an Activation (A) by which osteoblastic lineage cells start to secrete collagenase, which removes the thin layer of unmineralised bone typical of a resting bone surface. This exposes the mineralised bone underneath to the multinucleated, mobile osteoclasts. During osteoclastic bone Resorption (R), Howship's lacunae are excavated to a maximum depth of 50–60 μm. A short reversal phase, where the cement line is formed, follows, and then bone Formation (F) normally begins. If coupling has taken place, osteoblasts produce osteoid (collagen and ground substance) at a rate of 0.5–1.0 μm per day (Melsen et al. 1978). When the osteoid thickness has reached approximately 12–15 μm, mineralisation begins from the bottom (mineralisation front). At the termination of each remodelling process, the bone surface is again covered by an extremely thin layer of non-mineralised bone and a layer of flat lining cells. The bone is again converted into a resting surface.

The remodelling process, which is designed to renew old bone of inferior quality and to replace bone with microfractures, has its own cost – the balance in adults is normally negative (i.e. after each remodelling process there is a reduced mass). There is thus an unavoidable loss of bone mass with age. The remodelling process also has another cost: it causes disruption of the trabecular network with age (Kleerekoper et al. 1985; Pesch et al. 1980). Because the balance is negative, there is a thinning of the trabecular structures in the network, and this makes fortuitous osteoclastic perforations possible (Parfitt 1984B, 1987; Parfitt et al. 1983). As the normal resorption depth is approximately 50μm, one resorption site covering more than half of the circumference of a trabecula, or two resorption sites, one on each side of a trabecula, could easily perforate a trabecular structure with a diameter of 100–120 μm ($100-120\mu$m is the normal thickness of horizontal trabeculae in the vertebral body of elderly individuals (Mosekilde 1988, 1989)). In an adult skeleton of normal size, an activation of a remodelling process starts every

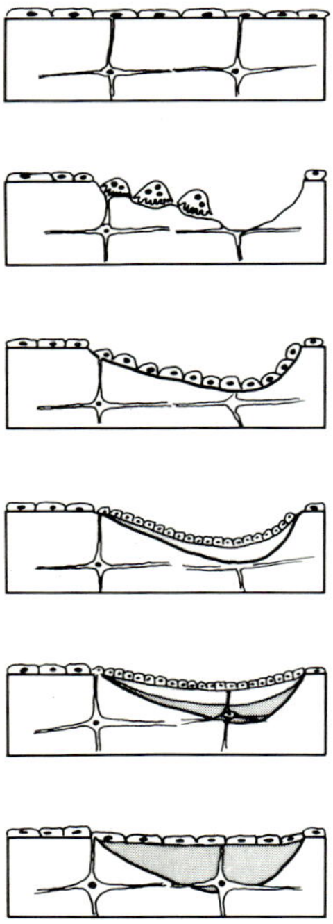

Fig. 2. The different phases in the remodelling process. Flat lining cells are seen on the resting surface. Large, multi-nucleated osteoclasts during the bone resorption. Cuboid osteoblasts are seen during bone formation.

10 seconds. The process lasts some 4–8 months, which means that a total of 1–2 million remodelling sites are active at any time (Parfitt 1988).

In summary, the three processes – growth, modelling, and remodelling – although different in working methods are all governed by the same cells and are all dependent on intrinsic factors as well as on mechanical stimuli. This means that the form, size, density, and strength of the skeleton at any age are dependent on its mechanical usage.

4 Changes in the Vertebral Trabecular Network During Aging

In younger individuals the vertebral trabecular network is dense and well connected (Fig. 3). On a vertical section through a vertebral body, three zones can be identified in this network; An upper, sub-endplate zone with an anisotropic trabecular network consisting of 200 – 220μm thick vertical columns and slightly thinner (180 – 200μm) horizontal struts. This network is regular and dense. In the middle of the vertebral body is a zone with irregular, vertical columns and plates supported by thin, horizontal struts. The irregularity of this zone is mainly caused by the large veins which pass through the vertebral body (basi-vertebral veins). The third zone, towards the lower endplate, is again regular with a dense network similar to the upper zone. The density of the whole trabecular network is identical for young men and young women. The size of the vertebral bodies is larger for men then women, and this factor causes the higher peak bone mass in men.

Haemopoietic marrow fills the space between the trabeculae and has an internal hydraulic effect (Karazian and Graves 1977). The marrow remains haemopoietic throughout life, although with age there is often a slight increase in fat cells. Several of the bone cells involved in the remodelling process are directly recruited from the haemopoietic blood marrow. Therefore, the close connection between red marrow and bone tissue in the vertebral bodies seems to be responsible for the high turnover (remodelling) at this site.

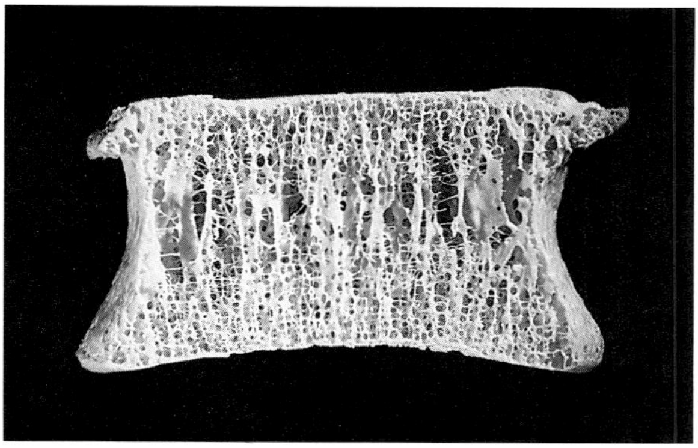

Fig. 3. Vertical section through a vertebral body from a young individual (plate-strut network). The network is anisotropic and clearly arranged in three different horizontal zones.

The decline in trabecular bone mass, the decline in trabecular thickness, and the perforations in the network are all caused by the remodelling process – and are, therefore, normal, age-related changes (Fig. 4). The described pattern, with selective thinning and perforation of horizontal struts, is seen in both men and women during normal aging (Mosekilde 1988, 1989). Around the menopause, the activation frequency increases (Brockstedt et al. 1993) and at the same time resorption depth might increase (Eriksen et al. 1985). These two factors accelerate the age-related changes and are thought to increase the number of trabecular perforations in the network. Furthermore, once trabecular structures have been perforated they are no longer loaded and they will, therefore, be resorbed rapidly by osteoclasts (Mosekilde 1990)(Fig.5).

The biomechanical consequences of the disruption of the loadbearing vertebral network are far-reaching. As the strength of a single trabecula is proportional to its radius squared, thinning of the vertical structures has

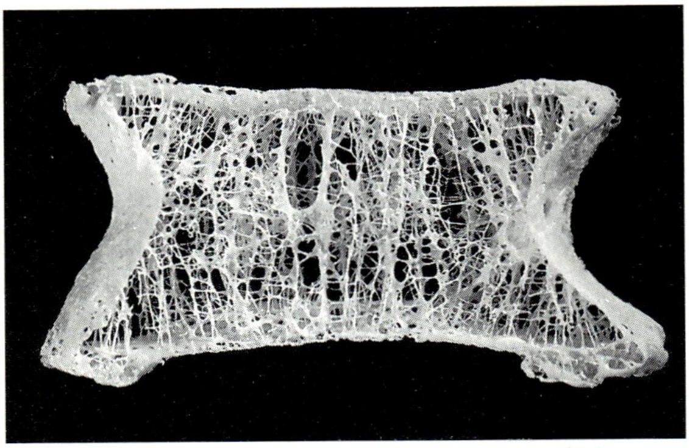

Fig. 4. Vertical section through a vertebral body from an elderly individual (strut-strut network). The network is highly anisotropic and there are many perforations of the horizontal struts.

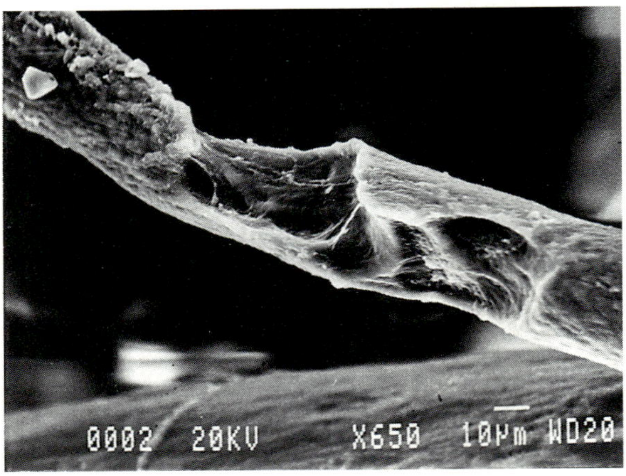

Fig. 5. Scanning electron microscopy (SEM) photograph of an unloaded horizontal trabecular strut. There are deep osteoclastic resorption pits. (magnification × 650).

a tremendous influence on their strength. The situation is similar concerning the length between supporting, horizontal struts: the compressive strength of the network is proportional to the square of the distance between the supporting struts (Bell et al. 1967). Furthermore, at a certain stage, when several horizontal struts have disappeared, the slenderness ratio of the long, unsupported, vertical trabeculae reaches a critical value, and the structure fails due to buckling rather than compression: when this stage is reached, there is a dramatic loss of strength (Bell et al. 1967). The loss of bone mass

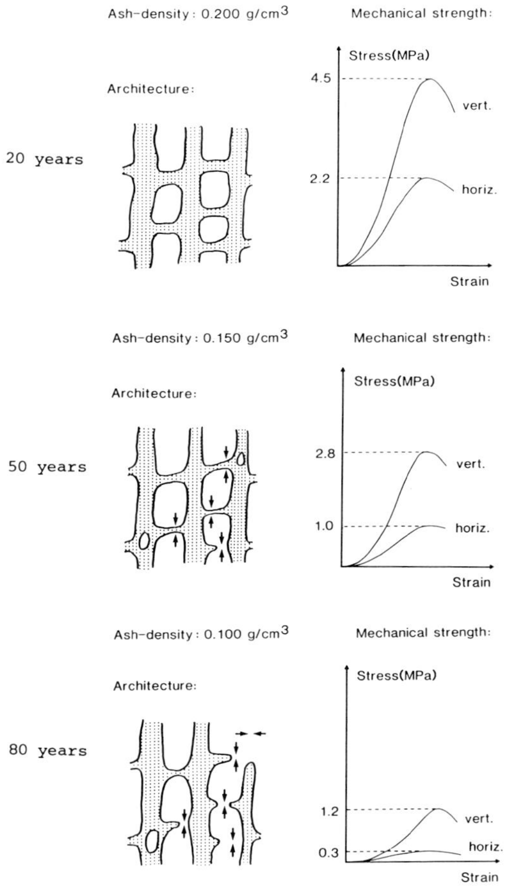

Fig. 6. Drawing indicating typical age-related changes in vertebral bone density, architecture, and strength.

during aging is therefore accompanied by a more pronounced loss of strength (Mosekilde et al. 1987)(Fig. 6).

Descriptions of changes in trabecular connectivity *with age* have been made by use of contact radiographs of 2.5 mm thick vertical sections (Atkinson 1967), X-ray images of 100μm thick horizontal sections (Pesch et al. 1980), and automatic computer analyses of both vertical and horizontal sections (Bergot et al. 1987). Most of these studies have demonstrated an age-related disappearance of horizontal struts in the network, and this was also proved and visualized by the use of 300μm thick, plastic-embedded, vertical sections investigated in polarized light (Mosekilde 1988). By use of this technique, it was also possible to show that although thinning and perforation of the horizontal struts occurred in both men and women, the destruction of the network was more pronounced in women.

Refined two-dimensional analyses (Kleerekoper et al. 1985; Vesterby et al. 1989; Garrahan et al. 1990) and three-dimensional analyses and reconstruction (Feldkamp et al. 1989) of human trabecular bone have recently been made. Both the star volume method (Vesterby et al. 1989) and the node-strut analysis (Garrahan et al. 1990) have shown pronounced disruption of the trabecular network during normal aging. Neither method has been able to give any measurements of the trabecular thicknesses or distances, however. The use of high-resolution microfocus computed tomography (Feldkamp et al. 1989) provided a tool for obtaining both two-dimensional and histomorphometric data and measurements of 3D connectivity (Euler number/tissue volume). A similar, but destructive, method has been developed by Hodgskinson and Currey (1990) and refined by Odgaard (Odgaard et al. 1990).

The importance of such methods comes into perspective when it is recognised that not only is bone strength highly dependent on bone architecture and on connectivity but also that loss of connectivity in the loadbearing network seems irreversible. Remodelling sites on disconnected, unloaded structures show no sign of bone formation (uncoupling) (Mosekilde 1990)(Figs. 7 and 8). Strain or stress applied to bone seems essential to enable osteoblasts to form new bone on existing surfaces (Parfitt 1984A, 1988). The osteoblast will therefore be unable to refill the gaps in the network under normal circumstances. Most therapeutic regimens, being primarily anti-resorptive, cannot facilitate this process either – but whether powerful anabolic agents like parathyroid hormone (PTH) will be able to do so remains unresolved. Consequently, a description of the mechanism by which the structures are perforated, the extent of the perforations, and how different therapeutic regimens affect loss of connectivity is important.

Again, currently available in vivo measurements can only measure changes in bone mass and bone density but cannot detect the important architectural changes during aging, the menopause, or different therapeutic regimens.

5 Computer Modelling

5.1 Computer Modelling of the Vertebral Trabecular Lattice

A "non-invasive", quick method for elucidating the importance of connectivity and the regularity of such a loadbearing network is by use of computer models for simulation of different, normally-occurring events (Reeve 1986; Lacy et al. 1994)). Realistic models based on human, biological material might also provide some insight into the capability of therapeutic agents to delay disruption of the network or eventually to restore connectivity (Jensen et al. 1990).

In such a study, an idealized, structural model of vertebral trabecular bone was used (Jensen et al. 1990). The architecture and lattice characteris-

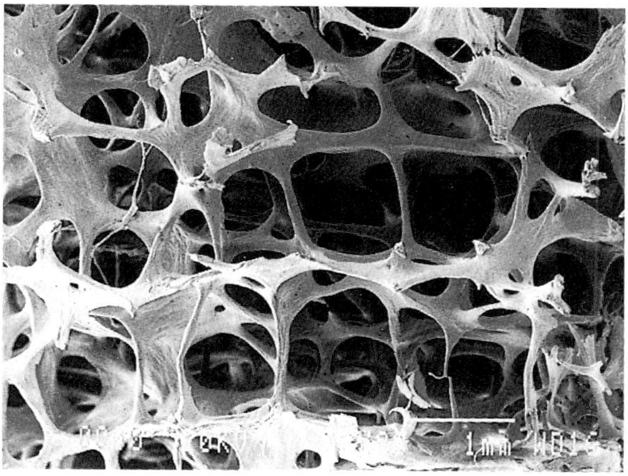

Fig. 7. SEM photograph of the vertebral trabecular network from a 58 year old women. A perforation of a horizontal strut in the network is visible. (magnification $\times$ 20).

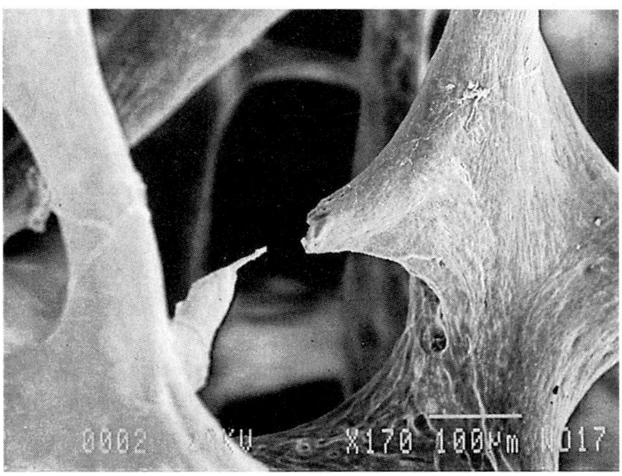

Fig. 8. SEM photograph from the same specimen as in Fig. 7. The osteoclastic perforation is clearly demonstrated, with osteoclastic footprints on one end. (magnification $\times$ 170).

tics were based on bone samples taken from the central part of human vertebral bodies (age span 30-90 years). With trabecular thicknesses and distances typical for individuals aged 40, 60, and 80 years, the model could account for the age-related changes in stiffness, stress, and bone volume found in experimental data. It was demonstrated that apparent stiffness varied by a factor of 5-10 from a perfect cubic lattice to a lattice of maximal irregularity. A considerable change in biomechanical competence was also demonstrated when the bone material was slightly redistributed between vertical and horizontal trabeculae in the network. In both situations, the changes were induced without changing the total amount of trabecular bone mass. The study concluded that measured bone mass or bone density should not be the sole indicator of trabecular bone biomechanical competence and that a detailed description of the architecture should also be provided. The model did not focus on changes in biomechanical competence in relation to perforations in the network, nor did it take into account the remodelling process and the temporal changes in this.

A combined model using this idealized network-model and a model of bone dynamics could provide data concerning changes not only in bone mass but also in bone quality during normal aging, the menopause, the development of osteoporosis, and during different treatment regimens. This would save time with regard to both pre-clinical and clinical studies and provide more insight into the work pattern of the remodelling process and how it might be manipulated in the prevention and treatment of osteoporosis.

5.2 A Simulation Model of the Remodelling Process

We have recently developed a model of the bone remodelling process. In this model we have used input parameters which have all been obtained from human studies. The input parameters were: activation frequency; resorption period; reversal period; formation period; initial trabecular thickness; resorption depth; "static" formation balance; extent of the remodelling site; and, number of trabeculae. The output variables in this simulation model were: the number of active remodelling sites; the bone mass lost relative to the initial mass (%); the mean trabecular thickness; and, the number of perforations in the network (Thomsen et al. in press).

The simulation model was a stochastic map which was discrete in time and continuous in state. The time between two iterations in the model was 1 day. This model of bone remodelling is represented in (Fig. 9) as flow charts.

The age-related changes in the amount of bone tissue in this model were represented by the bone formation balance b, which was slightly negative. Thus, for each remodelling process the trabecular thickness would be increased with b positive; and with b negative the trabecular thickness would therefore decline.

Our model was based on the model developed by Reeve in 1986. However, while Reeve used a plate-strut structural model characteristic of the iliac crest, we have used a strut-strut model in order to simulate the vertebral network found in elderly individuals (Mosekilde 1988). In Reeve's model, perforations mainly occurred as a consequence of two remodelling processes taking place at the same time – one on each surface of a trabecular plate. In our model, perforations were caused by the same remodelling process encircling a whole trabecular strut and thereby causing a virtual doubling of

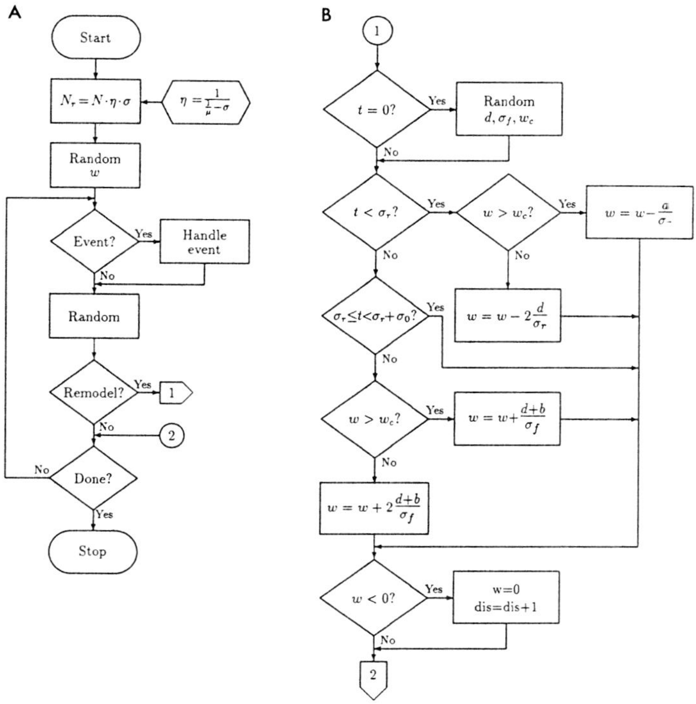

Fig. 9. Flow charts for the simulation model of the remodelling process. (A) Describes the overall remodelling process for the whole "network", and (B) describes the individual remodelling cycle on a given trabecula.

the resorption depth. Investigation of vertebral trabecular bone samples by scanning electron microscopy shows that this mechanism totally dominates after the age 40-50 years (Mosekilde 1990).

We have used the model to simulate the remodelling activity around the menopause. We used two slightly different concepts: (A) A doubling of the activation frequency; (B) a doubling of the activation frequency paralleled by an increased resorption depth (from 50μm to 70μm). The menopause was defined as an event (a period of change in the main input parameters) in the simulation model.

The model was simulated for 20 years with an event (the menopause) of 5 years duration. The result of a simulation of the two different menopause scenarios (A) and (B) can be seen in Fig 10. The parameters used in the simulation are listed in Table 1.

The model clearly confirmed a continuous loss of bone mass and trabecular thickness with age. During the menopause (A), an increase in number of trabecular perforations was detected, causing the described irreversible loss of bone mass. Simultaneously, the increase in activation frequency caused

Table 1. Input parameters for simulating the menopause without increased resorption depth (A) and with increased resorption depth (B). Age at simulation start 48 years, and age at menopause start 53 years. Duration of menopause: 5 years. The critical thickness is the trabecular thickness at which the remodelling site precisely encircles the trabecula.

	Normal	Menopause
Resorption period (σ_r), days[a]	42	42
Reversal period (σ_0), days[a]	9	9
Formation period (σ_f), days[a]	145 ± 45	145 ± 45
Trabecular thickness (w), μm[b]	135 ± 24	–
Resorption depth (d), μm[a]	50 ± 15	50 ± 15 (A)
Resorption depth (d), μm[c]		70 ± 15 (B)
Critical thickness (w_c), μm	$(130 \pm 50)/\pi$	$(130 \pm 50)/\pi$
Δ B.BMU (b), μm	-2	-2
Activation frequency (μ)[a], days^{-1}	1/1096	1/548
Number of trabeculae (N)	32000	32000

[a]Eriksen 1986
[b]Mosekilde 1989
[c]Eriksen et al. 1985

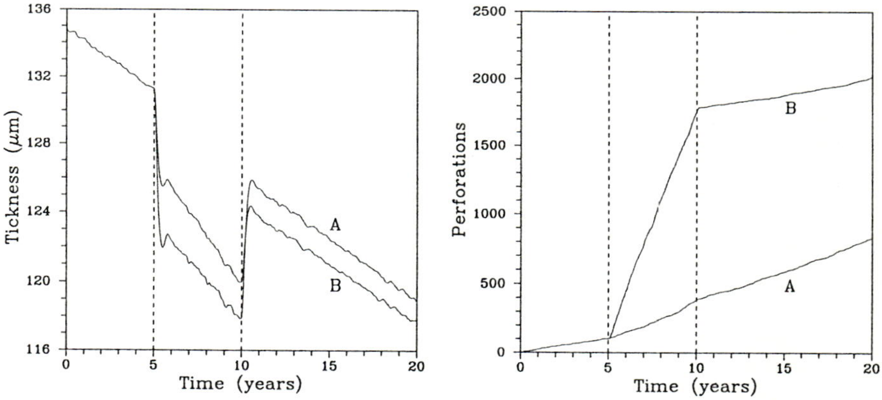

Fig. 10. Simulation of the menopause:(A) The activation frequency μ is doubled. (B) The activation frequency μ is doubled and the resorption depth d is increased from 50 to 70 μm.

an increase in the reversible loss of bone mass (thinning of trabecular structures). Using simulation (B) for the menopause, there was a more pronounced relative mass loss, representing both the reversible and the irreversible loss of bone. The number of perforations increased more than in simulation (A) during this simulation of the menopause where increased resorption depth accompanied the increased activation frequency.

The simulation model (Thomsen et al. in press) mirrored very closely normal age-related and also menopause-related changes in the human trabecular network.

Although the presented model could simulate clinical situations very closely, it should be borne in mind that many assumptions have been made: 1. we have used remodelling data from the iliac crest and applied them to vertebral trabecular bone. Although we know that the trabecular bone is lost at the same rate in the iliac crest and the vertebral bodies (Mosekilde and Mosekilde 1988), the remodelling characteristics might not be identical; the trabecular structures in the vertebral bodies are thinner than in the iliac crest, and resorption depth might be dependent on trabecular thickness (Eriksen et al. 1985); the trabeculae in the vertebral bodies might also be subject to a totally different loading pattern than in the iliac crest and thereby be subject to different strain patterns. 2. In the presented model we have taken only horizontal, supporting struts into consideration as these have been shown to exert much larger dynamic changes with age and during the menopause than the vertical columns (Mosekilde 1988). It is intended, though, to combine this model of bone remodelling with a three- dimensional structural model in order to predict the effects of changes in the remodelling process on the bone strength.

6 Conclusion

The dynamics of bone remodelling play a major role for the age-related and menopause-related changes in bone mass, structure, and strength. "Non-invasive" measurements of these changes are difficult to obtain. Therefore, the presented simulation models of the trabecular microarchitecture and of the remodelling process could be very useful in future studies concerning treatment of osteoporosis.

References

1 Atkinson P.J.: Variation in trabecular structure of vertebrae with age. Calcif. Tissue Res. 1: 24-32, 1967.

2 Baron R., Vignery A., Lang R.: Reversal phase and osteopenia: Defective coupling of resorption to formation in the pathogenesis of osteoporosis. In: Osteoporosis, Recent Advances in Pathogenesis and Treatment. (H.F. DeLuca, H.M. Frost, W.S.S. Jee, C.C. Johnston Jr. and A.M. Parfitt eds.), Baltimore University Press, pp. 311-320, 1981.

3 Baron R.: Molecular mechanisms of bone resorption by the osteoclast. The Anatomical Record 224: 317–324, 1989.

4 Bell G.H., Dunbar O., Beck J.S., Gibb A.: Variations in strength of vertebrae with age and their relation to osteoporosis. Calc. Tiss. Res. 1: 75-86, 1967.

5 Bergot C., Prêteux F., Laval-Jeantet A.-M.: Quantitative image analysis of thin sagittal and transversal slices from autopsy specimens from L3 vertebrae. In: Osteoporosis 1987. (C. Christiansen et al. eds.), Osteo Press, Copenhagen, pp. 338–340, 1987.

6 Boyde A. and Jones S.J.: Early scanning electron microscopic studies of hard tissue resorption: Their relation to current concepts reviewed. Scan. Microsc. Vol. 1,1: 369-381, 1987.

7 Brockstedt H., Kassem M., Eriksen E.F., Mosekilde Le., Melsen F.: Age- and sex- related changes in the iliac cortical bone mass and remodelling. Bone 14: 681–691, 1993.

8 Chambers T.J.: The pathobiology of the osteoclast. J. Clin. Pathol. 38: 241–252, 1985.

9 Chambers T.J., Darby J.A., Fuller K.: Mammalian collagenase predisposes bone surfaces to osteoclastic resorption. Cell. Tissue Res. 241: 671-675, 1985.

10 Chambers T.J. and Fuller K.: Bone cells predispose bone surfaces to resorption by exposure of mineral to osteoclastic contact. J. Cell Sci. 76: 155-165, 1985.

11 Eastell R., Mosekilde Li., Hodgson S.F., Riggs B.L.: Proportion of human vertebral body bone that is cancellous. J. Bone Min. Res. 5,12: 1237-1241, 1990.

12 Elmardi A.S., Katchburian M.V., Katchburian E.: Electron microscopy of developing calvaria reveals images that suggest that osteoclasts engulf and destroy osteocytes during bone resorption. Calcif. Tiss. Int. 46: 239-245, 1990.

13 Eriksen E.F., Melsen F., Mosekilde Le.: Reconstruction of the resorptive site in iliac trabecular bone. A kinetic model for bone resorption in 20 normal individuals. Metab. Bone Dis. Relat. Res. 5: 235-242, 1984.

14 Eriksen E.F., Mosekilde Le., Melsen F.: Trabecular bone resorption depth decreases with age: Differences between normal males and females. Bone 6: 141-146, 1985.

15 Eriksen E.F.: Normal and pathological remodelling of human trabecular bone: Three dimensional reconstruction of the remodelling sequence in normals and in metabolic bone disease. Endocrine Rev. 7,4: 379-408, 1986.

16 Feldkamp L.E., Goldstein S.A., Parfitt A.M. et al.: The direct examination of three- dimensional bone architecture in vitro by computed tomography. J. Bone Min. Res. 4: 3-11, 1989.

17 Frost H.M.: Bone remodelling dynamics (C.R. Lam ed.), C.C. Thomas, Springfield, IL, pp. 65-75, 1963.

18 Frost H.M.: Dynamics of bone remodelling. In: Bone Biodynamics (H.M. Frost ed.) Little Brown & Co., Boston, USA, pp 315-331, 1964.

19 Frost H.M.: Editorial: Tetracycline based histological analysis of bone remodelling. Calcif. Tissue Res. 3: 211-237, 1969

20 Frost H.M.: The spinal osteoporoses. Mechanisms of pathogenesis and pathophysiology. Clin. Endocrin. Metab. 2,2: 257-275, 1973.

21 Frost H.M.: A determinant of bone architecture. The minimum effective strain. Clin. Orthop. Rel. Res. 175: 286-292, 1983A.

22 Frost H.M.: The skeletal intermediary organization. Metab. Bone Dis. Relat. Res. 4: 281-290, 1983B.

23 Frost H.M.: Editorial: The mechanostat: a proposed pathogenic mechanism of osteoporosis and the bone mass effects of mechanical and nonmechanical agents. Bone Miner. 2: 73-85, 1987.

24 Frost H.M.: Editorial: Vital biomechanics: Proposed general concepts for skeletal adaptions to mechanical usage. Calcif. Tissue Int. 42: 145-156, 1988.

25 Garrahan N.J., Croucher P.I., Wright C., Compston J.E.: A computerised technique for the quantitative assessment of resorption cavities in trabecular bone. Bone 11: 241-245, 1990.

26 Gilsanz V., Gibbens D.T., Carlson M., Boechat M.I., Cann C.E., Schulz E.E.: Peak trabecular vertebral density: A comparison of adolescent and adult females. Calcif. Tissue Int. 43: 260-262, 1988.

27 Hodgskinson R. and Currey J.D.: Effects of structural variation on Young's modulus of non-human cancellous bone. Proc. Instn. Mech. Engrs. 204: 43-52, 1990.

28 Jensen K.S., Mosekilde Li., Mosekilde Le.: A model of vertebral trabecular bone architecture and its mechanical properties. Bone 11: 417-423, 1990.

29 Karazian L. and Graves G.A.: Compressive strength characteristics of the human vertebral centrum. Spine 21: 1-14, 1977.

30 Kleerekoper M., Villanueva A.R., Stanciu J., Rao D.S., Parfitt A.M.: The role of three dimensional trabecular microstructure in the pathogenesis of vertebral compression fractures. Calcif. Tissue Int. 37: 594-597, 1985.

31 Kragstrup J. and Melsen F.: Three-dimensional morphology of trabecular bone osteons reconstructed from serial sections. Metab. Bone Dis. Relat. Res. 5: 127-130, 1983.

32 Lacy M.E., Bevan J.A., Boyce R.W., Geddes A.D.: Antiresorptive druge and trabecular bone turnover: validation and testing of a computer model. Calcif. Tissue Int. 54: 179-185, 1994.

33 Marcus R., Kosek J., Pfefferbaum A., Horning S.: Age-related loss of trabecular bone in premenopausal women: A biopsy study. Calcif. Tissue Int. 35: 406-409, 1983.

34 Melsen F., Melsen B., Mosekilde Le., Bergmann S.: Histomorphometric analysis of normal bone from the iliac crest. Acta Path. Microbiol. Scand. Sect. A, 86: 70- 81, 1978.

35 Mosekilde Li.: Age related changes in vertebral trabecular bone architecture - Assessed by a new method. Bone 9: 247-250, 1988.

36 Mosekilde Li.: Sex differences in age-related loss of vertebral trabecular bone mass and structure - biomechanical consequences. Bone 10: 425-432, 1989.

37 Mosekilde Li.: Consequences of the remodelling process for vertebral trabecular bone structure - A scanning electron microscopy study (uncoupling of unloaded structures). Bone Miner. 10: 13-35, 1990.

38 Mosekilde Li. and Mosekilde Le.: Iliac crest trabecular bone volume as a predictor for vertebral compressive strength, ash density and trabecular bone volume in normal individuals. Bone 9: 195-199, 1988.

39 Mosekilde Li., Mosekilde Le., Danielsen C.C.: Biomechanical competence of vertebral trabecular bone in relation to ash density and age in normal individuals. Bone 8: 79-85, 1987.

40 Odgaard A., Andersen K., Melsen F., Gundersen H.J.G.: A direct method for fast three-dimensional serial reconstruction. J. Microsc. 159: 335-342, 1990.

41 Parfitt A.M.: The cellular basis of bone remodelling: The quantum concept reexamined in light of recent advances in the cell biology of bone. Calcif. Tissue Int. 36: S37-45, 1984A.

42 Parfitt A.M.: Age-related structural changes in trabecular and cortical bone: Cellular mechanisms and biomechanical consequences. Calcif. Tissue Int. 36: 37-45, 1984B.

43 Parfitt A.M.: Trabecular bone architecture in the pathogenesis and prevention of fracture. Am. J. Med. 82, 1B: 68-72, 1987.

44 Parfitt A.M.: Bone remodelling: Relationship to the amount and structure of bone, and the pathogenesis and prevention of fractures. In: Osteoporosis: Etiology, diagnosis and management. (B.L. Riggs and L.J. Melton III eds.), Raven Press, New York, pp. 45-93, 1988.

45 Parfitt A.M., Mathews H.E., Villanueva A.R., Kleerekoper M., Frame B., Rao D.S.: Relationships between surface, volume, and thickness of iliac trabecular bone in aging and in osteoporosis. J. Clin. Inv. 72, 4: 1396-1409, 1983.
46 Pesch H.-J., Scharf H.-P., Lauer G., Seibold H.: Der altersabhängige Verbundbau der Lendenwirbelkörper. Virchows Arch. A. (Pathol. Anat.) 386: 21-41, 1980.
47 Radin E.L.: Mechanical aspects of fractures and their treatment. In: Osteoporosis, Recent Advances in Pathogenesis and Treatment (H.F. DeLuca, H.M. Frost, W.S.S. Jee, C.C. Johnston Jr. and A.M. Parfitt eds.), University Park Press, Baltimore, pp. 191-199, 1983.
48 Reeve J.: A stochastic analysis of iliac trabecular bone dynamics. Clin. Orthop. Rel. Res. 213: 264-278, 1986.
49 Riggs B.L., Wahner H.W., Melton L.J.III, Richelson L.S., Judd H.L., Offord K.P.: Rates of bone loss in the appendicular and axial skeletons of women: Evidence of substantial vertebral bone loss before menopause. J. Clin. Invest.77: 1487-1491, 1986.
50 Thomsen J.S., Mosekilde Li., Boyce R.W., Mosekilde E.: Stochastic simulation of vertebral trabecular bone remodell-ing. Bone, in press.
51 Townsend P.R., Rose R.M., Radin E.L.: Buckling studies of single human trabeculae. J. Biomech. 8: 199-201, 1975A.
52 Townsend P.R., Raux P., Rose R.M., Miegel R.E., Radin E.L.: The distribution and anisotropy of the stiffness of cancellous bone in the human patella. J. Biomech. 8: 363-367, 1975B.
53 Vesterby A., Gundersen H.J.G., Melsen F.: Star volume of marrow space and trabeculae of the first lumbar vertebra: Sampling efficiency and biological variation. Bone 10: 7-13, 1989.

Modelling Heart Rate Variability Due to Respiration and Baroreflex

Henrik Seidel and Hanspeter Herzel

Abstract

This chapter is concerned with the interactions of heart beat, respiration, and blood pressure oscillations. We develop a nonlinear model which describes the essential parts of the baroreceptor loop: blood pressure wave, baroreceptor activity, cardio-respiratory center, and regulation of heart rate and vascular resistance via the autonomic nerves. Special attention is paid to the phase response properties of the cardiac pacemaker.

The model accounts for heart rate regulation, respiratory sinus arrhythmia (RSA), and resonant interaction of respiration and blood pressure oscillations (Mayer waves). External stimulations can induce toroidal oscillations, resonances, period-doubling, and chaotic behavior.

1 Introduction

In industrialized countries, one of the most frequent causes of death is a severe malfunction of the cardiovascular system. Yet, our present understanding of the dynamic regimes of this system and its interaction with respiration is very limited. This is partly due to the fact that many different mechanisms contribute to the control of circulation. They operate on time scales from seconds (baroreceptors, chemoreceptors) to minutes (renin-angiotensin) and even hours (renal fluid volume). Direct measurements of quantities involved in the control of the cardio-vascular system are very difficult (autonomic nervous system) or even impossible (medullary circulation centers) without serious intervention in the system. Hence, we have to deal with hidden or, at least, partially hidden variables.

For several reasons, information about the dynamics of these variables is of considerable interest. There is, for example, experimental evidence suggesting that a high sympathetic activity during myocardial infarction greatly increases the probability of a fatal cardiac arrhythmia [26]. Several diseases (e.g.

diabetes) can cause dysfunction of the autonomic nervous system. Anaesthetics may influence autonomic nerves or the vascular peripheral resistance and lead to circulation insufficiency. Failure of blood pressure control can cause hypertension and fainting or may facilitate heart arrhythmia and sudden death. Therefore, it is desirable to have a model that can relate the "hidden" variables to easily measurable data such as heart rate variability (HRV) and respiration.

The heart rate is not steady even when the level of physiological activity is constant or during deep sleep when there is no conscious psychic influence. Various rhythmic variations are observed. Coupling between the respiratory cycle and the heart beat causes oscillations of the heart rate with a frequency of approximately 0.25 Hz (4 s), referred to as respiratory sinus arrhythmia (RSA). The amplitude of RSA decreases with age. Furthermore, oscillations in heart rate appear at periods of approximately 10 s (Mayer waves) and, sometimes, 30 s. Their origin is imperfectly understood. Over a wide range, the heart rate exhibits a $1/f$ spectrum, the cause of which is also unknown [26]. Some of these rhythms are generated or, at least, influenced by circulation control mechanisms and their interaction with respiration. Their analysis by means of a model may allow conclusions about internal variables such as sympathetic and parasympathetic activity and, hence, can be used for diagnostic and prognostic purposes.

In this chapter, we attempt to establish a mathematical model which may contribute to the explanation of qualitative properties of circulation dynamics. Physiological systems are very complex, with numerous nonnegligible, interacting quantities. Only in the rarest cases it is possible to find simple mathematical expressions for these systems. Often only qualitative connections between physiological variables are known, and many parameters have to be guessed. Therefore, detailed modelling is nearly impossible, and we have to look for a compromise: the model must be complex enough to reconstruct essential features of the system. On the other hand, it must be simple enough to be useful for conceptualization.

The system treated in this chapter contains elements, where even qualitative properties are still disputed. This is particularly true for medullary interactions. For that reason, our model cannot and does not claim quantitative agreement with the physiological experiments. We want to investigate the qualitative behavior of the model and to compare it with experimental results. We hope that it can be a help in answering some as yet open physiological questions.

2 Heart, Respiration, and Baroreflex: Basic Dynamical Features

Before we present our model, we want to discuss some of its basic mechanisms. This discussion is very introductory, and the reader is referred to standard physiological literature [40, 3] for further information.

2.1 Heart

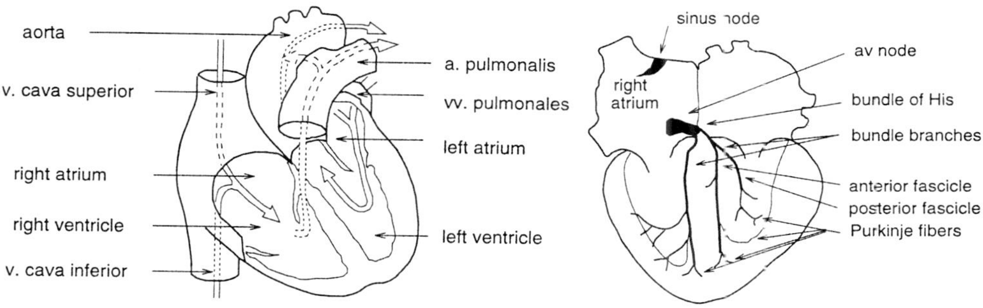

Fig. 1. The heart and its stimulus conducting system (adapted from [39])

The heart can be considered a complex oscillator. It contains several elements of specialized myocardial tissue forming a stimulus conducting system (Fig. 1). The sinus node, the atrio-ventricular node, and the Purkinje system are capable of oscillating independently. There is a hierarchy among these pacemakers with respect to their firing rates. If the heart is normal, the sinus node determines the heart rate (ca. 70 min^{-1} at rest). After the excitation has passed the AV-node, it quickly spreads along the bundle of His, to the Purkinje fibers, and then over the whole heart, resulting in a contraction. The frequency of the atrio-ventricular node is somewhat less (40–60 min^{-1}). Normally, the AV-node gets entrained by the sinus node. Only if the conduction from the sinus node towards the AV-node fails or is weakened (AV-block), the AV-node becomes the active pacemaker of the heart. If neither impulses from the sino-atrial nor from the atrio-ventricular node can reach the ventricles, the Purkinje system takes over rhythm generation (ca. 20 min^{-1}).

The myocardial tissue shows a post-systolic refractory behavior, i.e. each heart beat is followed by a certain time span characterized by the inexcitability of the cells. Hence, usually one impulse cannot excite a myocardial cell twice by wandering around and reentering. However, if conduction times are pathologically increased, e.g. after an infarction, the impulse may reenter a previously excited area after the end of the refractory period, leading to an extra-systolic heart beat or, even worse, to circulating waves or fibrillation.

Another cause of arrhythmia are *ectopic pacemakers*. They consist of degenerated self-oscillating cardiac muscle tissue. Sinus node and ectopic pacemakers interact and influence each other. Hence, they form a system of two coupled oscillators which is capable of evoking complex rhythms as entrainment, period doubling, and chaos [31, 4, 25, 20, 22, 5, 33, 19]

The heart is influenced by the autonomic nervous system in order to react to different stress situations. Three main effects are distinguished. High sympathetic activity increases, and high parasympathetic activity decreases the frequency of the sinus node (chronotropic effect), the contractility of the heart (inotropic effect), and the velocity of atrio-ventricular conduction (dromotropic effect).

2.2 Respiration

In the respiratory system, the rhythm generating pacemaker and the executing organ are spatially separated. Usually a central pattern generator is assumed to be situated in the *medulla oblangata.* Numerous physiological experiments have been performed in order to localize exactly the source of the respiratory rhythm. Until now, however, only preliminary concepts about the rhythm generation exist. A number of mathematical models have been established. These models are based on several groups of neurons mutually inhibiting or exciting [15, 14, 18, 50, 28, 44, 43, 16, 36] and are still rather speculative.

The impulses generated in the medulla oblangata are carried by the *phrenic nerve* to the diaphragm. Feedback is given by thoracic stress receptors and chemoreceptors sensitive to the blood concentrations of CO_2, H^+, and O_2. Numerous authors have tried to model these feedback loops. Some of the models include metabolic changes during muscular activity and contain more than one hundred equations [32].

2.3 Baroreceptors

Baroreceptors, sometimes also called *pressure receptors*, are neuronal cells wound around blood vessels. They respond with a change in their firing frequency to a strain in their cell membrane. The vascular walls expand at high pressures, enabling the baroreceptors to sense the blood pressure.

Upon closer examination, it turns out that the firing frequency of the baroreceptors depends both on the pressure and on its time derivative. Fig. 2 shows typical responses of baroreceptors to pressure waves. A review of baroreceptor models is given in [41].

Baroreceptors are located along the aorta. The most important of them are situated in the *aortic arch* directly at the heart and at the *sinus caroticus* in the neck. Their impulses are led by the *aortic nerve* and the *carotid sinus nerve* towards the brain. Both nerves cause a decrease in blood pressure when

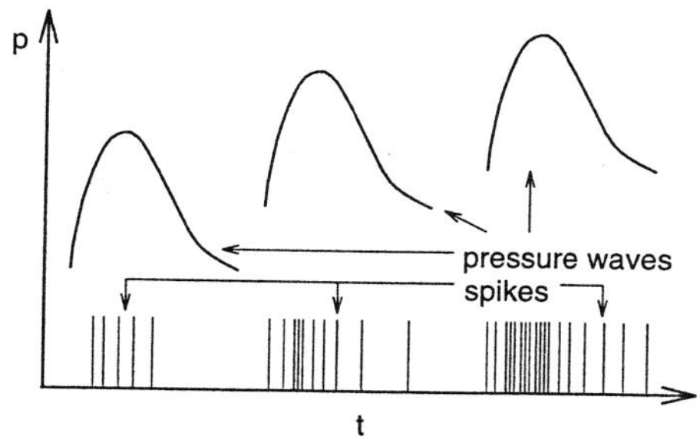

Fig. 2. Responses of baroreceptors to blood pressure waves of different strenghts (schematic).

stimulated. Thus, it it possible to understand how the baroreceptors manage their task of regulating the blood pressure: when the pressure increases, the vascular walls expand, the firing rates of the baroreceptors and, consequently, of the aortic and the carotid sinus nerve rise, causing a drop in blood pressure. Without any time delay in that control loop, the blood pressure would approach a constant value. However, as a consequence of the finite conduction times and further time delays, complicated rhythms can occur.

3 The Baroreceptor Loop

In figure 3, the structure of the baroreceptor loop is depicted. In this section we introduce the remaining parts of that loop.

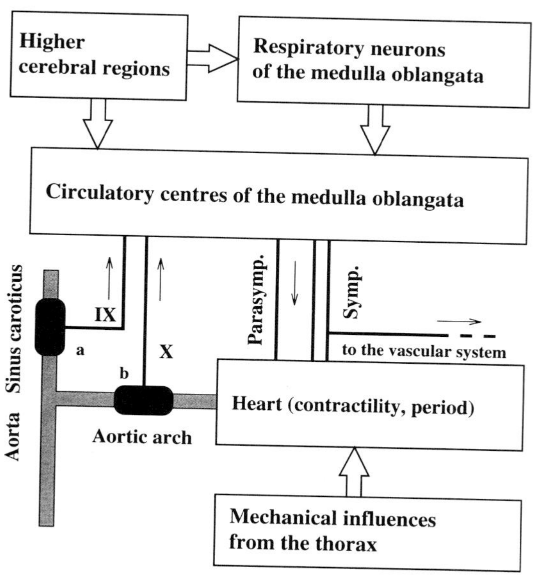

Fig. 3. Schematic representation of the baroreceptor control loop

3.1 Medullary Circulation Centers

The *medullary circulation centers* mix several inputs. Afferents from the baro- and chemoreceptors, impulses from higher cerebral regions, and from respiratory neurons are combined. The result of this process is led by the *sympathetic* and the *parasympathetic nerves* to the effector organs.

The knowledge about information processing in the medulla is very limited. The following points are generally accepted:

1. The sympathetic activity is reduced by an increased activity of the aortic nerve or the carotid sinus nerve.
2. The parasympathetic activity is enhanced by an increased activity of the aortic nerve or the carotid sinus nerve.
3. The phase of the respiratory cycle determines the strength of the influences mentioned above.

Statements going beyond this are hard to defend. In the physiological literature, one often finds contradicting opinions [29, 23, 42, 11, 13, 9, 17, 6, 36, 37]. A commonly accepted quantitative transfer function from the efferent to the afferent activities does not exist.

For these reasons, it is very difficult to model the medulla. Hence, it seems reasonable to limit our description to a minimal set of plausible equations representing the medulla at least qualitatively.

3.2 Sympathetic and Parasympathetic Nerves, Heart and Vascular System

In addition to voluntary movements and to the hormonal system, the autonomic nervous system controls the reaction of an organism to a change of its environmental or internal state. The autonomic system's afferent parts are the antagonistic *sympathetic* and *parasympathetic* nerves. Feedback is given by *visceral afferents*. Usually, the sympathetic nerves have an activating effect, whereas the parasympathetic nerves are inhibiting.

Large parts of the parasympathetic system are located in the left or right *vagus*. The common vagal synaptic transmitter is *acetylcholine*, the sympathetic nerves usually use *noradrenaline*. In consequence of their mythelization, the vagal axons have a very high impulse velocity (up to 100 m/s) in comparison with sympathetic neurons (ca. 1 m/s). Furthermore, the effect of acetylcholine is faster than that of noradrenaline. Acetylcholine is decomposed by *cholinesterases* in the synaptic gap. Noradrenaline needs to be removed from the synaptic gap and broken down at another location. This process takes much longer than the decomposition of acetylcholine. Hence, sympathetic dynamics are slower than their parasympathetic counterpart, and we will treat the two influences as mathematically distinct.

Now we want to turn to the actions of the autonomic nerves on several elements of the baroreceptor control loop. The blood vessels are influenced nearly exclusively by the sympathetic system. An elevated sympathetic activity decreases the elasticity of the vascular walls and thereby produces an increase in blood pressure.

The heart is supplied both by the sympathetic and the parasympathetic system. The sympathetic nerves innervate the whole heart. A high sympathetic activity increases the contractility and the heart rate by opening calcium channels. The vagus innervates mainly the sinus node and the atria. Its left part is responsible for the AV-node, the right one for the sinus node and the right atrium. The parasympathetic effect on the contractility is restricted to the atria. Since the main work of the heart is done by the ventricles, we shall neglect the vagal influence on the contractility in order not to complicate the model. The parasympathetic effect on the heart rate is based on the activity of potassium channels.

As a consequence of the slow sympathetic dynamics, the distribution of sympathetic impulses over a heart cycle is not very important. The noradrenaline concentration acts like a buffer.

This is completely different for the parasympathetic influence on the heart. The acetylcholine concentration follows the vagal activity nearly immediately. The effect of acetylcholine on the time course of the sino-atrial action potential strongly depends on the phase of the heart cycle. We shall use a *phase-response curve* to describe this behavior.

The respiration acts on the other elements of the baroreceptor loop by medullary respiratory neurons and by mechanical influences of the thorax on the heart. According to [9], the latter causes only approximately ten percent of the overall interaction between heart and respiration. In a first approach to modelling the system, we shall neglect mechanical influences.

4 The Model

Let us now try to formulate a nonlinear model of the baroreceptor control loop that accounts for processes within a heart cycle. A popular beat-to-beat-model was published some years ago by De Boer and Karemaker [7] (another one is given, e.g., by Baselli et al. [2]). However, their model has a number of severe disadvantages:

1. Respiration is taken into account only through its mechanical influence. As mentioned above, according to [9], the nervous influence by respiratory neurons is much stronger than the mechanical one.
2. Contrary to experimental evidence, a direct sympathetic influence on the contractility of the heart does not exist in that model.
3. In the physiological system several time delays occur. These are accounted for by considering the preceding six heart beats. This implies that there is no absolute time scale in the model – the duration of the heart period being used as the time step. Consequently, the simulated time delays

decrease to the same extent as the heart rate increases. This may distort the dynamics.

4. The effect of a vagal stimulation of the sinus node depends on the phase of the heart cycle [51]. Since the smallest time scale used by de Boer and Karemaker is the heart period, this essential effect cannot be taken into account in the model.
5. The model has been examined only in its linearized form.

We shall attempt to develop a model that overcomes these problems.

4.1 Modelling Neuronal Activity

In order to model neuronal activity on time scales of milliseconds, several complications have to be overcome. Typical firing rates of the fibers of the autonomic nervous system are in the order of 4–5 s^{-1}. Consequently, it seems not to be possible to describe neuronal activity only by firing rates. On the other hand, with current physiological knowledge, it is not possible to model such a complex system as circulation control on the basis of single spikes, especially given the lack of knowledge concerning the medullary processes.

To find a way out of this situation, one could try to use firing rates (particularly for the medullary transfer function) and to switch to single spikes for elements that are highly sensitive to the exact moment of occurrence of an impulse. A probability distribution for the occurrence of a spike at a given firing frequency could be used to determine the moment of the next spike.

In our model we adopt another formulation. Usually, a nerve consists of numerous axons. We assume that these fibers correlate in their firing frequency on greater time scales, but not in the occurrence of single spikes. Then it is possible to define an activity ν of single fibers for the transition to an infinite number N of axons by

$$\nu = \lim_{\delta t \to 0} \frac{1}{\delta t} \lim_{N \to \infty} \frac{N(\delta t)}{N} \tag{1}$$

where $N(\delta t)$ is the number of fibers firing in the time interval δt. Let τ be the duration of an impulse and f the mean firing frequency of a fiber. Then the limit is well approximated for $N \gg 1/(\tau f)$. With $\tau \approx 1$ ms and $f \approx 5\ \mathrm{s}^{-1}$, this means $N \gg 200$. In the following we assume that the number of fibers is high enough for this approximation to be valid.

4.2 Modelling Baroreceptors

In order to describe the response of the baroreceptor's activity to the blood pressure wave, we use the simple equation

$$\nu_{\mathrm{b}} = k_1 \left(p - p^{(0)}\right) + k_2 \frac{\mathrm{d}p}{\mathrm{d}t}\ , \tag{2}$$

originally proposed by Warner [47]. More complicated models are available (see the review given by Taher et al. [41]). However, it is not a good idea to put much effort into the modelling of one particular element of the system, when this accuracy may be lost because of insufficient knowledge in other parts. By applying (2) we neglect that

1. baroreceptors run into saturation at high pressures,
2. baroreceptors show adaption to a changed mean pressure,
3. there is a slight asymmetry in k_2 in dependence on the direction of the pressure change,
4. different baroreceptors have different inputs (the wave form changes along the aorta) and outputs (the coefficients k_1 and k_2 may vary for different baroreceptors),
5. the conduction times towards the medulla depend on the location of the baroreceptors.

Some of these phenomena may be included as the physiology of the medulla becomes known in more detail.

4.3 Modelling Sympathetic and Parasympathetic Activity

The description of the response of sympathetic and parasympathetic activity to the afferents from the baroreceptors and to respiratory influences is the most critical point of the model. As already mentioned, very little is known about medullary processes. For that reason, we shall keep the equations as simple as possible and assume a linear dependence of the sympathetic and parasympathetic activity on the activity of the baroreceptors.

One can find different, often contradictory opinions about the influence of respiratory neurons on the autonomic nerves. Eckberg [11, 10, 9, 12] found augmented sympathetic and vagal activities during expiration, though he points to the fact that other authors draw opposite conclusions.

Fortunately, concerning our model, it is not important during which phase in the respiratory cycle autonomic activity is augmented or reduced so long as we restrict the interaction between the respiration and baroreceptor loops to a single location only. Since we shall consider only medullary neuronal influences of respiration, this holds true in our case. An incorrectly chosen activating phase of the respiratory cycle corresponds to a shift of respiration relative to circulatory rhythms. Their dynamics remains unchanged, however. If we wanted to include mechanical interaction, we would have to use exact phase relations of gating.

In accordance with the above considerations, we shall describe the activities of the sympathetic and vagal activities, respectively, by the equations

$$\nu_{\mathrm{s}} = \max\left(0, \nu_{\mathrm{s}}^{(0)} - k_{\mathrm{s}}^{\mathrm{b}}\nu_{\mathrm{b}} + k_{\mathrm{s}}^{\mathrm{r}}|\sin\left(\pi f_{\mathrm{r}} t + \Delta\phi_{\mathrm{s}}^{\mathrm{r}}\right)|\right) \qquad (3)$$

$$\nu_{\mathrm{p}} = \max\left(0, \nu_{\mathrm{p}}^{(0)} + k_{\mathrm{p}}^{\mathrm{b}}\nu_{\mathrm{b}} + k_{\mathrm{p}}^{\mathrm{r}}|\sin\left(\pi f_{\mathrm{r}}t + \Delta\phi_{\mathrm{p}}^{\mathrm{r}}\right)|\right) \tag{4}$$

where superscripts of the proportionality constants k are mere labels. The sympathetic activity ν_{s} is composed of a resting-tone $\nu_{\mathrm{s}}^{(0)}$, an inhibitory influence $k_{\mathrm{s}}^{\mathrm{b}}\nu_{\mathrm{b}}$ from the baroreceptors, and an additional modulation by respiratory neurons. These three components are assumed to be linearly integrated by neurons of the circulation centers. When the inhibition is very strong, this integration can give a negative result, implying a complete suppression of sympathetic activity. We take this into account by means of the maximum function.

The description of the parasympathetic activity is performed analogously, with an activating effect of the baroreceptors.

4.4 Modelling the Heart

Our representation of the heart includes a variety of different phenomena:

4.4.1 Dromotropic Effect

The contractility of the heart is externally influenced by the sympathetic nerves. We neglect parasympatheticly induced contractility changes of the atria. Because of the slow dynamics of noradrenaline, it is necessary to introduce a cardiac noradrenaline concentration c_{cNa}

$$\frac{\mathrm{d}c_{\mathrm{cNa}}}{\mathrm{d}t} = -\frac{c_{\mathrm{cNa}}}{\tau_{\mathrm{cNa}}} + k_{c_{\mathrm{cNa}}}^{\mathrm{s}}\nu_{\mathrm{s}}\left(t - \theta_{\mathrm{cNa}}\right) , \tag{5}$$

where $\nu_{\mathrm{s}}(t)$ is the sympathetic activity at the medulla at time t. A time delay is taken into account by θ_{cNa}. This delay is caused by finite conduction velocity and the time that noradrenaline needs to produce an effect on the heart.

Furthermore, at high heart rates fatigue may occur and decrease the contractility. In addition, the potassium conductivity does not return to its resting-value. This has an accelerating effect on the next repolarisation and thereby decreases the contractility S_i. For these reasons, we use the equations

$$S_i' = S^{(0)} + k_S^c c_{\mathrm{cNa}} + k_S^t T_{i-1} \tag{6}$$

$$S_i = S_i' + \left(\hat{S} - S_i'\right)\frac{S_i'^{n_S}}{S_i'^{n_S} + \hat{S}^{n_S}} \tag{7}$$

to describe the contractility S_i of the ith heart beat. T_{i-1} denotes the duration of the heart cycle immediately preceding the systole. By equation (7), at high values the contractility is saturated. The fraction in the second term of that equation is a threshold function. Figure 4 shows this function for some exponents n_S.

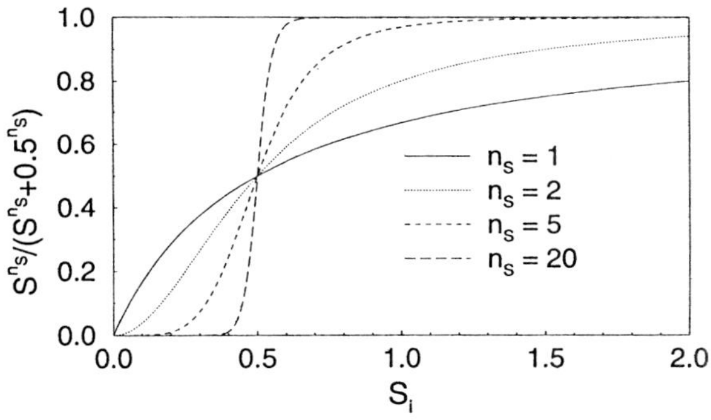

Fig. 4. Threshold function used for the description of saturation.

4.4.2 Chronotropic Effect

We want to model the heart on time scales less than the heart period. With this purpose we introduce a phase φ of the sinus node, satisfying the equation

$$\frac{\mathrm{d}\varphi}{\mathrm{d}t} = \frac{1}{T^{(0)}} f_\mathrm{s} f_\mathrm{p} \,. \tag{8}$$

The factors f_s and f_p represent the sympathetic and parasympathetic influence on the phase velocity. They are given by

$$f_\mathrm{s} = 1 + k_\varphi^{\mathrm{cNa}} \left(c_{\mathrm{cNa}} + (\hat{c}_{\mathrm{cNa}} - c_{\mathrm{cNa}}) \frac{c_{\mathrm{cNa}}^{n_{\mathrm{cNa}}}}{\hat{c}_{\mathrm{cNa}}^{n_{\mathrm{cNa}}} + c_{\mathrm{cNa}}^{n_{\mathrm{cNa}}}} \right) \tag{9}$$

$$f_\mathrm{p} = 1 - k_\varphi^{\mathrm{p}} \left(\nu_{\mathrm{p},\theta_\mathrm{p}} + \left(\hat{\nu}_\mathrm{p} - \nu_{\mathrm{p},\theta_\mathrm{p}}\right) \frac{\nu_{\mathrm{p},\theta_\mathrm{p}}^{n_\mathrm{p}}}{\hat{\nu}_\mathrm{p}^{n_\mathrm{p}} + \nu_{\mathrm{p},\theta_\mathrm{p}}^{n_\mathrm{p}}} \right) F(\varphi) \tag{10}$$

where $\nu_{\mathrm{p},\theta_\mathrm{p}} \equiv \nu_\mathrm{p}(t - \theta_\mathrm{p})$. Saturation is achieved analogously to equation (7).

When the phase φ reaches the value 1, a new heart beat is stimulated (*integrate-and-fire-model*). The phase is reset to 0 and the contractility S_i is added to the diastolic pressure to give the new systolic pressure.

The function F in equation (10) requires some explanation. We already mentioned that the effect of vagal impulses on the sinus node depends on the phase at which they occur during the heart cycle. For single stimuli, this can be described by a phase-response curve [30, 20, 24, 5, 19, 45, 46]. The theory of phase-response curves usually assumes instantaneous relaxation to a limit cycle after a perturbation. This is approximately attained if the relaxation time is negligible in comparison with the time interval between spikes. However, since we use vagal activities instead of single spikes, there are no longer discrete parasympathetic events, and the phase-response concept is not applicable in its original form. Nevertheless, we want to take into account that an increase of vagal activity has another effect at the beginning of a heart cycle than at the end. Therefore, we introduce a *phase-effectiveness curve*. In order to explain the meaning of that curve, let us first assume the activity to be a single rectangular impulse starting at t_s, having a duration of θ and an amplitude a with $\theta a = b \equiv \text{const}$. Let T be the undisturbed period of the system and t_0 the time of the last point of intersection of the trajectory with the $(\phi=0)$-plane. Then, the effect of an impulse can be calculated using the phase-effectiveness curve by

$$\Delta\phi = \int_{t_s}^{t_s+\theta} aF\left(\frac{t-t_0}{T}\right)\, \mathrm{d}t\,. \tag{11}$$

For continuous F, this can be written as

$$\Delta\phi = bF\left(\frac{t_\xi}{T}\right)\,, \tag{12}$$

where $t_\xi \in [t_0+t_s, t_0+t_s+\theta]$. Hence, for $\theta \to 0$ the phase-effectiveness should approach the phase-response curve

$$\Delta\phi = f_b\,(\phi) \tag{13}$$

except for a proportionality constant. In other words, if one knows the phase-response curve of a system, it is reasonable to set the phase-effectiveness curve proportional to that curve.

Of course, even the phase-effectiveness curve is only an approximation of the real processes. Each perturbation causes a deviation from the attractor. Consequently, the distance between the system's phase space position and the attractor depends on the stimulation history. Therefore, to be exact, F should be a function of the whole history of the system. Unfortunately, this would destroy our hope of getting a simple description of the system by means of phase-response concepts.

For that reason, we have to accept the inaccuracies resulting from the use of a phase-effectiveness curve. In order to estimate the error, one must know the complete high-dimensional system, as it is described, e.g., for the sinus node by the DiFrancesco-Noble-equations [8, 34]. Such an analysis is difficult and would require much time and effort. Presently, we restrict ourselves to the investigation of the capabilities of the model and try to compare its output with experimental results.

The phase-effectiveness curve that we have chosen is of the form

$$F(\varphi) = \varphi^{1.3}\,(\varphi - 0.45)\,\frac{(1-\varphi)^3}{(1-0.8)^3 + (1-\varphi)^3}\,, \tag{14}$$

approximating a phase-response curve derived numerically from a mathematical model of the sinus node by Reiner et al. [35] (fig. 5).

The functional form of equation (8) is similar to a model of Rosenblueth and Simeone [38]. They proposed that the dependence of the heart rate on autonomic activity can be expressed by

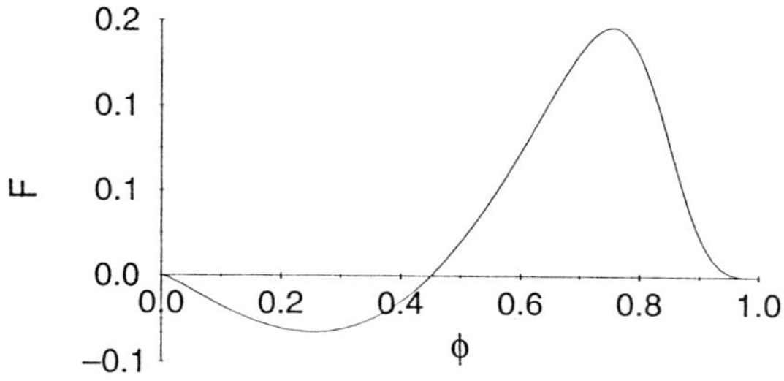

Fig. 5. Graph of the phase-effectiveness curve used in the model, see equation (14).

$$R = R^{(0)} f_{\mathrm{s}} f_{\mathrm{p}} = R^{(0)} \left[1 + \frac{\nu_{\mathrm{s}}}{R^{(0)}(k_1 + k_2 \nu_{\mathrm{s}})}\right] \left[1 - \frac{\nu_{\mathrm{p}}}{R^{(0)}(k_3 + k_4 \nu_{\mathrm{p}})}\right] . \quad (15)$$

Katona et al. [27] have shown the compatibility of this equation with another model proposed by Warner and Russel [48]. In both models, the total effect of vagal and sympathetic activity on the heart rate can be represented by a product of two factors, each depending only on the activity of one of the nerves.

4.4.3 Modelling the Vascular System

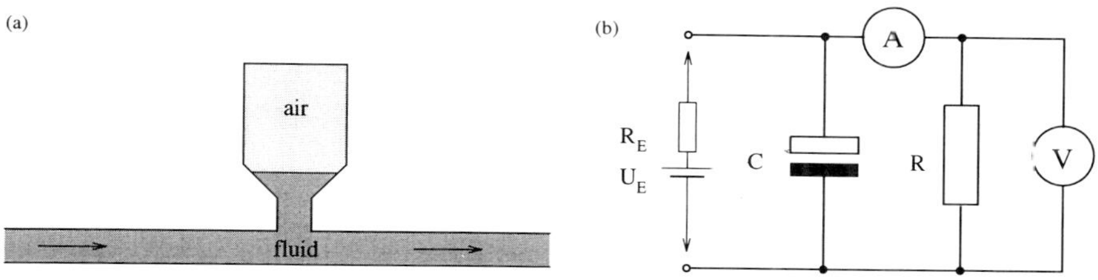

Fig. 6. (a) Origin of the *Windkessel* model, (b) Electric analogue

We use a *windkessel* model (fig. 6) to represent the vascular system. An elevated sympathetic activity causes a constriction of the blood vessels and, consequently, leads to an augmented peripheral resistance or, in other words, to an increased time constant τ of the *windkessel* model. Similarly to (5), we introduce another noradrenaline concentration at the blood vessels

$$\frac{\mathrm{d}c_{\mathrm{vNa}}}{\mathrm{d}t} = -\frac{c_{\mathrm{vNa}}}{\tau_{\mathrm{vNa}}} + k^{\mathrm{s}}_{c\mathrm{vNa}} \nu_{\mathrm{s}} (t - \theta_{\mathrm{vNa}}) . \quad (16)$$

Then, the *windkessel* time constant is given by

$$\tau_{\mathrm{v}} = \tau_{\mathrm{v}}^{(0)} + \bar{\tau}_{\mathrm{v}} \left(c_{\mathrm{vNa}} + (\hat{c}_{\mathrm{vNa}} - c_{\mathrm{vNa}}) \frac{c_{\mathrm{vNa}}^{n_{\mathrm{vNa}}}}{\hat{c}_{\mathrm{vNa}}^{n_{\mathrm{vNa}}} + c_{\mathrm{vNa}}^{n_{\mathrm{vNa}}}} \right) , \quad (17)$$

where $\tau_{\rm V}^{(0)}$ and $\bar{\tau}_{\rm v}$ are constants.

Finally, we need a representation for the blood pressure itself. We divide each pulse wave into two parts. During the systolic increase of pressure, we use

$$p_{\rm I} = d_{i-1} + S_i \, \frac{t - t_i}{\tau_{\rm sys}} \exp\left\{1 - \frac{t - t_i}{\tau_{\rm sys}}\right\} , \tag{18}$$

where d_{i-1} is the diastolic pressure immediately preceding the onset of systole at time t_i. Phase I has the constant duration $\tau_{\rm sys}$. For the diastolic part of the pulse wave we write

$$\frac{{\rm d}p_{\rm II}}{{\rm d}t} = -\frac{p_{\rm II}}{\tau_{\rm v}(t)} \tag{19}$$

according to the *windkessel* model.

5 Simulations

Let us hereafter present some of the results obtained from simulations with the model proposed in the preceding section.

5.1 Interaction of Mayer Waves and RSA

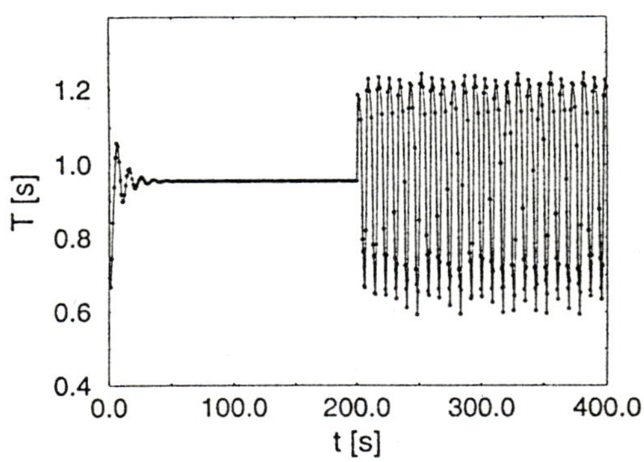

Fig. 7. Time series of the heart period. The first 200 seconds show a damped oscillation towards a constant period. The transition to a large amplitude cycle after the onset of stimulation can be seen (see 5.3.).

Mayer waves are blood pressure fluctuations with periods in the range of six to twenty seconds with a mean of ten seconds. They are quite common; their period and stability, however, can vary widely among individuals.

Our simulations show that the sympathetic part of the blood pressure control loop represents an oscillator with frequencies in the range mentioned above. Usually these oscillations are damped; for strong feedback or long time

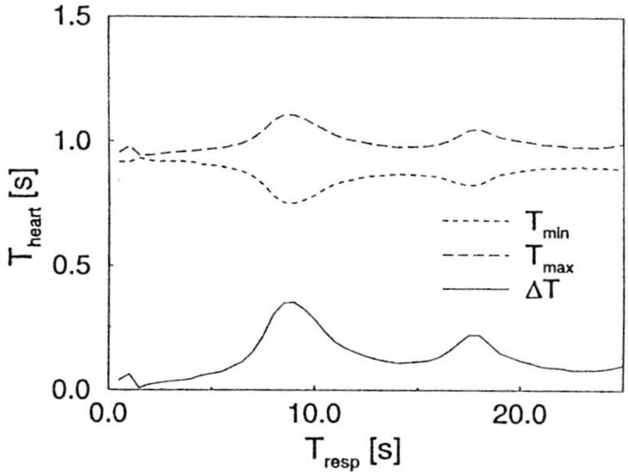

Fig. 8. Resonance between respiration and blood pressure control loop. T_{resp} and T_{heart} are the respiratory and the heart period, respectively.

delays, however, sustained oscillations are also possible. In figure 7 one can see a damped rhythm with a period of about nine seconds at the beginning of the record before the onset of stimulation.

The influence of respiratory neurons on the sympathetic and parasympathetic activity causes variations in blood pressure and heart period, called *respiratory sinus arrhythmia* (RSA, see fig. 9). If the respiratory frequency is approximately equal to the frequency of the sympathetic control loop, resonance effects occur. Figure 8 shows the maximum and minimum heart period during RSA, as well as the RSA's amplitude for varying breathing periods.

5.2 Response to Vagal Stimulation by Single Impulses

In some calculations, we have simulated the behavior of the system which is stimulated vagally both by single and repeated impulses. First we shall consider the system's response to single stimuli given near the medulla. We increase the parasypathetic activity ν_p for a stimulus duration τ_{stim} by a stimulus amplitude A_{stim}.

Figure 11 shows the period changes of the disturbed heart cycle T_i and the two following cycles T_{i+1} and T_{i+2} for several stimulus strengths, where θ is the time delay between the onset of systole of cycle i and the application of the stimulus, and T_0 is the heart period without additional stimulation. The phase-response curves of figure 11 are similar to those obtained experimentally by Yang et al. [51]. Both in our simulation and in Yang's experiments, the influence of respiration was suppressed.

5.3 Response to Repeated Vagal Stimulation

Finally, we would like to investigate the system's response to repeated vagal stimulation. Each heart cycle is stimulated after a time delay θ following the beginning of systole (see fig. 10). Some *return maps* for several time delays

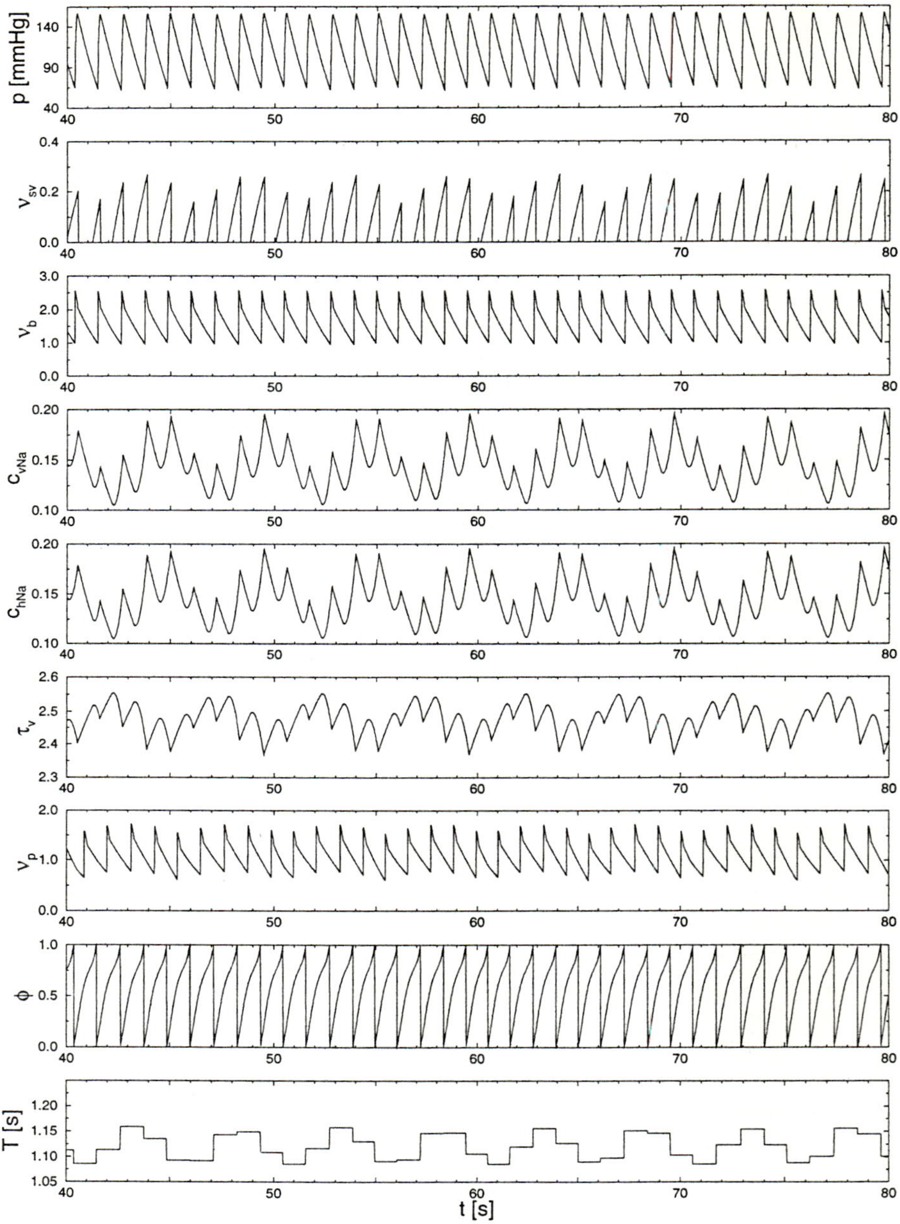

Fig. 9. Simulation of respiratory sinus arrhythmia ($f_r = 0.2\,s^{-1}$).

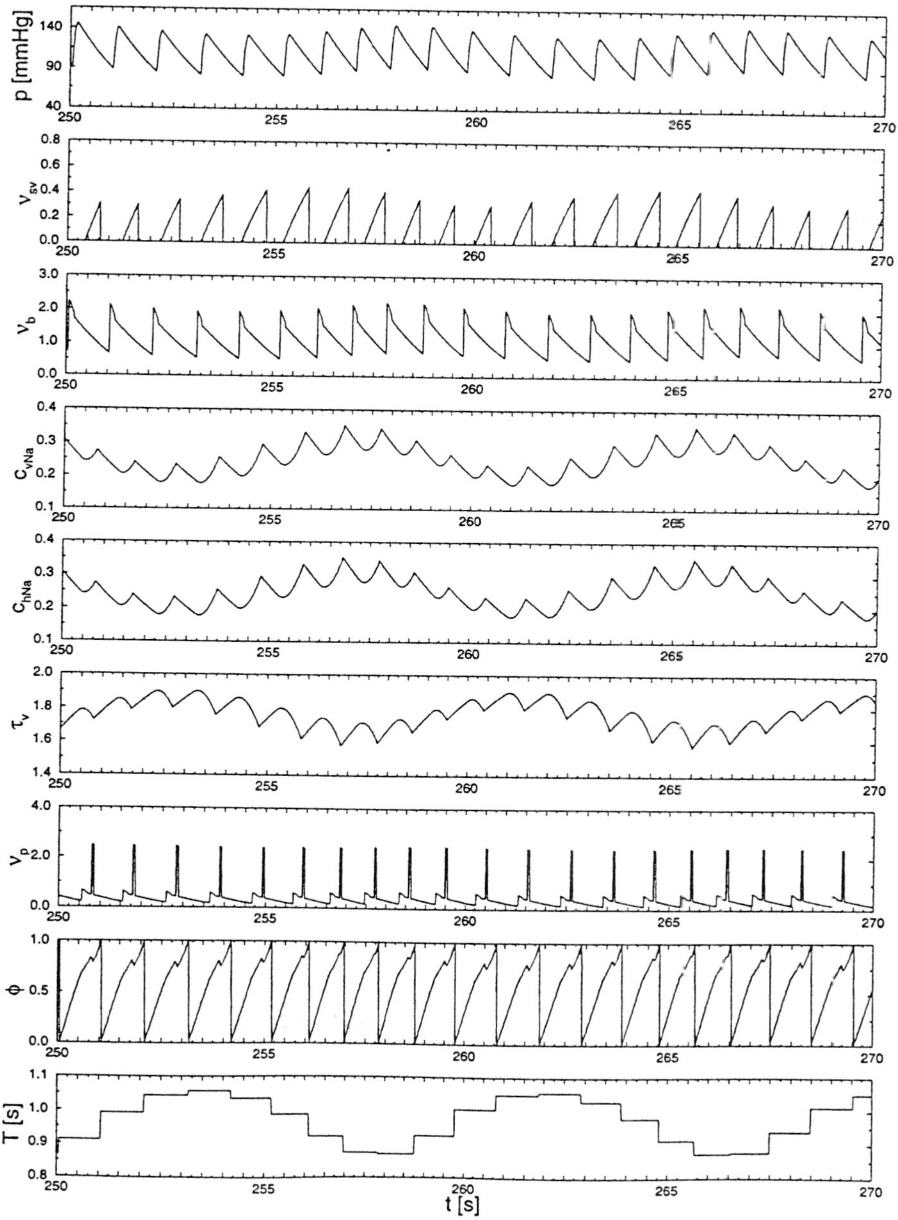

Fig. 10. Repeated stimulation (constant time delay $\theta = 0.2$ between systolic onset and stimulus), no respiratory influence

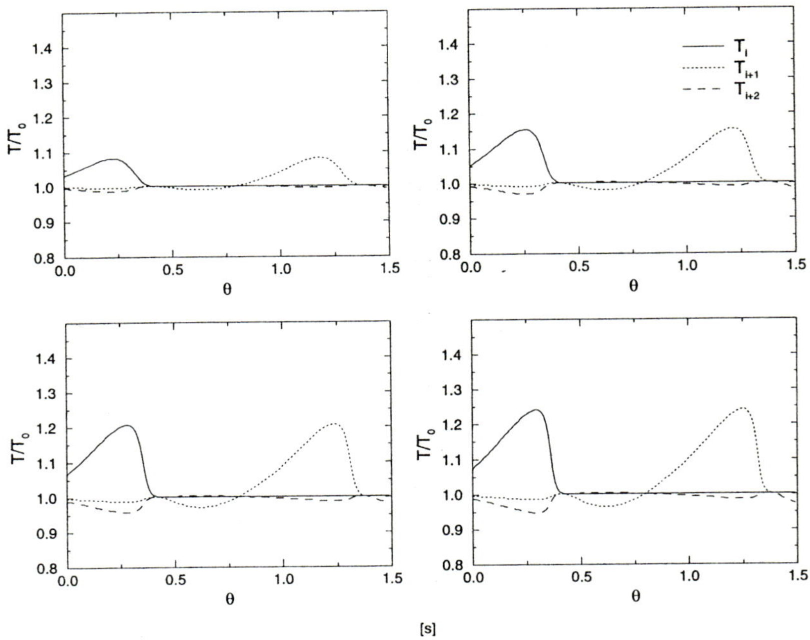

Fig. 11. Phase-response curves for several stimulus strengths (1.0, 2.0, 3.0, and 4.0), see text. The influence of respiration is not taken into account here.

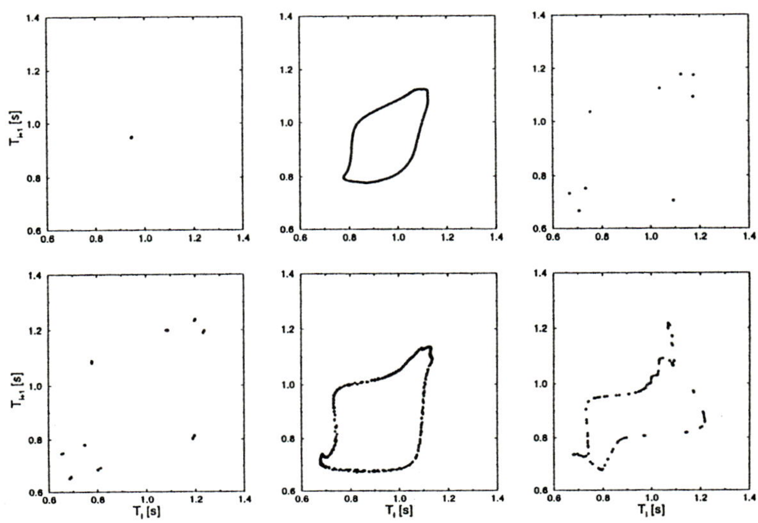

Fig. 12. Return map $T_{i+1}(T_i)$ for several time delays $(\theta, A_{\mathrm{stim}}, \tau_{\mathrm{stim}}) = (0.1,\ 2.0,\ 0.1)\ (0.23,\ 5.0,\ 0.05)\ (0.2,\ 2.0,\ 0.1)\ (0.25,\ 2.0,\ 0.1)\ (0.15,\ 2.0,\ 0.1)\ (1.0,\ 2.0,\ 0.1)$.

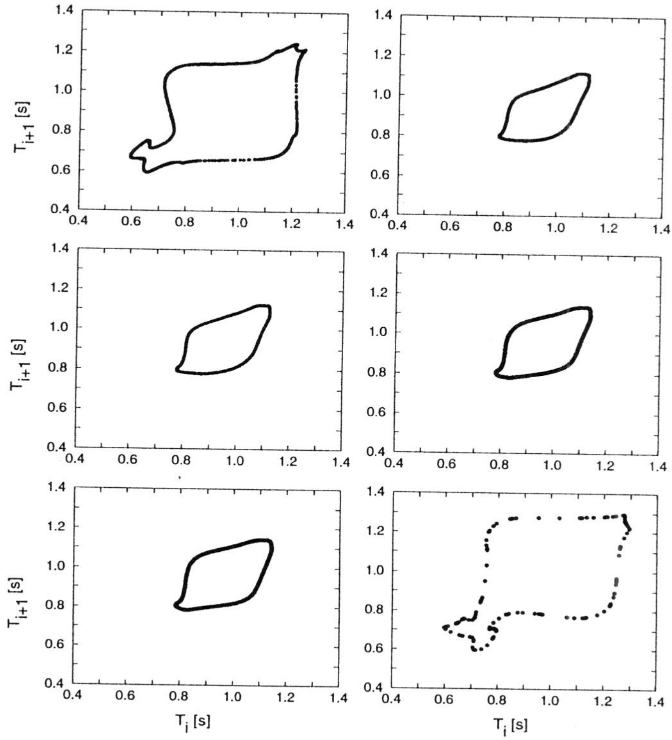

Fig. 13. Some tori for several time delays ($\theta = 0.20, 0.23, 0.24, 0.25, 0.26, 0.28$ s) $A_{\mathrm{stim}} = 5.0$, τ_{stim}=0.05.

are depicted in figure 12. θ turns out to be an essential bifurcation parameter. Changes in the time delay can result in stable nodes, periodic attractors, and tori, the latter sometimes being quite wrinkled.

Periodicity is observed only when the total period is in the range of the period of the baroreceptor loop or of one of its multiples. We found periods of 9–12, 18, and 35. Period doubling was also observed (transition from period 9 to period 18, see fig 12). The complicated shapes of some of the tori indicate chaotic dynamics for certain time delays. In figure 13 and 14 some of these tori and the corresponding Fourier spectra are depicted. The first of these spectra shows an augmented background, indicative of chaos. A comprehensive bifurcation analysis of the model will be left for further investigations.

A strong influence of the time delay on circulation rhythms has also been found in stimulation experiments on anaesthetized dogs [49]. This emphasizes that phase-response concepts are required for adequate modelling.

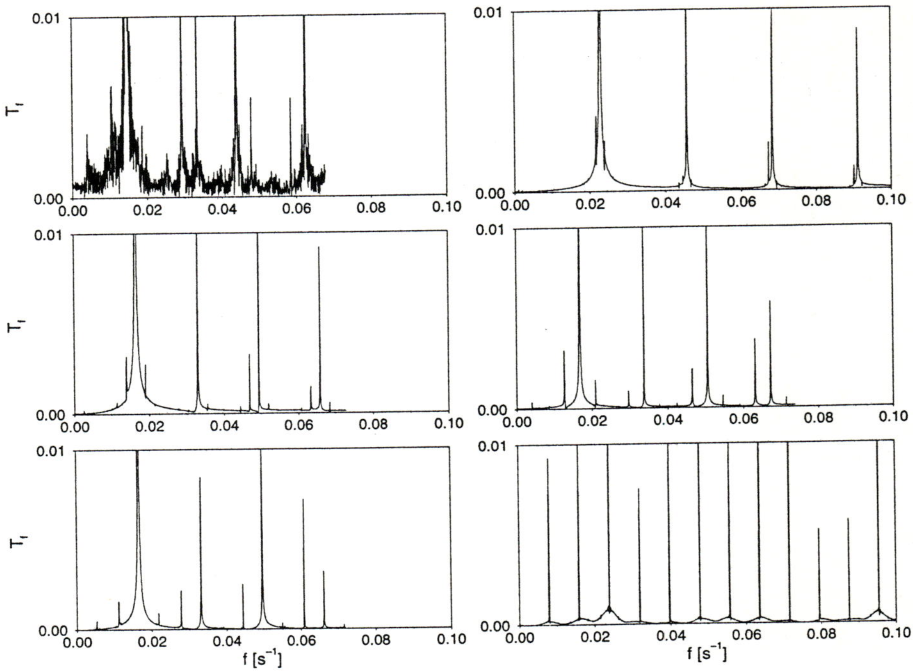

Fig. 14. Fourier spectra for the tori of fig. 13.

6 Discussion

We have introduced a model for the dynamics of the blood pressure control loop. It was designed to explain basic dynamical features of the system on time scales from milliseconds to minutes. Changes of the concentrations of CO_2, O_2, and H^+ are not included; neither are metabolic impacts on the system and influences from higher cerebral regions. The respiratory pattern generator is assumed to have a constant period. Mechanical interactions between respiration and heart beat are neglected.

In spite of these simplifications we are able to reproduce a number of experimental findings as such Mayer waves, respiratory sinus arrhythmia, resonance effects between respiration and Mayer waves, phase dependence of the impact of vagal activity on the heart, and the occurrence of complex dynamics at repeated vagal stimulation.

Appendices

A.1 Simplified Delay Model of Blood Pressure Control

In section 5.1 we saw that the circulation control exhibits a resonance around $0.1\ \mathrm{s}^{-1}$. We shall now show that this rhythm can also be obtained by a strongly simplified model. This model includes only the sympathetic part of blood pressure control, synonymous with cut vagi. The influence of respiration is neglected as well, and we are going to use an averaged blood pressure without the wave form induced by heart beats.

Let us assume that an augmented cardiac noradrenaline concentration leads to a blood pressure increase by way of a strengthened contractility

$$\frac{\mathrm{d}\tilde{p}}{\mathrm{d}t} = \alpha\left(\tilde{c} - c_0\right) . \tag{20}$$

On the pre-condition that the sympathetic activity never vanishes completely, so that we can delete the maximum function, one can derive the equation

$$\frac{\mathrm{d}\tilde{c}}{\mathrm{d}t} = -\beta\tilde{c} + \gamma - \sigma\left(\tilde{p}(t-\theta) - p_0\right) \tag{21}$$

from our model. The substitution $(\tilde{c}-c_0) \rightarrow c$ and $(\tilde{p}-p_0) \rightarrow p$ yields the two-dimensional linear differential delay equation system

$$\frac{\mathrm{d}p}{\mathrm{d}t} = \alpha c \tag{22}$$

$$\frac{\mathrm{d}c}{\mathrm{d}t} = -\beta c + \gamma - \sigma p(t-\theta) \tag{23}$$

or

$$\ddot{c}(t) + \beta\dot{c}(t) + \alpha\sigma c(t-\theta) = 0 . \tag{24}$$

Equation (24) exhibits a Hopf bifurcation for increasing time delay (see, e.g., [1]). In order to calculate the bifurcation point, we use $c = \hat{c}\exp(\mathrm{i}\omega t)$. The conditions for a vanishing imaginary part of ω are

$$\omega^4 + \omega^2\beta^2 - \alpha^2\sigma^2 = 0 \tag{25}$$

$$\tan(\omega\theta) = \frac{\beta}{\omega} . \tag{26}$$

The frequency of the system is determined by $\alpha\sigma$ and β according to

$$\omega = \frac{\beta}{\sqrt{2}}\sqrt{\sqrt{1 + \left(\frac{2\alpha\sigma}{\beta^2}\right)^2} - 1} . \tag{27}$$

With $\alpha = 30$, $\beta = 0.5$, and $\sigma = 0.2$, approximating our global model, the period of the blood pressure variations is in the order of ten seconds. The linear model allows, however, no analysis of the resulting limit cycle.

If the damping β is small or the feedback $\alpha\sigma$ is strong ($\alpha\sigma/\beta^2 \gg 1$), equation (26) can be expanded to yield

$$\theta \approx \frac{\beta}{\omega^2} = \frac{\beta T^2}{4\pi^2} \, . \tag{28}$$

For a strong damping or a weak feedback ($\alpha\sigma/\beta^2 \ll 1$) equation (27) gives

$$\frac{\beta}{\omega} \approx \frac{\beta^2}{\alpha\sigma} \gg 1 \, . \tag{29}$$

According to equation (26) the lowest frequency of the system is given by $\omega\theta \lessapprox \pi/2$ or, in other words,

$$\theta \lessapprox \frac{T}{4} \, . \tag{30}$$

Hence, for strong damping or weak feedback we get sustained oscillations if the time delay is greater than a quarter period of the system's smallest frequency.

According to figure 15, for a period of ten seconds one can expect stable sustained oscillations at time delays $\theta \gtrapprox 1.5\,\mathrm{s}$.

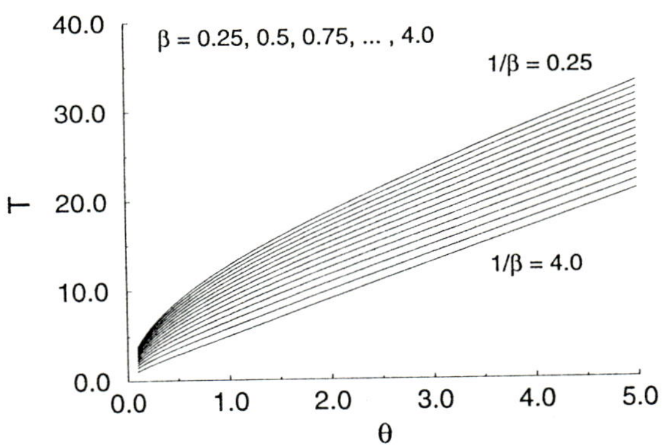

Fig. 15. Bifurcation line $\theta(T)$ for several strengths of damping

To sum up, one can say that even the extremely simplified approximation introduced in this section can account for the occurrence of sustained oscillations and resonance phenomena at periods around ten seconds. We want to emphasize that this result does not exclude the generation of these rhythms at other locations (e.g. in the *formatio recticularis*).

A.2 Sample Parameter Set

$p^{(0)}$	50.0 mmHg	θ_{p}	0.5 s
k_1	0.02 1/mmHg	$S^{(0)}$	25 mmHg
k_2	0.00125 s/mmHg	k_S^{c}	40 mmHg
$\nu_{\mathrm{s}}^{(0)}$	0.8	k_S^t	10 mmHg/s
$k_{\mathrm{s}}^{\mathrm{b}}$	0.7	n_S	2.5
$k_{\mathrm{s}}^{\mathrm{r}}$	0.1	$\hat{S}$	70.0 mmHg
f_{r}	0.2 1/s	$T^{(0)}$	1.1 s
$\Delta\phi_{\mathrm{s}}^{\mathrm{r}}$	0.0	$k_{\varphi}^{\mathrm{cNa}}$	1.6
$\nu_{\mathrm{p}}^{(0)}$	0.0	$\hat{c}_{\mathrm{cNa}}$	2.0
$k_{\mathrm{p}}^{\mathrm{b}}$	0.3	n_{cNa}	2.0
$k_{\mathrm{p}}^{\mathrm{r}}$	0.1	k_{φ}^{p}	5.8
$\Delta\phi_{\mathrm{p}}^{\mathrm{r}}$	0.0	$\hat{\nu}_{\mathrm{p}}$	2.5
τ_{cNa}	2.0 s	n_{p}	2.0
$k_{c_{\mathrm{cNa}}}^{\mathrm{s}}$	1.2	$\tau_{\mathrm{v}}^{(0)}$	2.2 s
θ_{cNa}	1.65 s	$\bar{\tau}_{\mathrm{v}}$	1.5 s
τ_{vNa}	2.0 s	$\hat{c}_{\mathrm{vNa}}$	10.0
$k_{c_{\mathrm{vNa}}}^{\mathrm{s}}$	1.2	n_{vNa}	1.5
θ_{vNa}	1.65 s	τ_{sys}	0.125 s

References

1. U. an der Heiden, A. Longtin, M. C. Mackey, J. G. Milton, and R. Scholl: Oscillatory modes in a nonlinear second-order differential equation with delay. Journal of Dynamics and Differential Equations **2**, 423–449 (1990)
2. G. Baselli, S. Cerutti, S. Civardi, A. Malliani, and M. Pagani: Cardiovascular variability signals: Towards the identification of a closed-loop model of the neural control mechanisms. IEEE Trans. Biomed. Eng. **35**, 1033–1046 (1988)
3. R. M. Berne, N. Sperelakis, and S. R. Geiger, editors: Handbook of Physiology, volume I. American Physiological Society 1979
4. T. R. Chay and Y. S. Lee: Phase resetting and bifurcation in the ventricular myocardium. J. Physiol. **47**, 641–651 (1985)
5. M. Courtmanche, L. Glass, M. Rosengarten, and A. L. Goldberger: Beyond pure parasystole: promises and problems in modelling complex arrhythmias. Am. J. Physiol. **257**, H693–H706 (1989)
6. M. R. Cowie and J. M. Rawles: A modified method of quantifying the carotid baroreceptor-heart rate reflex in man: the effect of age and blood pressure. Clin. Sci. **77**, 223–228 (1989)
7. R. W. de Boer, J. M. Karemaker, and J. Strackee: Hemodynamic fluctuations and baroreflex sensitivity in humans: a beat-to-beat model. Am. J. Physiol. **253**, 680–689 (1987)
8. D. DiFrancesco and D. Noble: A model of cardiac electrical activity incorporating ionic pumps and concentration changes. Phil. Trans. R. Soc. Lond. B **307**, 353–398 (1985)
9. D. L. Eckberg: Human sinus arrhythmia as an index of vagal cardiac outflow. J. Appl. Physiol. **54**, 961–966 (1983)

10. D. L. Eckberg, Y. T. Kifle, and V. L. Roberts: Phase relationship between normal human respiration and baroreflex responsiveness. J. Physiol. **304**, 489–502 (1980)
11. D. L. Eckberg and C. R. Orshan: Respiratory and baroreflex interactions in man. J. Clin. Inv. **59**, 780–785 (1977)
12. D. L. Eckberg, R. F. Rea, O. K. Andersson, T. Hedner, J. Pernow, J. M. Lundberg, and B. G. Wallin: Baroreflex modulation of sympathetic activity and sympathetic neurotransmitters in humans. Acta Physiol. Scand. **133**, 221–231 (1988)
13. R. B. Felder and M. D. Thames: Interaction between cardiac receptors and sinoaortic baroreceptors in the control of efferent cardiac sympathetic nerve activity during myocardial ischemia in dogs. Circ. Res. **45**, 728–736 (1979)
14. J. L. Feldman and J. D. Cowan: Large-scale activity in neuronal nets I: A model for the brainstem respiratory oscillator. Biol. Cybern. **17**, 29–38 (1975)
15. J. L. Feldman and J. D. Cowan: Large-scale activity in neuronal nets II: A model for the brainstem respiratory oscillator. Biol. Cybern. **17**, 39–51 (1975)
16. J. L. Feldman et al.: Neurogenesis of respiratory rhythm and pattern: emerging concepts. Am. J. Physiol. **259**, R879–R886 (1990)
17. F. M. Fouad, R. C. Tarazi, C. M. Ferrario, S. Fighaly, and C. Alicandri: Assessment of parasympathetic control of heart rate by a noninvasive method. Am. J. Physiol. **246**, H838–H842 (1984)
18. S. Geman and M. Miller: Computer simulation of brainstem respiratory activity. J. Appl. Physiol. **41**, 931–938 (1976)
19. L. Glass: Cardiac arrhythmias and circle maps – a classical problem. Chaos **1**, 13–19 (1991)
20. L. Glass, A. L. Goldberger, and J. Bélair: Dynamics of pure parasystole. Am. J. Physiol. **251**, H841–H847 (1986)
21. L. Glass and M. C. Mackey: From Clocks to Chaos. Princeton University 1988
22. A. L. Goldberger, D. R. Rigney, J. Meitus, E. M. Antman, and S. Greenwald: Nonlinear dynamics in sudden cardiac death syndrome: heartrate oscillations and bifurcations. Experientia **44**, 983–987 (1988)
23. N. M. Greene and R. G. Bachand: Vagal component of the chronotropic response to baroreceptor stimulation in man. Am. Heart J. **82**, 22–27 (1971)
24. M. R. Guevara, A. Shrier, and L. Glass: Phase-locked rhythms in periodically stimulated heart cell aggregates. Am. J. Physiol. **254**, H1–H10 (1988)
25. N. Ikeda, S. Yoshizawa, and T. Sato: Difference equation model of ventricular parasystole as an interaction between cardiac pacemakers based on the phase response curve. J. theor. Biol. **103**, 439–465 (1983)
26. Danial T. Kaplan and Mario Talajic: Dynamics of heart rate. Chaos **1**, 251–256 (1991)
27. P. G. Katona, P. J. Martin, and J. Felix: Neuronal control of heart rate: A conciliation of models. IEEE Trans. Biomed. Eng. **23**, 164–166 (1976)
28. T. Kawahara: Coupled Van der Pol oscillators - a model of excitatory and inhibitary neuronal interactions. Biol. Cybern. **39**, 37–43 (1980)
29. P. Lindgren and J. Manning: Decrease in cardiac activity by carotid sinus baroreceptor reflex. Acta physiol. scand. **63**, 401–408 (1965)
30. L. G. Michael, R. Guevara and A. Shrier: Phase locking, period-doubling bifurcations, and irregular dynamics in periodically stimulated cardiac cells. Science **214**, 1350–1353 (1981)
31. G. K. Moe, J. Jalife, W. J. Mueller, and B. Moe: A mathematical model of parasystole and its application to clinical arrhythmias. Circulation **56**, 968–979 (1977)

32. E. Mosekilde and J. I. Jensen: Dynamic simulation of human ventilatory regulation. System Dynamics Conf. Albany, New York 1981.
33. L. Glass, P. Hunter and A. McCulloch, editors: The Theory of Heart. Springer-Verlag New York 1991.
34. D. Noble and S. J. Noble: A model of sino-atrial node electrical activity based on a modification of the DiFrancesco-Noble (1984) equations. Proc. R. Soc. Lond. B **222**, 295–304 (1984)
35. V. S. Reiner and C. Antzelevitch: Phase resetting and annihilation in a mathemetical model of sinus node. Am. J. Physiol. **249**, H1143–H1153 (1985)
36. D. W. Richter and K. M. Spyer: Cardiorespiratory control. In A. D. Loewy and K. M. Spyer, editors: Central Regulation of Autonomic Functions. Oxford University Press 1990, pp. 189–207
37. D. W. Richter, K. M. Spyer, M. P. Gilbey, E. E. Lawson, C. R. Bainton, and Z. Wilhelm: On the existence of a common cardiorespiratory network. In H.-P. Koepchen and T. Huopaniemi, editors: Cardiorespiratory and Motor Coordination. Springer-Verlag Berlin, 1991, pp. 118–130
38. A. Rosenblueth and F. A. Simeone: The interrelations of vagal and accelerator effects on the cardiac rate. Am. J. Physiol. **110**, 42–55 (1934)
39. R. F. Schmidt and G. Thews, editors: Physiologie des Menschen. Springer-Verlag Berlin, 22nd edition, 1985.
40. R. F. Schmidt and G. Thews, editors: Human Physiology. Springer-Verlag Berlin, 3rd edition, 1989.
41. M. F. Taher et al.: Baroreceptor responses derived from a fundamental concept. Ann. Biom. Eng. **16**, 429–443 (1988)
42. S. F. Vatner, C. B. Higgens, D. Franklin, and E. Braunwald: Extent of carotid sinus regulation of the myocardial contractile state in conscious dogs. J. Clin. Inv. **51**, 995–1008 (1972)
43. C. von Euler: On the central pattern generator for the basic breathing rhythmicity. J. Appl. Physiol. **55**, 1647–1659 (1983)
44. C. von Euler: On the origin and pattern control of breathing rhythmicity in mammals. In A. Roberts and B. Roberts, editors: Neuronal origin of rhythmic movements. Society for Experimental Biology, Society for Experimental Biology 1983, pp. 469–485
45. Z. Wanzhen, L. Glass, and A. Shrier: Evolution of rhythms during periodic stimulation of embryonic chick heart cell aggregates. Circ. Res. **69**, 1022–1033 (1991)
46. Z. Wanzhen, L. Glass, and A. Shrier: The topology of phase response curves induced by single and paired stimuli in spontaneously oscillating chick heart cell aggregates. J. Biol. Rhythms **7**, 89–104 (1992)
47. H. R. Warner: The frequency-dependent nature of blood pressure regulation by the carotid sinus studied with an electric analog. Circ. Res. **6**, 35–40 (1958)
48. H. R. Warner and R. O. Russel: Effect of combined sympathetic and vagal stimulation on heart rate in the dog. Circ. Res. **24**, 567–573 (1969)
49. H. Warzel, H.-U. Eckhardt, and U. Hopstock: Effects of carotid sinus nerve stimulation at different times in the respiratory and cardiac cycles on variability of heart rate and blood pressure of normotensive and renal hypertensive dogs. J. Auton. Nerv. Syst. **26**, 121–127 (1989)
50. R. J. Wyman: Neuronal generation of the breathing rhythm. Ann. Rev. Physiol. **39**, 417–448 (1977)
51. T. Yang, M. D. Jacobstein, and M. N. Levy: Synchronisation of automatic cells in S-A node during vagal stimulation in dogs. Am. J. Physiol. **246**, H585–H591 (1984)

A Dynamical Approach to Normal and Parkinsonian Tremor

Anne Beuter and Anne de Geoffroy

Abstract

Many phenomena in motor control recur at regular or almost regular intervals (gait, tremor, fibrillations) while others occur irregularly (cerebellar gait, fasciculations, myoclonus). Subtle changes in the dynamics of motor control rhythms (i.e., going from regular to irregular or from irregular to regular rhythms) can be of major clinical importance in the early detection of neurodegenerative diseases. Here, we examine the effect of lesions located in the basal ganglia on the amplitude and frequency of tremor. As one of the major reentrant subcortical loop modulating the output of the cerebral cortex through thalamo-cortical projections, the basal ganglia play a major role in sensorimotor integration. Some lesions affecting these neural structures are associated with Parkinson's disease and produce qualitative changes in tremor dynamics. These tremor dynamics may represent the missing link between the anatomical level (i.e., the lesion) and the behavioral level (i.e., the pathophysiology).

1 Introduction

Tremor is a sustained involuntary and irregular oscillatory motion of a body part observable in all human beings in a relaxed position (resting tremor), during the maintenance of a posture (static tremor), and during the execution of movements (kinetic tremor). Some forms of tremor such as physiological tremor or enhanced physiological tremor (a physiological tremor augmented by fatigue, emotion or shivering) are perfectly normal. However, other forms of tremor such as essential, cerebellar, or parkinsonian tremors are a sign of brain malfunction. There is no easy way to classify the different tremors partly because their causes are still relatively unknown.

Tremor has been recorded with a variety of instruments based on the measurement of position, velocity, acceleration, force or electromyography

(EMG). These recordings have been performed in different joints including the metacarpo-phalangeal joint of the thumb, index or middle finger, the back of the hand, the elbow, the ankle, etc. Because of the complexity of tremor mechanisms involved and the variety of methodologies used, results have been somewhat inconsistent and sometimes even contradictory. As a consequence progress in understanding tremor experimentally have been relatively slow. This situation may also have somewhat limited the attempts made to model parkinsonian tremor (Gurfinkel and Osovets, 1973; Beuter, Belair, Labrie, 1993).

This chapter starts with a review of the physiology of normal and parkinsonian tremor. Then, some qualitative changes observed in the dynamics of normal and pathological tremor are presented and illustrated using data recorded in our laboratory. Finally the chapter ends by presenting some of the difficulties encountered in the analysis of tremor dynamics.

2 Characteristics of Normal Physiological Tremor

Today it is agreed that physiological tremor does not have one but multiple causes. According to Marsden (1984) one of the leaders in this field, these causes include the ballisto-cardiogram, muscle properties, motoneurone firing, spindle feedback, as well as supraspinal and pharmacological influences. In the fingers, these oscillations have an average amplitude of 1 to 2 minutes of arc and are a mixture of frequencies between 5 and 15 Hz (Halliday and Redfearn, 1956). The term physiological tremor is defined by Marsden, Meadows, Lange and Watson (1969) as the "rhythmical oscillations of the unsupported actively outstretched fingers" (p.648). Physiological tremor is apparently associated with attempts to maintain a constant posture, attempts to execute smooth movements, and attempts to maintain a constant force of muscle contraction (Marsden et al, 1969).

Oscillations at Rest

Tremor at rest is barely visible with the naked eye in most human beings but fluctuations in position can be detected with precision by using position measurements instruments such as laser systems (Beuter, Cordo and de Geoffroy, 1994). When looking at the raw data it becomes clear that physiological parameters such as breathing and cardiac activity play an important role in generating tremor in normal subjects. Figure 1 illustrates the resting tremor of a normal subject and will be discussed below.

The influence of hemodynamics on tremor has been described by Brumlik (1962), Buskirk and Fink (1962) and Marsden et al. (1969) More recently, Lakie, Walsh and Wright (1982) have shown that the heart beat represents a significant proportion of resting tremor. Using an averager, they found that a large component of resting tremor is phase-locked to the electrocardiogram (ECG). However, when the muscles around the radiocarpal joint are contracted as in postural tremor, the cardiac component represents a smaller proportion of the total tremor recorded with an accelerometer (Gurfinkel, Sotnikova, Tereshkov, Fomin, and Shik, 1971). This is coherent with the observation that muscles in relaxed limbs have little or no EMG discharge (Lakie, Walsh and Wright, 1986). In the absence of strong muscle activity, resting tremor is related to the heart beat as shown by a change in the envelope of the tremor waveform in the presence of an occasional extrasystole (Lakie, Walsh and Wright, 1983) or in subjects with sinus arrhythmia or atrial fibrillation (Brumlik, 1962, Boshes, 1966).

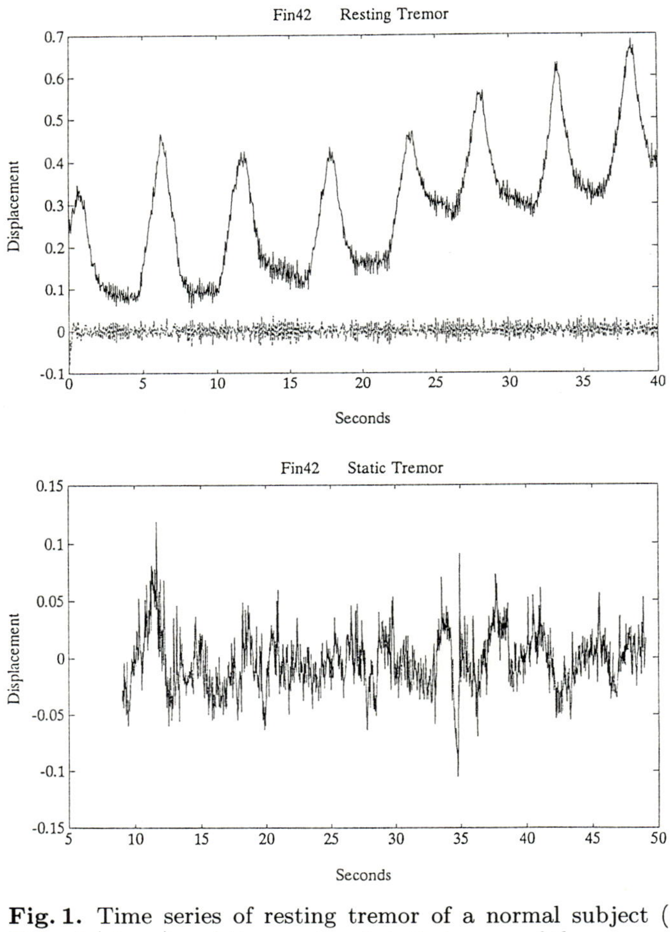

Fig. 1. Time series of resting tremor of a normal subject (#42) unfiltered and filtered (above) and time series of static tremor of the same subject (below).

With increasing inertia added to the limb, the relationship to the heart beat is observed with added clarity but the frequency of tremor is reduced. The cardiac component slows correspondingly but the demodulated EMG peak frequency is not significantly affected at least in horizontal finger movements (Lakie, Walsh and Wright, 1986). The heart beat/tremor relationship persists when a cuff is applied to the upper arm (Lakie et al., 1983) or at the wrist above the arterial pressure (Marsden et al., 1969). However, arterial occlusion at the elbow diminishes tremor progressively (Halliday and Redfearn, 1954).

Another way to explore resting tremor of the hands is to apply light taps to the finger or to stimulate the motor point of wrist muscles (at 1 Hz or less). Light taps induce decrementing transients at a frequency similar to the frequency of spontaneous tremor (around 8 Hz in the study of Lakie, Walsh and Wright, 1982) and stimulation of the extensor digitorum communis produces oscillatory transients which are essentially similar to those produced by taps (Lakie, Walsh and Wright, 1986).

Finally, tremor at rest is not suppressed by neuromuscular blocking drugs administered during anesthesia which according to Lakie et al. (1986) clearly suggests that "the resonant behaviour of the relaxed wrist is not due to neurological circuits but is a passive property of the tissues" (p. 674).

While mechanical properties of the limbs appear to play a major role in resting tremor, other contributing factors exist as well. In a series of studies, Ito (1961a and b) showed that tremor frequency increases with muscle tone which itself depends on the position of the subject (sitting, lying, etc.). They use the Settle Down Time (SDT) which is the time required for the beta component (higher frequencies) to disappear from the tremor record (about 5 min). The SDT is supposed to be an indicator of the rate of tranquilization. Their study demonstrates that tremor is influenced by various drugs and emotions.

In summary, it appears that tremor at rest is mainly of mechanical origin when muscle relaxation can be readily achieved. As indicated by Marsden (1984) limbs possess inertia and stiffness and act as passive mass-spring systems which oscillate freely at their natural frequencies. The greater the mass, the lower the natural resonant frequency and the more damped the oscillations will be. The resonant frequency of the finger is about 25 Hz and that of the hand is about 9 Hz (Stiles and Randall, 1967; Joyce and Rack, 1974). Spontaneous resting tremor has the characteristics of a damped system subjected to minor mechanical disturbances. The ballistocardiogram is one of these disturbances, breathing activity is another one, and these disturbances cause the system to ring at its resonant frequency.

As can be seen on Figure 1 above, the resting tremor of a 62 year old, right handed normal subject (#42, block 1) contains oscillations related to breathing visible on the unfiltered data. Tremor was recorded using a position laser sensor at a sampling frequency of 100 Hz while the subject was

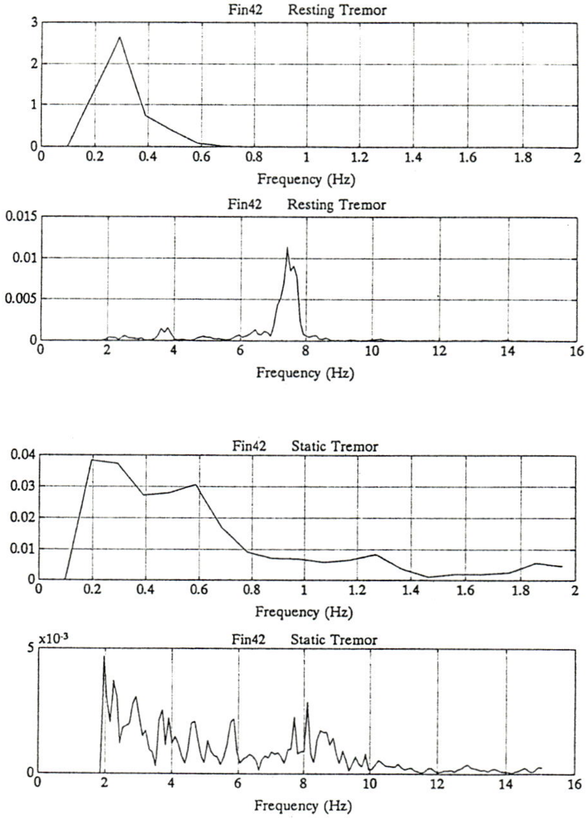

Fig. 2. FFT of the time series presented in Figure 1 for resting tremor (above) and static tremor (below) of a normal subject (#42). Each spectrum is divided into a first section corresponding to frequencies less than 2 Hz and a second section corresponding to frequencies between 2 and 16 Hz

relaxed with his eyes closed. Spectral analysis of the resting tremor of this subject in the 2 to 16 Hz range reveals a peak at 7.4 Hz having a power of 0.0085 and a secondary peak at 3.7 Hz with a power of 0.002 (Figure 2 above). Power spectra in this and in subsequent figures are calculated by first detrending the data (i.e., by removing the best straight line fit to the data using a least squares approach), then sectioning the data into four 1024 segments. Successive segments are Hanning windowed, then transformed with a 1024 point Fast Fourier Transform (FFT) and summed. The peaks in the FFT correspond to local maxima which are the largest local maximum in the 0.5 Hz neighbourhood and are greater than the mean for the one Hertz neighbourhood. In subject #42, the peak at 3.7 Hz appears to represent a subharmonic of the higher peak (7.4Hz). In the range below 2 Hz, a large peak at 0.3 Hz corresponds to breathing.

Oscillations During Maintenance of a Posture: Static Tremor

Physiological (static) tremor is a normal phenomenon which is evident when we attempt to hold a position. The amplitude of static tremor varies widely between subjects. Figure 1 below illustrates the static tremor of a normal subject (#42, block 2) extending (but not hyperextending) the index finger. Tremor was recorded using a position laser sensor at a sampling frequency of 100 Hz while the subject was watching his finger position and a reference line on the screen of an oscilloscope. Oscillations related to breathing are no longer visible on the time series but can be seen on the power spectrum around 0.4 Hz. Spectral analysis of the static tremor of this subject in the 2 to 16 Hz range reveals a peak at 8.4 Hz having a power of 0.002 and secondary peaks at 5.9, 4.7, 3.9, 2.7 and 1.9 Hz with powers between 0.003 and 0.001 (Figure 2 below).

The influence of physiological oscillations (such as breathing, ECG, etc.) is not limited to resting tremor. However, as muscle contraction increases the contribution of these physiological oscillations decreases. Padsha and Stein (1973) examined postural tremor and their spectral analysis revealed a major peak at 0.3 Hz, a secondary peak at 1 Hz and a broad peak at about 10 Hz. They showed that the peak at 0.3 Hz correlates with breathing, the peak at 1 Hz correlates with the ECG and the peak at 10 Hz correlated with the EMG of the finger extensor muscles. They propose that breathing accounts for about 23% of the finger tremor, the ECG and EMG account for 10% and 11% of the tremor respectively. Incidently, Marsden et al. (1969) had previously found a similar contribution of the heart by using cross spectral and coherence analyses of tremor between the two hands. Finally the remaining 58% in Padsha and Stein's study (1973) are due to non-linear effects of the causes mentioned and to other causes as well. These authors calculated the time delay between breathing, the R-wave of the ECG and finger tremor. They found that the delay between the peak in the R wave and the peak in the tremor wave was about 343 ms (sd=55ms) which may correspond to the time needed for the pressure wave produced in the heart to reach the finger. The delay between the breathing and finger tremor varied across subjects and the delay between EMG peaks and tremor was about 50 to 60 ms which may correspond to the motor units contraction time in finger muscles (Padsha and Stein, 1973).

As we have seen in resting tremor, various manipulations have been used to characterise postural tremor. Since the work of Robson in 1959, several researchers have investigated the natural mechanical resonance of limbs oscillating at a joint by adding a load which increases inertia and causes resonant properties to decrease (Marsden, 1984). Joyce and Rack (1974) and more recently Homberg, Hefter, Reiners and Freund (1987) used loads to change the mechanical properties of the limb and examined the peak tremor frequency. Fox and Randall (1970) examined the forearm acceleration and the biceps electromyogram in normal subjects while loads up to 10 pounds were applied at the wrist. Adding weights systematically decreased the peak of the acceleration record and systematically increased the peak of the EMG record. Since cross-correlation analysis revealed no significant coherence at the tremor frequency between acceleration and EMG, the authors concluded that physiological tremor frequency is determined primarily by the natural frequency of the system rather than neural properties of the forearm position control system. Later, Stiles (1976) performed a study in which normal subjects were asked to continuously extend their hand for 15 to 45 min and observed that the RMS displacement amplitude of the tremor recorded with an accelerometer reached values on the order of 100-1000 times control. With time they observed an increase in tremor amplitude and a decrease in tremor frequency. Spectral analysis of the tremor records revealed a consistent relation between extensor EMG modulation amplitude at the tremor frequency and the RMS displacement amplitude for tremor records with large RMS displacement (fatigue) but not with small ones (control). They concluded that small amplitude tremor is largely determined by mechanical factors while larger ones and determined both by mechanical and neural feedback factors (Stiles, 1976).

In the frequency spectrum of tremor recorded with the outstretched hand using an accelerometer, there is a peak of activity in the range of 8 to 12 Hz which appears to be unrelated to passive resonance properties of the limb (Marsden, 1978) since finger resonance frequency is much higher (25 Hz). Today a number of investigators believe that postural tremor reflects motor unit activity eventhough a hemodynamic component is also contained in it (Gurfinkel et al., 1971). As indicated by Marsden (1984) static tremor appears to be the result of a complex interaction between several factors including:

"(1) the natural resonance of the limb, which is dependent upon the inertia and stiffness of the system;

(2) an interaction between the initial firing rates of motor neurones and the low-pass filter properties of the muscle system. The latter cuts out any tendency to oscillate when motor neurones fire at frequencies above about 15 Hz, and

(3) the size principle of motor unit activation. As force is increased, those units already engaged increase their firing rate, so contribute less and less to tremor as a result of the filtering properties of muscle. However, newer and larger motor units are recruited at the lower frequency, so superimposing an oscillation roughly proportional to the force of contraction." (p.53, Marsden, 1984).

Other factors such as supraspinal influences (e.g., visual feedback) and pharmacological influences (e.g., circulating catecholamines) can also modulate static tremor intensity.

In 1984, Allum stated that the emphasis in research was on the role of neural mechanisms in generating tremor. The role of muscle mechanics was restricted to influencing the amount of muscle stretch during tremor "rather than being, per se, tremogenic" (p.135, 1984).

The mechanical recording of tremor consists of the sum of multiple factors in varying proportions depending upon the amplitude and nature of the tremor (Marsden, 1984). Thus, it is important that experimenters report precisely the conditions of tremor recording, muscles involved, load, task, force required in order to avoid getting ambiguous (Marsden, 1978) or even contradictory results. In other words, the resonant properties may dominate or be dominated by the tremor depending on these conditions.

3 Tremor in Patients with Parkinson's Disease

Parkinson's Disease (PD) is a chronic progressive neuro-degenerative disorder characterized by tremor, rigidity and bradykinesia (Nutt, 1992). Loss of midbrain dopamine neurons is a major neurochemical characteristic of patients with PD. The most severe neuronal death always occurs in the substantia nigra. Replacement with a dopamine precursor (L-dopa) through pharmacotherapy can reverse parkinsonism but clinical use of L-dopa is complicated by several pharmacokinetic factors and long term L-dopa therapy almost invariably results in adverse effects. Thus, researchers are continuously exploring new ways to treat the disease (e.g., transplantation of dopamine neurons) and the symptoms (e.g., controlled release preparations, thalamic stimulation).

There are several types of tremulous movements in PD including resting, postural and kinetic tremors, cogwheeling and clonus phenomena (Findley, Gresty and Halmagyi, 1981). Resting tremor is a dominant feature of PD which has a characteristic frequency around 4-5 Hz. Usually, it begins distally in the extremities on one side and later involves both sides. One question that has been debated at length is whether tremor in PD is a self-sustained oscillation in a peripheral feedback loop or originates from the action of an oscillator located in the Central Nervous System (CNS). After recording force and EMG in patients during imposed joint movements, Rack and Ross (1986) concluded that there exists a continuum between the contribution of peripheral reflexes and oscillatory mechanisms located in the CNS. But the debate is not over yet (see Burne, 1987; Allum, Dietz and Freund, 1978).

Resting Tremor in PD

As can be seen on Figure 3 above, the resting tremor on the less affected side of a patient with PD (#03, block 2) contains oscillations related to breathing visible on the unfiltered data. This patient is 61 years old, he is right handed,

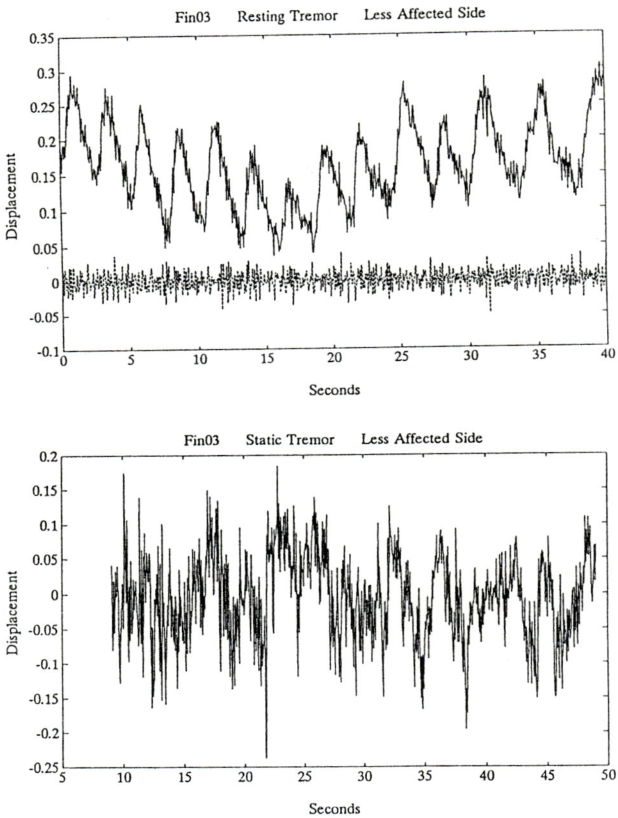

Fig. 3. Time series of resting tremor of a patient with Parkinson's disease (#03) unfiltered and filtered (above) and time series of static tremor of the same subject (below). These data are recorded on the <u>less</u> affected side.

has had PD seven years at the time of testing and is considered to be in the stage III of the Hoehn and Yahr (1967) classification. Spectral analysis of the resting tremor of this subject in the 2 to 16 Hz range reveals a peak at 3.6 Hz having a power of 0.004 and secondary peaks at 2.4 and 7.2 Hz with powers of 0.003 (Figure 5 above). In this subject the peak at 7.2 Hz appears to represent a harmonic of the peak at 3.6 Hz. In the range below 2 Hz, a large peak at 0.4 Hz corresponds to breathing.

On the more affected side (right side) the breathing rhythm is obscured by a large amplitude resting tremor (Figure 4 above). Frequency analysis of the 2 to 16 Hz interval reveals an important peak at 5.9 Hz with a power of 2.25 (Figure 6 above). This peak is more than 500 times larger on the more affected side than on the less affected side. A secondary peak with a power of 0.4 is noted around 5 Hz. The peak observed in frequencies less than 2 Hz and corresponding to breathing is 0.3 which is considerably lower in power than the comparable tremor peak in the less affected side (Figure 5 above). The RMS is, as expected, larger (0.35) on the more affected side than on the less affected side (0.17). These values are calculated using the unfiltered data. To filter the data we removed the lower frequency component by taking an FFT of the signal, then by removing all frequencies below 1.65 Hz and doing an inverse FFT to convert the signal back in the time domain. With the filtered data the RMS values become 0.163 on the more affected side and 0.012 on the less affected side. Clearly it is preferable to filter out the low component of the data in order to quantify adequately resting tremor amplitude.

Static Tremor in PD

Clinicians frequently observe that patients with PD have a large amplitude resting tremor which decreases or even disappears during the maintenance of a posture. The static tremor of a patient with PD (#03, block 2) is presented in Figure 3 below, for the less affected side and in Figure 4 below, for the more affected side. Oscillations related to breathing are no longer visible on the time series but peaks are clearly present on the FFT around 0.3 and 0.4 Hz. Spectral analysis of the static tremor of subject #03 in the 2 to 16 Hz range reveals multiple peaks at 2.05 and 6 Hz having powers around 0.025 (Figure 5 below) for the less affected side. The pattern is similar on the more affected side (Figure 6 below) but the multiple peaks are 2.25 and 4.1 Hz. The RMS of static tremor are 0.06 and 0.07 for the patient and 0.02 for the normal subject.

Problems Linked to the Study of Tremor Dynamics

Despite a considerable number of investigations on tremor, progress have been relatively slow. While one problem is linked to the methodologies used as indicated above, another one is linked to the consistent use of spectral analysis as the primary source of information about tremor frequency. Some investigators have pushed frequency analysis to its limits by using cross correlation and coherence analyses in order to extract meaningful information out of tremor amplitude and frequency but sometimes the results of different studies have lacked consistency. For example, how can we combine the 7 Hz frequency associated with the ballistocardiac impulse when it is known that finger resonance frequency is 25 Hz? Spectral analysis deals adequately with linear phenomena but these is ample evidence that human tremor is a highly nonlinear system. Some modulation of amplitude and frequency are not easily picked up from FFT or power spectra but may be of critical importance in the interpretation of the results. As indicated recently by Gresty and Buckwell (1990) spectral analysis of a tremor record can produce spectra with multiple components of significant amplitude and the problem is to

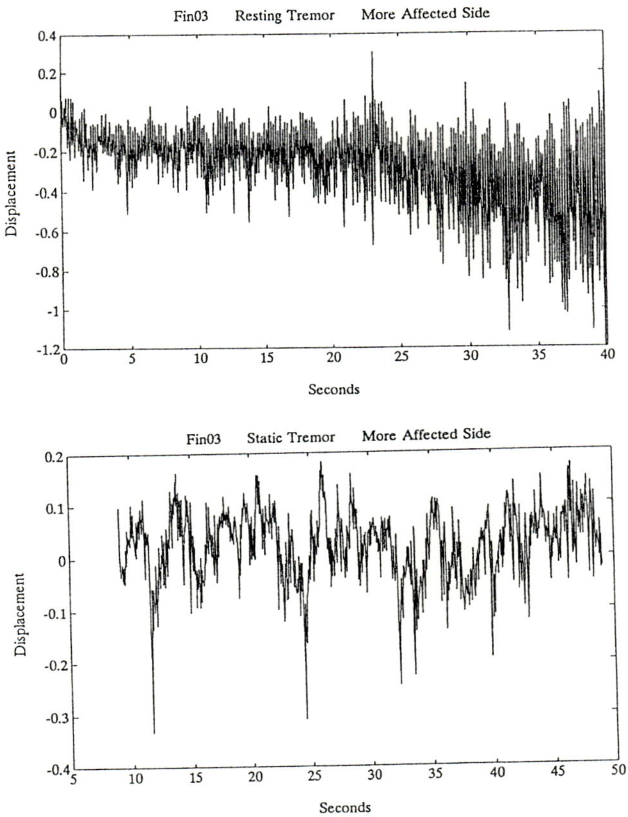

Fig. 4. Time series of resting tremor of a patient with Parkinson's disease (#03) unfiltered (above) and time series of static tremor of the same subject (below). These data are recorded on the more affected side.

determine whether the presence of several peaks represents the "the coexistence of separate tremor mechanisms or is a consequence of fluctuations in the frequency or amplitude of a single tremor" (p.976). The investigation of harmonics, subharmonics and sidebands has been neglected for a long time by tremor experts and this should be corrected if tremor mechanisms are ever to be understood. For example, when position and not acceleration is measured, there are tremor peaks at both 4 and 8 Hz in normal and parkinsonian subjects. The peak powers follow an inverse pattern between these two populations. The investigation of these peaks could provide the beginning of an answer (Beuter, de Geoffroy, Thomsen, 1994).

Another problem is that several phenomena described as pathological (such as the 4 Hz peak described above) are not really so, if the context of the comparison is taken under consideration. For example, the frequency of

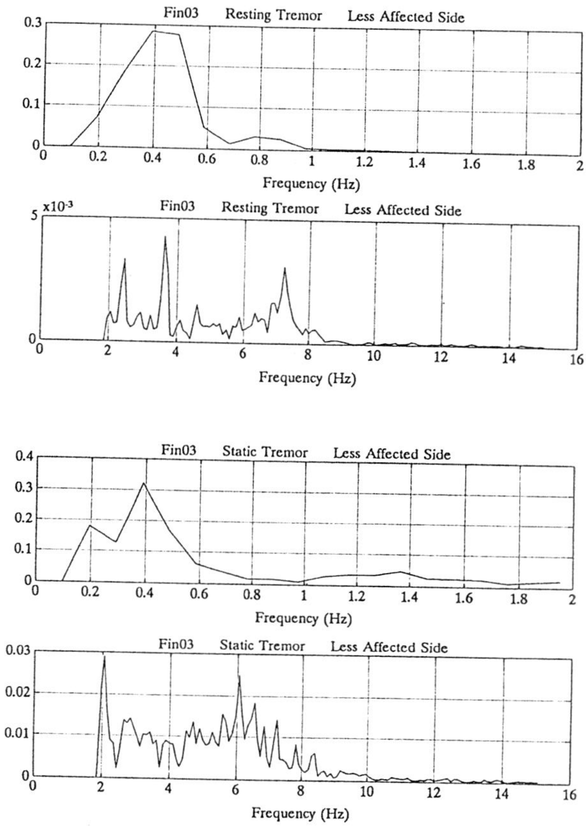

Fig. 5. FFT of the time series presented in Figure 3 for resting tremor (above) and static tremor (below) of a patient with Parkinson's disease (#03) corresponding to the less affected side. Each spectrum is divided into a first section corresponding to frequencies less than 2 Hz and a section corresponding to frequencies between 2 and 16 Hz

parkinsonian tremor is similar to that of control subjects when frequencies are compared at similar RMS displacements levels (Stiles and Pozos, 1976). Another example is related to paired motor unit discharges which have been considered for years as a sign of pathology. Today, we known that these paired discharges are present in a normal subject mimicking parkinsonian tremor. These few examples suggest that the distinction between normality and pathology cannot always be described by different mechanisms but rather by similar mechanisms whose parameter values are out of range. This is in fact the definition of a dynamical disease: a disease occurring in a normal physiological control system operating in an abnormal range of its parameter values (Mackey and Glass, 1977).

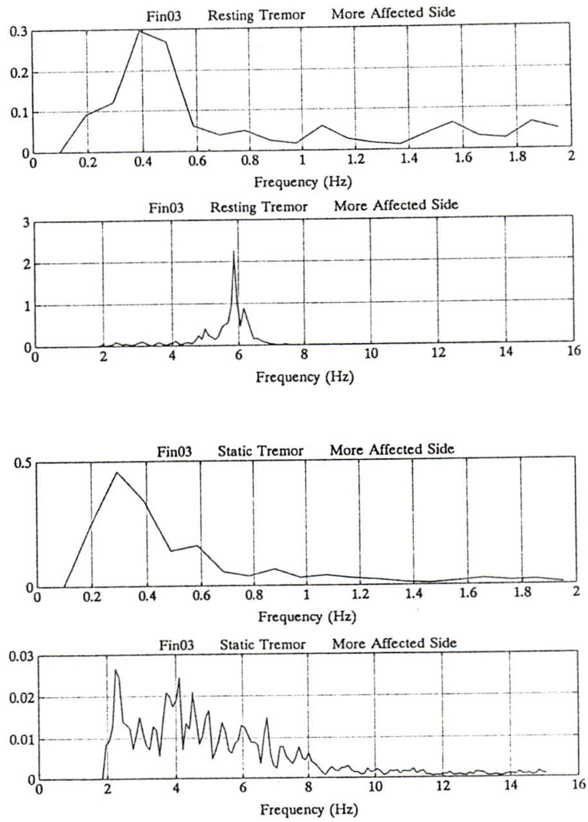

Fig. 6. FFT of the time series presented in Figure 4 for resting tremor (above) and static tremor (below) of a patient with Parkinson's disease (#03) corresponding to the less affected side. Each spectrum is divided into a first section corresponding to frequencies less than 2 Hz and a section corresponding to frequencies between 2 and 16 Hz

The on-off phenomenon affecting patients with PD also affects tremor and has interesting dynamics. These fluctuations in disability (especially and rigidity) follow different time scales. The short term variations occur over seconds or minutes, are seen in the untreated patient and are believed to be due to the disease itself (Marsden et al., 1982). In our laboratory, we have observed these short term fluctuations in tremor (Figure 7) and are now actively engaged in exploring these unpredictable changes by monitoring the overall autononic tone (heart rate, blood pressure, breathing). Generally tremor is considered in the framework of a central oscillator or as a peripheral feedback model depending on whether it is thought to originate from a group of bursting neurons or from the instability of large neuronal circuits (Wichmann and DeLong, 1993). In fact it appears that tremor dynamics follow a continuum

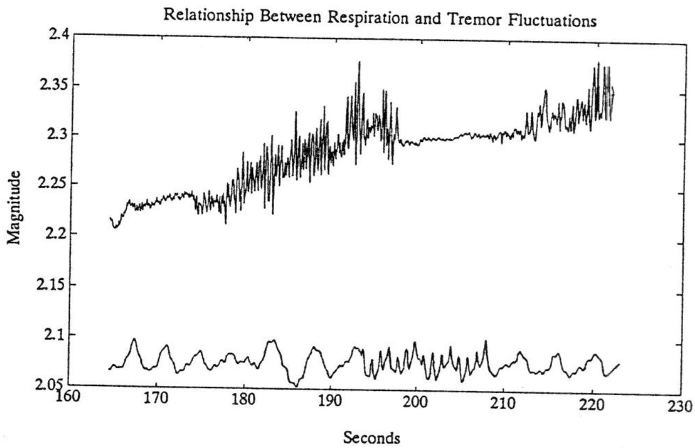

Fig. 7. Relationship between tremor fluctuations and breathing in a patient with PD (H) while performing a mental task.

which is changing over time The debate between centralists and peripheralists regarding the origin of tremor should now be replaced by the following question: what switches tremor on and off in PD? by what mechanism is the balance between the central and the peripheral contribution modified at a given time?

4 Modelling of Normal and Parkinsonian Tremor

In the past, a number of hypotheses have been proposed to explain parkinsonian tremor going from an abnormal rise in the threshold of the Renshaw cells (Gelfand, 1963), to an increased delay in the spinal feedback with subsequent conversion from negative to positive feedback (Gurfinkel and Osovets, 1973). Some investigators have suggested that parkinsonian tremor is caused by a central generator (Lamarre and Cordeau, 1967) which could be located in the inferior olive (Armstrong, 1974) while others have proposed that this pathological tremor appears in the spinal cord under the influence of abnormal supraspinal influences. The current state of knowledge suggests that both the CNS and the sensory feedback pathways participate in the generation of tremor. What is not clear however, is the exact nature of their respective contribution. Overall few attempts to model human tremor have been published in the literature. Interestingly, all models reviewed used a macrodescription of tremor to describe parkinsonian tremor. We review briefly these models below.

In 1962, Austin and Tsai proposed a fairly sophisticated model of parkinsonian tremor. In their model, the rate of change of tremor amplitude with respect to time or emotional excitation is proportional to the amount of tremor present. They hypothesized that parkinsonian tremor is based on an abnormal ratio of supraspinal facilitation to inhibition, acting downstream from the basal ganglia to the spinal cord. The abnormal level of excitation is due to the destruction of the inhibitory regions of the brain caused by PD. They use two ordinary differential equations describing the damping function of the tremor, with emotional factor assumed to come from supraspinal centers. They use adrenalin and sodium amytal intravenous injections to simulate emotional excitation and fatigue, which would lead to inhibition. Unfortunately the authors do not indicate how they recorded tremor nor what type of tremor they recorded (i.e., resting, postural, etc.). Furthermore, their model does not take under consideration the time delays inherent to any physiological feedback loop.

Later, Gurfinkel and Osovets (1973) investigated the mechanisms generating parkinsonian tremor. They are the first to mention the irregularities and instabilities present in parkinsonian tremor. They hypothesize that the stability in the maintenance of a posture is obtained by periodic changes in the strength of the muscle at the frequency of physiological tremor. The oscillations in strength have an amplitude that is a function of the deviation angle from the set position and the coefficient of amplification in a feedback loop. In their model the authors have neglected all delays present in the central and peripheral feedback loops. Although they noted the presence of transition between parkinsonian and physiological tremor with no intermediate frequencies they did not consider the possibility that both forms of tremor can be present in the same patient. Since the frequency of parkinsonian tremor is about half that of physiological tremor, they suggest the possibility that parametric resonance could explain their results. They explain the transition by a spontaneous rise in amplification in a feedback loop but because of their insufficient data available, they do not elaborate on this point.

In 1976, Stein and Oguztoreli examined the mechanical (i.e., how a muscle interacts with its load) and reflex (i.e., high gain in a reflex pathway) factors in the generation of tremor. They assume that physiological tremor is a basically linear phenomenon which has a small amplitude and use a second order system based on a mass spring system. The differential equations they used generate oscillations corresponding to the frequency of physiological tremor (8-12 Hz) using spinal pathways and muscle properties and oscillations corresponding to parkinsonian tremor (4-6Hz) using longer supraspinal reflexes. In addition the authors examine briefly the possibility that nonlinearities present in muscle receptors limit the magnitude of the oscillations. In this model at a specific gain the reflex oscillation becomes dominant while at a higher gain, it is the mechanical oscillation that dominates.

The model of parkinsonian tremor proposed by Fukumoto (1986) is the first model to include both an oscillator (4-6 Hz) located in the central nervous system and a peripheral feedback loop located in the stretch reflex (8-12 Hz). While the central loop is often believed to play a role in pathological tremor, the peripheral loop is often associated with physiological tremor. Fukumoto (1986) hypothesized that parkinsonian tremor is caused by a reduced contractive force maybe due to fatigue of intrafusal muscle fibers and modeled it with an autoregressive model. Eight independent variables were used and the manipulation of only one parameter (i.e., the intensity parameter of intrafusal fibre force) allowed the author to reproduce two different kinds of tremor (i.e., parkinsonian and physiological). Although the model is an interesting new contribution to the field, it is extremely limited by the fact that it is based on linear differential equations, and only deals with frequency while ignoring the morphology of the oscillation. There is ample evidence that tremor in general and parkinsonian tremor in particular are nonlinear phenomena. Parkinsonian tremor, for example, fluctuates markedly in intensity, the oscillations are not regular and it is not clear how the model could address these experimental observations.

The first model to incorporate nonlinear dynamics was proposed by Gantert, Honerkamp and Timmer (1992). They measured acceleration of the stretched hand and measured the correlation dimension to test for stochasticity, estimation of the Lyapunov exponent and an ARMA model to test for linearity. The calculation of the correlation dimension in case of normal physiological tremor shows that the dynamics are stochastic and the hand behaves as a linear damped oscillator driven by white noise. In patients with PD however, the dynamics of patients with PD was recognized as a deterministic system characterized by nonlinear deterministic dynamics.

Finally, the model presented by Beuter, Belair and Labrie (1993) focus on the analysis of tremor using nonlinear dynamics. They examine the influence of delays in two feedback loops from a physiological and analytical perspective and study the role of each delay on the stability of the finger position. The model was tested over a wide range of parameter values. This led the authors to explore the role of an increased gain and noise in one of the loops to match more accurately the performance of some patients with PD. They found that the influence of stochastic elements (i.e., noise) in the differential delay equations was found to contribute qualitatively to a more accurate reproduction of experimental traces in patients with PD.

In summary, applying nonlinear dynamics to a neuro-degenerative such as PD may provide a way to reveal the underlying simplicity of neural dynamics, that is, to discern patterns that would not be visible with linear analysis or the naked eye. Such an approach may also provide new modeling avenues as well as new diagnostic tools.

5 Conclusion

In conclusion, normal physiological tremor may be regarded as "a necessary consequence of the capacity of the neuromuscular system to respond briskly" (Marsden, 1978, p.15). It may even play a role in the organization of the nervous system (Llinas, 1984). If it is an indicator of nervous system integrity, then subtle changes may, if adequately detected and analysed, help progress toward a presymptomatic diagnosis of PD. This is especially important since we know that around 80 % of striatal dopamine has disappeared before the first symptoms of the disease appear (McGeer, McGeer and Suzuki, 1977).

References

Allum, J.H.J (1984) Segmental reflex, muscle mechanical and central mechanisms underlying human physiological tremor. In: Findley, L.J. and Capildeo, R. Movement Disorders: Tremor. Macmillan (London), 135-155.

Allum, H.J., Dietz, V. and Freund, H.J. (1978) Neuronal mechanisms underlying physiological tremor. Journal of Neurophysiology, 41, 3, 557-571.

Armstrong, D.M. (1974) Functional significance of connections of the inferior olive. Physiological Review, 54:358-417.

Beuter, A., Belair, J. and Labrie, C. (1993) Feedback and delays in neurological diseases: a modeling study using dynamical systems. Bulletin of Mathematical Biology, 55, 3, 525-541.

Beuter, A., Cordo, P. and de Geoffroy, A. (1994) The measurement of tremor using simple laser systems. Journal of Neuroscience Methods. In press, january 1994.

Beuter, A., de Geoffroy, A. and Thomsen, J.S. (1994) in preparation.

Boshes, B. (1966) Measurement of tremor. Journal of Neurosurgery, 24, 324-330.

Brumlik, J. (1962) On the nature of normal tremor. Neurology, 12, 159-179.

Burne, J.A. (1987) Reflex origin of Parkinsonian tremor. Experimental Neurology, 97, 327-339.

Buskirk V.C. and Fink, R.A. (1962) Physiologic tremor: An experimental study. Neurology, 12, 361-370.

Findley, L.J., Gresty, M.A. and Halmagyi, G.M. (1981) Tremor, the cogwheel phenomenon and clonus in Parkinson's disease. Journal of Neurology, Neurosurgery and Psychiatry, 44, 534-546.

Fox, J.R. and Randall, J.E. (1970) Relationship between forearm tremor and the biceps electromyogram, Journal of Applied Physiology, 29, 1, 103-108.

Fukumoto, I. (1986) Computer simulation of parkinsonian tremor. Journal of Biomedical Engineering, 8:49-55.

Gantert, C., Honerkamp, J. and Timmer, J. (1992) Analyzing the dynamics of hand tremor time series. Biological Cybernetics, 66:479-484.

Gelfand, I.M., Gurfinkel, V.S., Y.M., Kots, V.I. Krinskiy, M.L. Tsetlin and M.L. Shik (1963) Biofizika, 9, 6, 710.

Gresty, M. and Bruckwell, D. (1990) Spectral analysis of tremor: understanding the results. Journal of Neurology, Neurosurgery, and Psychiatry, 53, 976-981.

Gurfinkel, V.S. and Osovets, S.M. (1973) Mechanism of generation of oscillations in the tremor form of parkinsonism. Biofizika, 18, 4, 781-790.

Gurfinkel, V.S., Sotnikova, L.E., Tereshkov, O.D., Fomin, S.V. and Shik, M.L. (1971). An analysis of physiological tremor by means of a general purpose computer. In: Models of structural-functional organization of certain biological systems. Gelfand, I.M., Gurfinkel, V.S., Fomin, S.V. and Tsetlin, M.L. (eds), The MIT Press: Cambridge (MA).

Halliday, A.M. and Redfearn, J.W.T. (1956) An analysis of the frequencies of finger tremor in healthy subjects. Journal of Physiology, 134, 600-611.

Halliday, A.M. and Redfearn, J.W.T. (1954) The effect of ischemia on finger tremor. Journal of Physiology, 123, 23-24.

Hoehn, M.M. and Yahr, M.D. (1967) Parkinsonism: Onset, progression and mortality. Neurology, 17, 427-442.

Homberg, V., Hefter, H., Reiners, K. and Freund H.J. (1987) Differential effects of changes in mechanical limb properties on physiological and pathological tremor. Journal of Neurology, Neurosurgery, and Psychiatry, 50, 568-579.

Ito, K. (1961a) The influence of postural change on the frequency of human minor tremor. Yonago Acta Medica, 5, 1, 32-35.

Ito, K. (1961b) Settle down time of human minor tremor. Yonago Acta Medica, 5, 2, 32-35.

Joyce, G.C. and Rack, P.M.H. (1974) The effect of load and force on tremor at the normal human elbow joint. Journal of Physiology, 240, 375-296.

Lakie, M., Walsh, E.G. and Wright, G.W. (1986) Passive mechanical properties of the wrist and physiological tremor. Journal of Neurology, Neurosurgery, and Psychiatry, 49, 669-676.

Lakie, M., Walsh, E.G. and Wright, G.W. (1983) On tuning a physiological tremor. Journal of Physiology, 334, 32-33.

Lamarre, Y and Cordeau, J.P. (1967) Etude du mécanisme pathophysiologique responsable chez le singe d'un tremblement expérimental de type parkinsonian. Actualités Neurophysiologiques, 7:141-166.

Llinas, R.R. (1984) Possible role of tremor in the organisation of the nervous system. In: Findley, L.J. and Capildeo, R. Movement Disorders: Tremor. Macmillan (London), p.475-477.

Mackey, M.C. and Glass, L. (1977) Oscillation and chaos in physiology control systems. Sciences, 197, 287-289.

Marsden, C.D. (1978) The mechanisms of physiological tremor and their significance for pathological tremors. In: Physiological tremor, pathological tremors and clonus. Progress in Clinical Neurophysiology, 5, 1-16. Ed. J.E. Desmedt (Karger, Basel).

Marsden, C.D. (1984) Origins of normal and pathological tremor. In: Findley, L.J. and Capildeo, R. Movement Disorders: Tremor. Macmillan (London), p.37-84.

Mc Geer, P.L., Mc Geer, E.G. and Suzuki, J.L. (1977) Aging and extrapyramidal function. Archives of Neurology, 34, 33-35.

Marsden, C.D., Meadows, J.C., Lange, G.W. and Watson, R.S. (1969) The role of the ballistocardiac impulse in the genesis of physiological tremor. Brain, 92, 647-662.

Nutt, J.G. (1992) Therapy of Parkinson's disease. Neuroscience Facts, 3,2, 85-86.

Padsha, S.M. and Stein, R.B. (1973) The basis of tremor during a maintained posture. In: Control of Posture and Locomotion. (Eds) R.B. Stein, K.G. Pearson, R.S. Smith, J.B. Redford. Plenum Press: New York, 415-419.

Rack, P.M.H. and Ross, H.F. The role of reflexes in the resting tremor of Parkinson's disease. Brain, 109, 115-141.

Robson, J.G.(1959) The effect of loading upon the frequency of muscle tremor. Journal of Physiology, 149, 29-30.

Stiles, R.N. (1976) Frequency and displacement amplitude relations for normal hand tremor. Journal of Applied PHysiology, 40, 1, 44-54.

Stiles, R.N. and Randall, J.E. (1967) Mechanical factors in human tremor frequency. Journal of Applied Physiology, 23, 324-330.

Stiles, R.N. and Pozos, R.S. (1976) A mechanical reflex oscillator hypothesis for parkinsonian hand tremor. Journal of Applied Physiology, 40, 6, 990-998.

Wichmann,T. and Delong,M.R. (1993) Pathophysiology of Parkinsonian Motor Abnormalities. In Advances in Neurology, Parkinson's Disease: From Basic Research to Treatment. Narabayashi, H., Nagatsu, T., Yanagisawa, N. and Mizuno, Y. (eds), Raven Press: New York, 60, 53-61.

Part V

Complex Ecologies and Evolution

Dynamics of Complex Ecologies

Jacqueline M. McGlade

Abstract

In this chapter I address the problems of analysing real ecosystems using artificial complex ecologies. These artificial ecologies are biological models which are explicitly designed to capture the spatio-temporal dynamics of multiple forms of interaction and their evolution. I first examine general aspects of evolutionary dynamics and discuss the concept of *evolutionarily stable attractors*. I then go on to discuss the inclusion of space explicitly in continuum models and individual-based models. In particular I examine the possibility of data analysis and reconstruction techniques for ecology given that the dynamics of many complex ecologies are the result of a low-dimensional dynamical system.

1 Introduction

The fundamental aim of modelling the dynamics of complex ecologies is to better understand the processes and mechanisms underlying the evolution of patterns in ecosystems. Implied in this is the ability to detect multiple states of stability, ascertain the resilience of ecosystems to perturbations and understand the local and global interdependencies. However, most studies of pattern have been little more than looking at the effects of ordering processes operating on randomness.

In this chapter I will address the issue of interpreting local and global patterns arising from different processes using both analytic and numerical approaches. Although I will not discuss the more general problems of self-organisation (see [1, 2]), much of what is given below has an immediate bearing on this debate. In particular, it is clear that any conclusions as to the origin of differential patterns must include the effects of asymmetric interactions between individuals as well as those derived from simple spatial interactions.

There are three metaphors which can be used to model ecosystems: *equilibrium centered* in which there is constancy through time and there are no penalties with size; a *dynamical view* in which the existence of stability, resilience, adaptation and the role of instability are discussed, where the system has notable qualitative properties, spatial graininess is not averaged out and where local events can cascade; and an *evolutionary view* or organisational view where self-regulated change and self-evolved criticality are discussed alongside succession and fitness. What this shows is that the axiomatic basis of ecosystem modelling is not yet completely formed. Just as the Milesians felt that changes in nature were real and the Eleatic school considered that matter was unique and unchanging and hence patterns retained, so ecologists today are split between using non-equilibrium, dynamical models versus models which generate some sort of unchanging evolutionary pattern such as successional stable climaxes.

There is of course much empirical support for stable patterns in nature. For example there are observations such as Dyar's Rule (that successive instars of many insects' larvae differ in mass by a factor of 2), the $\frac{3}{2}$ Self-thinning Law of plant growth, Cope's Law (that large species tend to appear later in a group's phylogeny), Chossat's Rule (that animals die when total weight loss approaches half of the initial body mass) and Size Dependent Theory (that living biomass occurs in logarithm intervals of organism size across the biosphere). But there is also evidence of the chaotic dynamics and sudden changes in ecosystems which have led to mass extinctions or large-scale alterations in species diversity. One of the most important issues in analysing real ecosystems is thus our ability to characterise those processes which lead to systematic changes and those which are largely random.

In the first section I will extend the ideas of evolutionary dynamics and stability to a broad class of ecologies drawing on work given in [3]. This work will be used to show that evolutionary stable attractors and chaotic dynamics are perfectly compatible. I will then discuss the study of local dynamics in ecologies using continuum systems. The example I will use is a coupled map lattice model for studying growth of plant monocultures [4]. In particular I show how the model can be used to derive methods to distinguish between asymmetric competition and spatial effects. In the final section I analyse another type of spatial model which uses an individual-based approach to study a forest mosaic cycle. This is an example of an artificial ecology.

2 Dynamics and Evolution

First, I will consider a simple resource-prey-predator model commonly used in modelling ecosystems of the form:

$$\frac{x_1'}{x_1} = (1 + b_1) \exp(-\alpha \frac{x_1'}{x_3} - c_1 \frac{x_2}{1 + d_1 x_1}), \tag{1}$$

$$\frac{x_2'}{x_2} = (1 - d_2)\exp(c_2 x_1), \tag{2}$$

$$\frac{x_3'}{x_3} = (1 + b_3)\exp(-\frac{x_3}{k} - c_3 x_1). \tag{3}$$

Here x_1, x_2 and x_3 represent the prey, predator and resource, and the primed variables represent the corresponding numbers in the next period. Chaotic dynamics can be obtained using either one of two sets of parameter values of $\alpha = 0.3, 0.5; b_1 = 1.1, 1.1; b_3 = 1.8, 2.8; c_1 = 0.0025, 0.001; c_2 = 0.0015, 0.0005; c_3 = 0.005, 0.005; d_1 = 0.01, 0.01; d_2 = 0.05, 0.05$ and $k = 1000, 1000$.

If we now consider the effect of adding mutations within each group, then we can rewrite the equation above in the abstract form: $x_i' = X_i(x, p); (i = 1, ...s)$ showing the dependence of the dynamics on the phenotype parameter. In this example the number of species, s = 3.

The phenotype of a species group i is described by a vector p_i of real numbers. Each aspect of the phenotype is capable of continuous variation. So if p denotes the set of possible values that p_i can take, it can be considered to represent the ecosystem's phenotype. Within the framework used by [3] the phenotypes involved are only those affected by the set of parameters in the model. As in a real ecosystem, these parameters are constrained from within (e.g. through energy and resource limitation) and through cross-group (e.g. species interactions). So P, the set of values which p can take, is called the *phenotype constrained manifold.*

In running the model, mutant forms (e.g. y_i) are introduced into the prey equation as $(x_1 + y_1)/x_3$. The resource limiting term is replaced by exp $(-\alpha(x_1 + y_1)/\ x_3)$, which in a general form is given by $\exp(-e_{11}/e_{13})$ The dynamics can thus be written in the following way:

$$\frac{x_1'}{x_1} = (1 + b_1)\exp(-\alpha\frac{e_{11}}{e_{13}} - \frac{e_{12}}{1 + d_1 e_{11}}), \tag{4}$$

$$\frac{x_2'}{x_2} = (1 - d_2)\exp(e_{21}), \tag{5}$$

$$\frac{x_3'}{x_3} = (1 + b_3)\exp(-\frac{e_{33}}{k} - e_{31}) \tag{6}$$

where e_{11} is the population size of the prey, and $e_{13} = e_{33}$ the population size of the resource. These interaction parameters are a function of the phenotype distribution ξ_j of the jth species group and the phenotype p_i. Such a distribution $\xi_j = x_j dp_j$ records the phenotypic density of the species groups j population, i.e. the number of individuals in species group j whose phenotype lies in the volume dp_j based at p_j. For each p_i, $e_{ij}(\xi_j; p_i)$ can be either a scalar or vector quantity. In general the species groups may interact through different average properties such as abundance, density, mean strategy, average clustering, biomass, etc. Moreover, these general equations can be used to model size and age-based structures.

If $\xi = (\xi_1, .., \xi_s)$ gives the phenotype density of all species groups, by $e(\xi; p)$ we denote the matrix $(e_{ij}(\xi_j; p_i))_{i,j=1,...,s}$ and by $e_i(\xi; p)$ we denote $(e_{ij}(\xi_j; p_i))_{j=1,...,s}$. In the pure case the distribution ξ_j is $x_j \delta p_j$ where δp_j is the delta function on P_j concentrated at p_j, with only one phenotype present. The distribution of all species groups is given by the vector $(x_1 \delta p_1, ..., x_s \delta p_s)$, which can be denoted by $x\delta p$. Then the overall interaction $(e(x_j \delta p; p))$ is given by the matrix $(e_{ij}(x_j \delta p_j; p_i))$. The pure, unmutated equations in abstract form look like:

$$x_i' = X_i(x, e_i, p); (i = 1, ...s) \tag{7}$$

$$e_i = (e_{ij}(x_j \delta p_j; p_i))_{j=1,..,s}. \tag{8}$$

A small mutant population y with phenotype p' can thus be introduced as follows:

$$x_i' = X_i(x, e_i, p)(i = 1, ...s) \tag{9}$$

$$y_j' = X_j'(x, e_j', p')(j \in M) \tag{10}$$

$$e_i = (e_{ij}(x_j \delta p_j + y_j \delta p_j'; p_i))_{j=1,..,s} \tag{11}$$

$$e_i' = (e_{ij}(x_j \delta p_j + y_j \delta p_j'; p_i'))_{j=1,..,s} \tag{12}$$

where M is the set of i such that $p_i' \neq p_i$. This p' – *mutated* system is completely determined by the pure equations and the interactions.

We now suppose that this system has an attractor Λ; the advantage of the approach given above and in [3] is that it applies to all types including stationary, periodic, quasi-periodic and chaotic. The attractor of the pure system is said to be *strongly evolutionarily stable* if for all p' in P near p, Λ_0 is an attractor for the p'-mutated system. This means that for all p' in P near p a small invading mutant population y will die out and the system will relax back into its pure state. In other words the probability of very small invasions succeeding is very small and goes to zero with the size of the invasion.

We say that Λ is *evolutionarily stable* if it is evolutionarily stable to P' in P near p. This attractor is called an ESA, and the associated phenotype is the ESA value. There are interior ESAs and boundary ESAs depending on whether the phenotype constraint manifold is a smooth manifold near p or not. We characterise the evolutionary stability of an attractor Λ_p by the invasion exponent $\vartheta_p(p') = \vartheta_p(\Lambda_p, p')$. This measures the rate of growth of small invading populations with phenotype p'. A positive growth rate ϑ means that a small mutated population will be able to invade and either take over or co-exist with the original population. The principle of mutual exclusion holds in that if a population invades and withstands the dynamics of the existing population, it can take over and replace the original population. The magnitude of ϑ depends on the selective pressure and determines the speed at which the invasion initially occurs.

There are no ESAs for an unconstrained ecology. To find the ESAs when P is one-dimensional we choose a small positive number ε and look at the functions $f_+(p) = \vartheta_p(p+\varepsilon)$ and $f_-(p) = \vartheta_p(p-\varepsilon)$. These measure the selective advantage of $p+\varepsilon$ and $p-\varepsilon$ over p. If we plot f_- for the parameter values given above (see figure 1) we see in the first case (a) there is a single ESA value $p_* \approx 0.59$ and the corresponding ESA is chaotic. In the second case (b) there are two ESA values separated by an evolutionary repellor. One is a fixed point corresponding to 0.75, the other, which is chaotic, to a value of 1.1. Thus even with rather simple dynamics we should expect to see multiple ESAs. And as this example shows, evolutionary stability and chaotic dynamics are perfectly compatible.

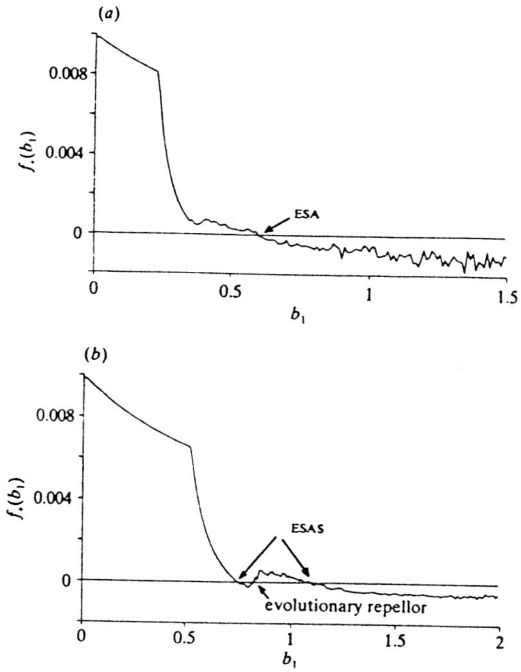

Fig. 1. The graphs of $f+$ for the two sets of parameter values (after [3]). The jagged parts of the graph are where the attractor is chaotic, caused by non-uniform convergence of the time-series for the ergodic measures of chaotic attractors.

3 Spatially Extended Continuum Ecologies

The next area I will address is that of space and its inclusion in ecosystem models. A number of ecological models represent continuum systems using reaction-diffusion equations (see [5]); others have used patch models (see [6]), and coupled map lattice models (see [7] [8]). In this section I will concentrate on artificial ecologies using coupled map lattice (CML) models. CMLs have been used extensively in other areas such as for looking at *coupled logistic maps* [9], *Hyperbolic CMLs*[10] and the *complex Ginzburg-Landau equation* [11], and are now increasingly being used in ecology.

In CMLs the variables can vary continuously on a discrete grid through space and time. Very briefly the formalism is as follows (see [9]): a lattice of points is considered, i.e.,

$$\Lambda(L) = \{k = (i, j)|1 \leq i, j \leq L\} \tag{13}$$

and a dynamical system defined as:

$$x_{n+1}(k) = F_\mu(x_n) + D\nabla^2 x_n(r). \tag{14}$$

Using CMLs a number of important results have emerged: first is the presence of *supertransients* [12]; these mean that spatiotemporal chaos can be understood as a very long transient. When supertransients are present, all periodic windows in the lattice give a spatially homogeneous solution to the CML equations. The length of the supertransient has a non-trivial dependence on parameters; when calculated as a function of a set of bifurcation parameters, the supertransient length shows a divergence at the transition of the chaotic domain. This has been termed dynamical behaviour 'at the edge of chaos', and has attracted a lot of interest because close to the transition point there is an immense opportunity for analysis of information about the system. This aspect is discussed further below in section 4 on individual-based models.

The example I wish to discuss in this section seeks to resolve the issue of why heterogeneous growth patterns occur in plant monocultures. Although simple in construction, plant growth models often imply processes of two kinds: i) neighbourhood/spatial effects and ii) the form of competition between individuals [13]. Two alternative views of how competition contributes to the generation of size hierarchies have been put forward. One is that competition among plants is usually asymmetric, i.e. larger plants are able to obtain a disproportional share of contested resource, suppressing the growth of smaller plants [13]. The other is that because plants do not grow in a uniform pattern, individuals vary in the degree of crowding they experience. So variation in growth rates, caused by differences in local neighbourhood conditions, leads to size differences [14].

In the artificial ecology constructed by [4], the basic model is a coupled map lattice, in which each cell has a continuous variable of plant mass which is updated every time step as a function of its own value and those of the cells in a specified neigbourhood. The lattice is toroidal, i.e. the boundaries are periodic. The plant mass is given as a proportion of the maximum possible plant size attainable by an uninhibited plant. A von Neumann neigbourhood is used, i.e. made up of five cells, so each cell has a fixed area equal to one fifth of the maximum area attainable by a plant. The model for plant growth is an extension of [15], where Δm_i of plant i from one time step to the next is given as:

$$\Delta m_i = (g(a_i - l_i) - bm_i^2)\Delta t \tag{15}$$

with $a_i = \mathrm{cm}_i^{2/3}$, a_i given by the '3/2 self-thinning rule', l_i the growing area lost by competitors, g the intrinsic growth rate of the plant, b and c constants and Δt the time step. The lost growing area depends on the competition: the maximum size per cell can thus be derived, by setting l_i and $\Delta m_i = 0$ and dividing over 5 sites to give:

$\alpha = (c/5)\sqrt{gc/b}$. If a plant overlaps into its four neighbouring sites equally the total overlap between two neighbours i and j is:

$$\Omega_{i,j} = \max(\frac{a_i - \alpha}{4}, 0) + \max(\frac{a_j - \alpha}{4}, 0) \tag{16}$$

and for the four-cell neighbourhood is:

$$\Omega_{i,j} = \sum_{j \in nhd} \left\{ \max(\frac{a_i - \alpha}{4}, 0) + \max(\frac{a_j - \alpha}{4}, 0) \right\}. \tag{17}$$

Four types of competition are examined: *absolute asymmetry*, i.e. the larger plant of two plants takes resources from the entire overlap area: equal-sized plants will share the area equally; *absolute symmetry*, i.e. plants divide the area equally; *relative symmetry* i.e. the overlap is weighted by the relative masses linearly for *relative symmetry* and quadratically for *relative asymmetry*.

A second model is also used to make comparisons between the CML and a circular neighbourhood model as developed by [14]. Seedlings are randomly distributed over the plot and growth is simulated using the same model as above. The area of a plant is translated into a circle of appropriate radius centred on the plant. The overlap between two interfering neighbouring plants is given as:

$$\Omega_{i,j} = p^2 \cos^{-1}(\psi_1) + q^2 \cos^{-1}(\psi_2) - qd \sin(\cos^{-1}(\psi_2)) \tag{18}$$

where

$$p = \sqrt{\frac{cm_i^{2/3}}{\pi}}, \quad q = \sqrt{\frac{cm_j^{2/3}}{\pi}}, \quad \psi_1 = \frac{p^2 + d^2 - q^2}{2pd} \quad \text{and} \quad \psi_2 = \frac{q^2 + d^2 - p^2}{2qd}$$

The two models are shown in figure 2. Parameter values for the growth equations were $g = 25\mathrm{gm}^2/\mathrm{day}; b = 0.00147\mathrm{g/day}, c = 0.00434\mathrm{m}^{-2}\mathrm{g}^{-2/3}$.

The CML model was run on grids 20 by 20, 50 by 50 and 100 by 100 cells with densities ranging from 0.1 to 1.0. the circular neighbourhood model was run on a square plot with dimensions of 0.1 m, with densities of 7,19,37,61,91,127,181,233, 291 and 355 as used by [14]. Statistical analysis of the numerical simulations included derivation of the mean mass μ over the whole lattice/plot, the coefficient of variation and the Gini coefficient (related to the Lorenz curve of cumulative frequencies).

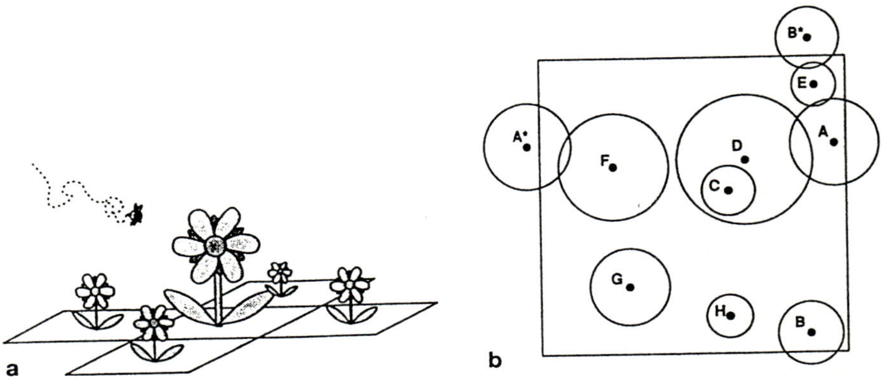

Fig. 2. a) Configuration of the lattice cell neighbourhood. b) Schematic representation of eight plants, A-H. A* and B* are 'virtual' generated to satisfy the toroidal boundary conditions.

The results show quite clearly that i) mean mass decreases as density increases; ii) growth is sigmoidal, so the early phase is exponential; variation in growth rates only occurs when competition sets in i.e. *after* the exponential stage; iii) the mean mass of plants is the same for all competition models except asymmetric competition; in this case the even spacing over the lattice is suppressed; higher variability at higher densities thus implies greater asymmetry in neighbourhood interactions. The key results are shown in figure 3, where the variability in both models increases from symmetric to asymmetric competition and figure 4 in which the variation in plant sizes can be seen.

Both models support the view that size hierarchies can be used as evidence to establish asymmetric competition as the dominant process of determining size variation in plant populations and communities over the alternative view that size hierarchies are evidence for neighbourhood effects.

The excessive computation needed in the circular model limits the extent of investigation within such as system. On the other hand the CML allows critical mechanisms to be quickly and thoroughly studied and general results extracted. I would therefore argue that the CML is a robust model with which to study such artificial ecologies.

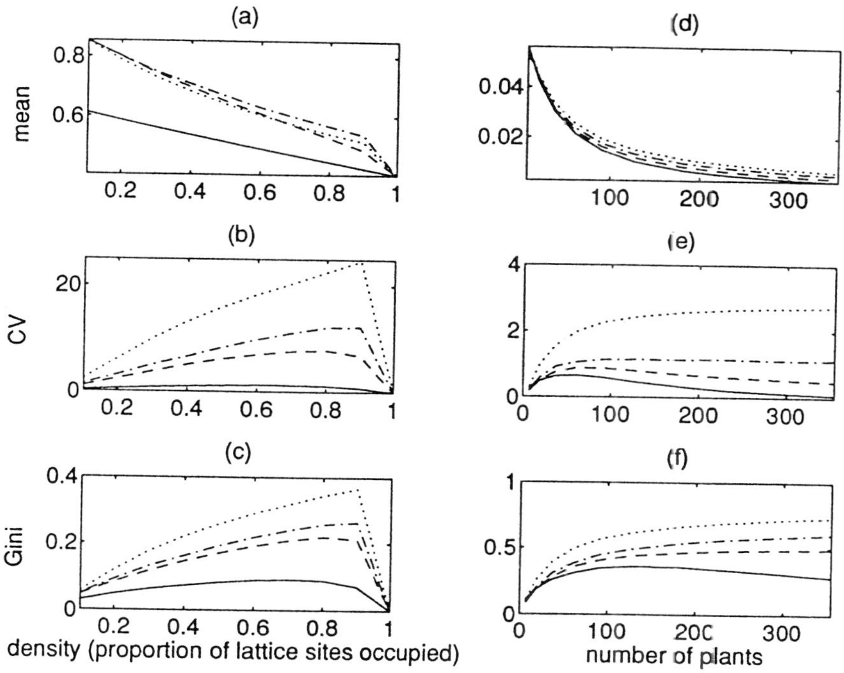

Fig. 3a–f. Results for the CML model after 200 days as a function of density. a) mean mass ; b) coefficient of variation of mass, c) Gini coefficient of mass. d-f: results from the circular neighbourhood model at time 100 days as a function of number of plants (equivalent to density). d) mean mass, e) coefficient of variation, f) Gini coefficient of mass (after[16]).

As shown earlier, the local dynamics for the *resource-prey-predator* model are chaotic. The corresponding diffusively coupled CML displays space-time intermittency, i.e. competition between two states; the first or background state acts as an absorbing state and is generally linearly stable. The active state is usually made up of localised objects which carry the disorder. So in the case of the *resource-prey-predator* model there is an initial state of zero prey with high levels of resource, followed by random nucleation of prey outbreaks and then followed by predator waves. These structures propagate across the lattice with a particular velocity [18]. In such a system, changes in the prey-predator booms and corresponding drop in resources can be used to develop a statistical theory for the lifetime and size of nucleated objects, and hence to a definition of spatial and temporal coherence lengths. These are crucial elements in any analysis of complex ecologies, because they help to determine the strategies needed to sample real ecosystems. This aspect of artificial ecologies is the focus of discussion in the next section, which looks at a different type of spatial model.

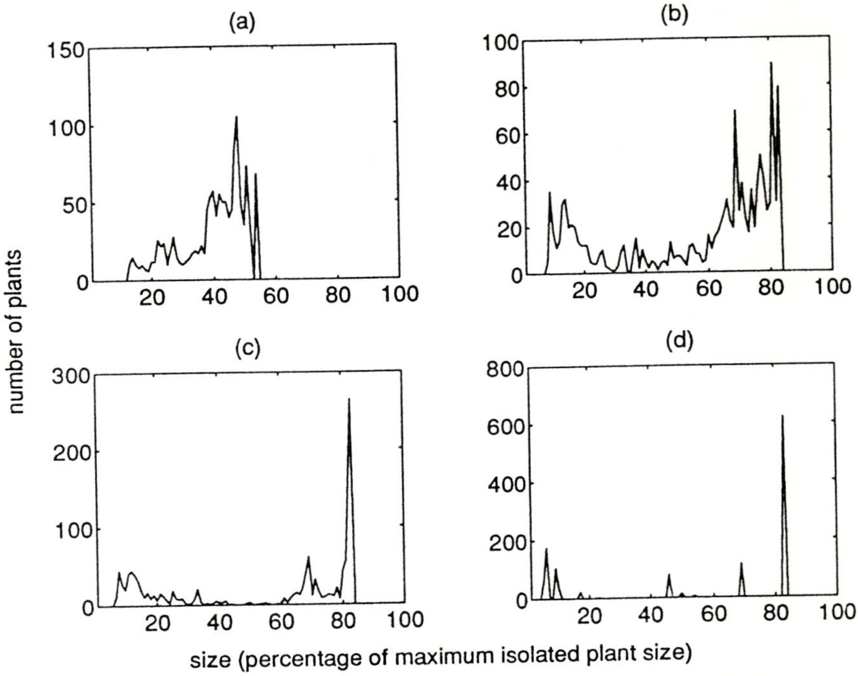

Fig. 4. Distribution of plant sizes at time 500 and density 0.5 for the CML, as a percentage of the maximum size of an isolated plant. a) absolute symmetry, b) relative symmetry, c) relative asymmetry, d) absolute asymmetry.

4 Individual-based Artificial Ecologies

In this section I will discuss two non-equilibrium individual-based artificial ecologies. The main ideas come from [16],[17],[18] and [19].

Individual-based cellular automata are now widely discussed in ecology [20];[21]. Examples of particular systems include rocky shore dynamics [22], grassland communities [23], host-parasitoid interactions [7] and predator-prey dynamics [24];[25]. However in many of these examples there is no real attempt to analyse the spatial patterns or to evaluate the relative importance of the mechanisms used to establish the set of behaviours or events. In the artificial ecologies described below I shall therefore outline some new approaches to quantifying and analysing the data that can be derived from such disordered spatial systems.

Artificial ecologies are similar to probabilistic cellular automata in that physical space is represented by a two-dimensional lattice of sites Ω. The biological communities which 'live' on the rectilinear grid are governed by simple probabilistic rules which act locally. Time, space and state variables are all treated discretely. Thus a site x can be in any one of a number of discrete states, and the configuration of these states e.g. $S = \{S_x\}_{x \in \Omega}$ defines the state of the system. This state at time t determines the future potential of the system. Behaviour of each site in the lattice of an artificial ecology is determined by its neighbourhood; this as compared to a probabilistic cellular automata (PCA) in which each site is updated independently of its neighbour. In this way artificial ecologies can be considered to be more biologically realistic.

In each artificial ecology there is a set of behaviours or events which cause a transformation of the state of the neighbourhood. These behaviours can be easily defined in terms of biological functions, e.g. death, birth/growth, predation, migration etc. as shown in the schema below:

The lattice can be updated either synchronously or asynchronously. In the former, the state of each neighbourhood determines the probability distribution on the set of behaviours: in the latter the state of each neighbourhood determines the rates at which the various behaviours occur in the neighbourhood. For the artificial ecologies used here, the lattice is updated synchronously. However, Huberman and Glance [26] have shown that in some systems different behaviours emerge from the two forms.

4.1 Forest Mosaic Cycle

The concept of the mosaic cycle was revived by Remmert [27] in his work on the middle European beech forest. Contrary to the idea in classical ecology that an ecosystem reaches a climax state or fixed equilibrium after a certain period, mosaic cycle theory asserts that an ecosystem is in a constant state of flux. The mosaic derives from patches or "stones" that cycle through a set of states.

The idea of a mosaic cycle has been used in studies of temperate forests [27–29], tropical and sub-alpine forests [30] and marine kelp forests [31]. The example described here is a summary of [16]. The model itself is based on [28] and [29], and is summarised in figure 5.

The dominant long-lived species is beech: gaps created by fallen trees are invaded by an early successional monoculture of birch which survives for approximately 50 years. A mixed forest then appears consisting of oak, cherry, ash and maple and survives for up to 150 years. Beech gradually succeeds this forest, initially as young thicket, followed then though thinning and growth of survivors to a mature stand lasting up to three centuries.

The cycle can be interrupted by losses of trees and understorey through disease, poor climatic conditions etc. and stages can be missed by invasion of later successional forms from neighbouring sites. Autosuccession (i.e. when a species replaces itself) does not occur in the case of beech because the soil nutrients have generally become too depleted for another beech tree to survive. Another important process is radiation death. Beech trees are particularly sensitive to solar radiation, and their smooth black bark splits open if ex-

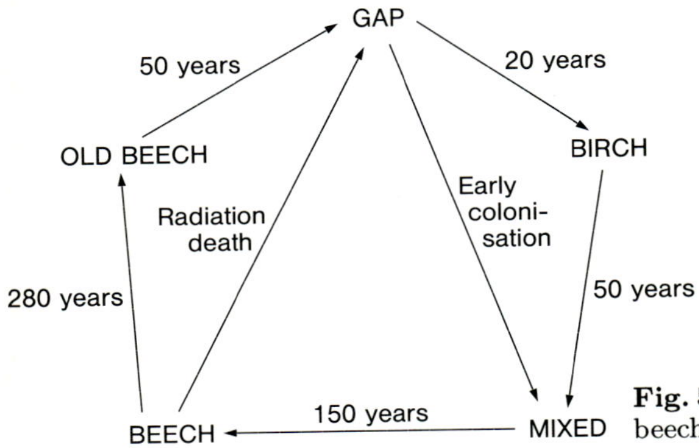

Fig. 5. Representation of the mid-European beech forest mosaic cycle.

posed to intense sunlight. Thus gaps appearing in the northern hemisphere to the south of a beech tree will generally lead to it death.

The ecology used is based on Wissel's model, and has 55 states, each corresponding to a 10 year period, giving an overall cycle of 550 years. A nine-cell Moore neighbourhood was used. At each iteration the lattice is updated synchronously and the states progress by 1, reverting to state 1 when they reach 55. Early colonisation is modelled by letting either gap state (1 or 2) move immediately to birch (3) if the neighbouring states are birch (3-8). If however one of the neighbouring states is mixed (9-22) then the gap will move to a mixed forest (9). There is an increased probability of die-back (in this case in the beech forest only) in states 51–54, given by P_0. Beech trees can also die through radiation death. To achieve this the lattice is given an orientation, and beech trees with gaps in cells to the south-west, south and south-east of their cells die with probability P_{sw}, P_s, P_{se}.

Wissel showed that the cyclic cellular automata rules together with the local neighbourhood effects were sufficient to produce a mosaic. The pattern was also insensitive to the parameter values used. He also claimed that no patches were formed in the absence of local radiation effects. However in the ecology described above [16] a second mechanism – memory – is also shown to be vital in the creation of spatial patterns.

Memory in this context means the way in which the past history of the system affects the present structure and behaviour. From the earlier example of plant monocultures, it was shown that for sessile organisms such as trees, current growth and size is strongly related to the effects induced by and on neighbours. The history of current spatial patterns also goes back well beyond the lifetime of the plants currently established because of the nutrient status of the soil, water levels, presence of fauna etc. all have an effect. For example, after a beech forest has been growing in an area for a number of centuries, the water table becomes substantially lowered and certain essential nutrients absent. This exhaustion of the soil and water is taken into account in the ecology by the absence of autosuccession in the long-lived beech stands.

If we take the longevity of different species in the system as the age of individuals in the beech stands, then the importance of memory in the system and its impact on the structure of the mosaic can be studied through analysis of the beech trees.

To do this, four models were constructed relating to *no memory/no radiation; no memory/radiation; memory/no radiation; memory/radiation*. For *no memory* the gap, birch, and mixed states are converted into single states with transition probabilities equal to the reciprocal of the expected lifetimes:

$$P_{\text{gap}} = \frac{1}{2}, P_{\text{birch}} = \frac{1}{5}, P_{\text{mixed}} = \frac{1}{15}, \text{ and}$$

$$P_{\text{beech}} = (33 - 10P_0 + 10P_0^2 - 5P_0^3 + P_0^4)^{-1} \tag{19}$$

to take account of early death in states 51 - 54.

No radiation is simply implemented by setting $P_w = P_s = P_e = 0$.

Contrary to what Wissel[29] claimed, i.e. that removal of radiation death would prevent patches forming, it is clear from our simulations that it is the memory mechanism which is fundamental (see figure 6). The top two patterns (*memory/radiation* and *memory/no radiation*) show mosaic patterns whilst the other two are randomly distributed.

A mean field approximation to the mosaic cycle can be given in the form of a Markov process on the set of configurations. Comparisons between analytical solutions of the mean field models and numerical results of the spatial ecology show that the root mean square errors increase when local radiation effects are included as expected, however what is more interesting is the amplification of these local spatial effects when memory is included. Thus memory will be a crucial component in any definition of characteristic length scales in the ecology.

Within the real beech forest a number of characteristic spatial scales are likely to occur. Wissel identified from his model and from comparisons with data from southern Europe a fundamental cell size for the lattice of $30m^2$. However, scales other than the cell size are emergent properties of the ecology. To study the dynamics of artificial ecologies it is very important to be able to determine not only the critical length scale, i.e. the minimum grid size which does not affect the dynamics, but also other scales (see [19]). For example, at micro-scales any deterministic dynamics will be dominated by noise arising from the interactions between cells; whilst at the infinite size length limit, the dynamics can become averaged out and will yield little relevant information about the ecology. However, at an intermediate scale it is possible to define a length scale at which the non-trivial determinism is maximised and where any population dynamics are likely to be deterministic, low-dimensional and chaotic. This will be the scale where each cell directly affects its neighbourhood, and where beyond this distance this influence is decoupled.

A method to determine the critical length scale is to analyse the deviations due to the spatial structure from the long-term mean dynamics [16]. The state

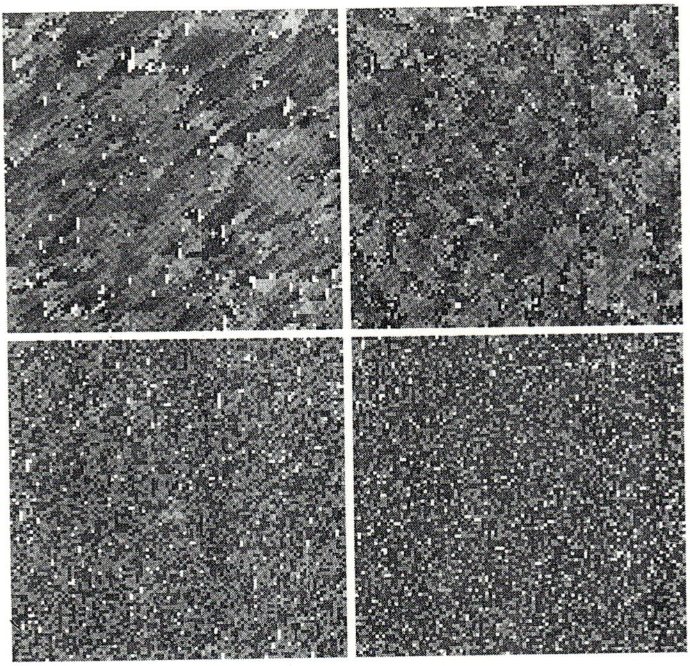

Fig. 6. Results of the ecology model at iteration 1000. Top left: memory/radiation; top right: memory/no radiation; bottom left: no memory/radiation; bottom right: no memory/no radiation. White - gap; pale grey - birch; mid-grey - mixed forest; dark grey to black - beech of increasing age.

of cell i at time t is given by x_i^t, which provides a measure on the state space represented by the lattice. The mean measure, x_μ, is found by averaging over a large $N \times N$ lattice and over a large time T (after transients have passed away). The deviation from this mean, or error, on a sub-grid $n \times n$ can therefore be evaluated for the ecology and compared to the spatial dynamics unaffected by the size of the grid. Using the central limit theorem [32] showed that the theoretical error varies inversely with the sub-grid size n. For the mosaic forest ecology the fitted $\frac{1}{n}$ curves meet the numerical 'curves' at less than 5×5 for the simple *no memory/no radiation* model and 20×20 for the *memory/radiation* model. This is considerably smaller than many discrete models, and means that it is not necessary or even desirable to run the model on a large grid, except in certain special applications where consideration of transience is important.

The intermediate length scale which will capture the maximum amount of information about the system and its dynamics relative to the quantity of data needed to describe it can be defined as having the following properties

[18]: non-overlapping windows on the grids are nearly independent, and the dynamics of any points within the window are significantly correlated.

This scale can be found using the time-series obtained by counting the total number of a certain species or successional stage inside a fixed window or grid of size $V = l \times l$. The intermediate scale l_s has the properties given above. Take a window of a certain size V where $l = \sqrt{V}$ is large compared to any correlation length. Divide this window into $m = V/V_s V_s - windows$. Then the time series for the large window is the sum of the time series for the l_s windows. If $Var(V)$ is the variance of the time series in an l-window then if V_s is large enough:

$$\frac{Var(V)}{V} = \frac{Var(V_S)}{(V_S)} + C_0 + C_1 V \tag{18}$$

where $C_0 + C_1 V$ is a small error term due to the fact that the time-series are finite. As the length of the time series T is increased $C_1 \to 0$. The crossover from small-scale behaviour to the scaling behaviour given by the above can be clearly seen in a plot of $Var(V)/V$. For the full mosaic ecology the scale l_s is about 150, but is significantly less for the *no memory* models. This implies that the memory mechanism causes the spatial coupling to extend over a larger distance, and so a larger grid is needed to see the dynamics as statistically stationary.

Characterising the patterns of aggregation is also an important aspect of understanding the dynamics of artificial ecologies. In the past biometricians have looked at distinguishing between random, aggregated and regular patterns (e.g. [33, 34]). From [16] a *clumping index* for state i, C_i is given based on the joint-count statistics of [35] and others. Using the clumping index, C_ρ for a random distribution of density ρ a *relative clumping index* C_i^R can be defined. This takes the value of 1 for random distributions and is greater or less than 1 for aggregated and regular patterns respectively.

The results showed that the spatial structures evolve towards clumped patterns. In the *memory/no radiation* case the density fluctuates, but the clumping index indicates a slow evolution throughout a long transience from a random to a clumped state. Comparison with the other models shows quite clearly that the memory mechanism is fundamental to the production of aggregated mosaic patches. The clumping index can also be used to demonstrate structural transience; for example the transience persists in the *no radiation* model for almost 1000 iterations.

Using an extension to the standard method of evaluating maximum positive Lyapunov exponents [16] for the full ecology, indicates that it is chaotic. Removal of *memory* removes the chaos; however, removal of *radiation* does not, suggesting again that memory has a fundamental effect on the dynamics of the ecology. Nevertheless, the *no memory* model has a large stochastic element and does not have a significantly higher predictability than the chaotic model. Despite the fact that the average Lyapunov exponent is simpler to evaluate and provides a measure of global predictability, it fails to detect the chaotic nature of the system.

Singular Value Decomposition (SVD) combined with embedding techniques can be used to examine the dimensionality of artificial ecologies [16]. In this case, the time-delayed matrix for the number of beech cells was used at each iteration. Results show that for the *memory* model the first four eigenvectors are fairly smooth whilst the later ones are clearly just noise, suggesting that the system is predominantly low-dimensional and can be captured by a 4-dimensional representation. The *no memory* model has a larger dimensionality of not less than 8, i.e. a simpler model does not necessarily correspond to a lower-dimensional dynamical system. These techniques can thus be used to identify and hence reconstruct data from ecologies which may appear highly complex.

Finally, the stability of the mosaic cycles can be studied by removing whole sub-regions of the lattice for different durations and over various scales. Using the various measures outlined above it can be seen that all but the smallest disturbances cause large oscillations in species distributions and considerably increase the levels of clumping. This is because the range of spatial coupling is temporarily increased. However, even large disturbances do not prevent the original dynamical equilibria from re-establishing itself. This suggests that the ecology is very resilient. The characterisation of stability in this ecology is that it is persistent, constant and resilient but not resistant to a range of perturbations.

5 Conclusion

From the above, it is possible to see that a number of approaches have been developed to analyse the spatial and temporal structures of complex ecologies. Each type of model allows different aspects of the artificial ecologies' to be looked at. Overall, however, one of the most important outcomes of these approaches and models is that by identifying that the dynamics of many of these ecologies are in fact the result of a low dimensional dynamical system, it should be possible to i) reconstruct the dynamics of unsampled species from the SVD analysis of the patterns observed i.e. if one species has a similar error structure to one that has been sampled or observed then it will be slaved to it and ii) distinguish long-term structural change i.e. change in the parameters governing the dynamics, from the natural dynamical variation of the ecosystem. The main idea for this comes from [36] and [19] who use recurrence plots to distinguish between changes in time-series. Here we want to find out if two time-series obtained from the same real or artificial ecology are significantly different, e.g. if something fundamentally has altered between obtaining samples. By counting the recurrences within one time series and comparing this with cross- recurrences that occur in both it should be possible to see if there are significant differences between the two.

Acknowledgments

I am indebted to my student Ruth Hendry, and colleagues Peter Bauer, Steve Gaito and Jake Weiner because three of the examples come from our joint work. It is also a pleasure to thank David Rand and Matthew Keeling for useful discussions. This research work is supported by the UK Natural Environmental Research Council.

References

1 P. Bak, C. Tang, K. Wiesenfeld "Self-organized criticality" Phys. Rev. A **38**, 364 (1988)
2 S. A. Kauffman "The origins of order. Self-organization and selection in evolution" (Oxford Univ. Press, New York 1993)
3 D. A. Rand, H. B. Wilson, J. M. McGlade "Dynamics and evolution: evolutionary stable attractors, invasion exponents and phenotype dynamics" Phil. Trans. R. Soc. Lond. B **343**, 261-283 (1994)
4 R. J. Hendry, J. M. McGlade, J. Weiner "A coupled map lattice model of the growth of plant monocultures" Warwick Preprint (1994)
5 J. D. Murray "Mathematical Biology" (Springer-Verlag, Berlin 1990)
6 S. Levin, T. M. Powell, J. Steele (eds.) "Patch Dynamics". Lecture Notes in Biomathematics **96**(1993)
7 M. P. Hassell, H. N. Comins, R. M. May "Spatial structure and chaos in insect population dynamics" Nature **353**, 255-258 (1991)
8 R. V. Solé, J. Bascompte, J. Valls "Noneqilibrium dynamics in lattice ecosystems: chaotic stability and dissipative structures" Chaos **2**,387-395 (1992)
9 K. Kaneko "Simulating physics with coupled map lattices" in *Formation, Dynamics and Statistics of Patterns*, eds. K. Kawasaku, M. Suzuki, A. Onuki, (World Scientific, Singapore 1990) vol. **1**
10 L.A.Buminovich, Ya. G. Sinai "Spacetime chaos in coupled map lattices" Nonlinearity **1**, 491-516 (1988)
11 T. Bohr, A. W. Pederson, M. H. Jensen, D. A. Rand "Vortex dynamics in a coupled map lattice" in *New Trends in Nonlinear Dynamics and Pattern Forming Phenomena* eds. P. Coullet, P. Huerre (Plenum, New York 1990)
12 J. P. Crutchfield, K. Kaneko "Are attractors relevant to turbulence?" Phys. Rev. Lett. **60**, 2715-2718(1988)
13 J. Weiner "Asymmetric competition in plant populations" TREE **5**,360-364 (1990)
14 G. B. Bonan "Analysis of neighbourhood competition among annual plants: implications of a plant growth model" Ecol. Modelling **65**, 123-136 (1993)
15 D. O. Aikman, A. R. Watkinson "A model for growth and self-thinning in even-aged monocultures of plants" Ann. Bot. **45**, 419-427 (1980)
16 R. Hendry, J. M. McGlade "Using probabilistic cellular automata to analyse mechanisms in a forest mosaic cycle" Warwick Preprint (1994)
17 J. M. McGlade "Alternative Ecologies" New Scientist, **137**, 14-16 (1993)
18 D. A. Rand "Measuring and characterising spatial patterns, dynamics and chaos in spatially-extended dynamical systems and ecologies" Proc. Roy. Soc. A (in press) (1994)

19 D. A. Rand, H. B .Wilson "Using spatio-temporal chaos and intermediate-scale determinism in artificial ecologies to quantify spatially-extended ecosystems" Warwick Preprint (1994)
20 R. Durrett "Crabgrass, measles and gipsy moths: an introduction to interacting particle systems" The Mathematical Intelligencer **10**, 37-47 (1988)
21 G. B. Ermentraut, L. Edelstein-Keshnet "Cellular automata approaches to biological modelling" J. theor. Biol. **160**, 97-133 (1993)
22 M. T. Burrows, S. J. Hawkins, B. J. Wilson "Patch dynamics in rocky shores: a deterministic cellular automata model" Preprint (1993)
23 K. A. Moloney, S. A. Levin, N. R. Chiarello, L. Buttel "Pattern and scale in a serpentine grassland" Theor. pop. Biol. **41**, 257-276 (1992)
24 A. M. DeRoos, E. Cauley, W. G. Wilson "Mobility versus density-limited predator-prey dynamics on different spatial scales" Proc. Roy. Soc. Lond. B **246**, 177-192 (1991)
25 W. G. Wilson, A. M. DeRoos, E. McCauley "Spatial instabilities with the diffusive Lotka-Volterra system: individual-based simulation results" Theor. pop. Biol. **43**, 91-127 (1993)
26 B. A. Huberman, N. S. Glance "Evolutionary games and computer simulations" Proc. Nat. Acad. Sci. U.S.A. **90**, 7716-7718 (1993)
27 H. Remmert "The mosaic-cycle concept of ecosystems - an overview", in *The mosaic-cycle concept of ecosystems*, ed. H. Remmert, Ecological Studies **85**, (Springer, New York 1991) pp. 1-21
28 C. Wissel "A model for the mosaic-cycle concept" in *The mosaic-cycle concept of ecosystems*, ed. H. Remmert, Ecological Studies **85**, (Springer, New York 1991) pp. 22-45
29 C. Wissel "Modelling the mosaic cycle of a middle European beech forest" Ecol. Modelling **63**, 29-43 (1992)
30 Müller-Dombois "The mosaic theory and the spatial dynamics of natural dieback and regeneration in the Pacific forest" in *The mosaic-cycle concept of ecosystems*, ed. H. Remmert, Ecological Studies **85**, (Springer, New York 1991)
31 Reise "Mosaic cycles in the marine benthos" in *The mosaic-cycle concept of ecosystems*, ed. H. Remmert, Ecological Studies **85**, (Springer, New York 1991)
32 Y. A. Rosanov "Probability theory: a concise course" (Dover, New York 1977)
33 M. Thomas "Some tests for randomness in plant populations" Biometrika **38**, 102-111 (1951)
34 J. F. Heltshe, T. A. Ritchey "Spatial pattern detection using quadrat samples" Biometrics **40**, 877-885 (1984)
35 P. A. P. Moran "The interpretation of statistical maps" Roy. Stat. Soc. J. B. **10**, 243-251 (1948)
36 J. P. Eckmann, S. O. Kamphorst, D.Ruelle "Recurrence plots of dynamical systems" Université de Geneve Preprint (1987)

A Self-Organized Critical Model for Evolution

H. Flyvbjerg, P. Bak, M. H. Jensen and K. Sneppen

Abstract

A simple mathematical model of biological macroevolution is presented. It describes an ecology of adapting, interacting species. Species evolve to maximize their individual fitness in their environment. The environment of any given species is affected by other evolving species; hence it is not constant in time. The ecology evolves to a "self-organized critical" state where periods of stasis alternate with avalanches of causally connected evolutionary changes. This characteristic intermittent behaviour of natural history, known as "punctuated equilibrium," thus finds a theoretical explanation as a self-organized critical phenomenon. In particular, large bursts of apparently simultaneous evolutionary activity require no external cause. They occur as the less frequent result of the very same dynamics that governs the more frequent small-scale evolutionary activity. Our results are compared with data from the fossil record collected by J. Sepkoski, Jr., and others.

1 Introduction

There is a good deal of evidence that biological evolution is not gradual, but episodic, with long periods of stasis interrupted by bursts of rapid activity. This intermittent pattern has been observed for the evolution of single species, as represented by their morphology. It has also been observed across taxa. In particular, Raup, Sepkoski, and Boyanian [1–4] have found a similar pattern for the distribution of extinction events, by studying fossil records. Gould and Eldredge [5] have coined the term *punctuated equilibrium* to describe the intermittent behavior of the evolution of single species; see [6] for a review and documentation of the phenomenon. We shall use the term to describe the intermittent nature of evolution in general.

Punctuated equilibrium is sometimes incorrectly presented as an explanation or a theory of the observed intermittency. Rather, it is a phenomenological principle, describing certain empirical features of the fossil record. The fundamental cause of evolutionary change is explained by Darwin's theory [7] which locates it to the natural selection operating by struggle among individual organisms for reproductive success. Darwin's theory may thus be thought of as the "atomic theory" for evolution. However, there is no theory deriving the consequences of Darwin's principles for macro-evolution. This is the challenge we are responding to here.

By studying the stratigraphic records of 19.897 fossil genera, Raup, Sepkoski and Boyanian found that not only do the extinction events occur in bursts within families, but different genera often show the same extinction profile. It thus appears that the evolution of different families "march to the same drummer." Some extinction events are regional [8], and the largest events are global. It has therefore been suggested that extinction events are caused by external forces, such as changing sea levels [9], worldwide climatic pulses [10], or meteorites [11]. Although this has perhaps caused some extinction, we shall nevertheless argue that punctuated equilibrium may well be the natural consequence of the dynamics of biology itself, with no need for external triggering mechanisms.

Indeed, large dynamical systems have a tendency to evolve, or *self-organize,* into a "critical" non-equilibrium state characterized by bursts, or *avalanches,* of dynamical activity of all sizes [12]. This behavior is known as self-organized criticality (SOC) and below we discuss how such behaviour may appear in an ecology driven by Darwinian evolution.

The present chapter is not the first one theorizing that the intermittency of biological evolution might be caused by self-organized critical behaviour. But theoretical investigations have been hampered by the difficulty of constructing even remotely realistic, yet tractable mathematical models. First, punctuated equilibria were observed by Bak, Chen, and Creutz [13] in the "Game of Life," a simple computer caricature of a society of living and dying individuals living on a two dimensional lattice. However it is not robust against small changes in rules, as it should be in order to represent real evolution. Later, Kauffman and Johnsen [14] in a very imaginative work studied the so-called "NKC-models" for co-evolving species evolving with periods of stasis interrupted by co-evolutionary avalanches. It was argued that the ecology as a whole was "most fit" at the critical point. However, as these models where driven they do not self-organize: some external tuning of the system, "divine intervention," was needed to obtain critical behaviour [15,16]. Finally, intermittent bursts indicating evolution to a critical state have been observed in computer simulations of reproductive organisms, first by Ray [17], and very recently by Adami [18].

This chapter is organized in the following way. In Section 2 we present our model. In Sections 3 and 4 we discuss analytically tractable versions of it. Readers less keen of mathematical details should skip these sections. In section 5 we discuss a version that is easily simulated. In section 6 we discuss pertinent data from the fossil record in the light of our results. In section 7 we discuss the characteristics of life in an ecology that has reached the self-organized critical state predicted by all version of the model. Section 8 contains the discussion and conclusions.

2 Modelling

Our investigation starts at the level of species. We consider the microevolution acting up to this level as decoupled from the macroevolution that we wish to understand. This decoupling, of course, is not a claim of falseness or irrelevance of microscopic mechanisms. It is an assumption that divides the problem into more manageable parts.

The basic picture that we have in mind is the evolutionary fitness landscape envisioned by Wright in his seminal work, the shifting balance theory of evolution, reviewed in [20]. The properties of a population are modified by means of mutation and differential selection towards higher fitness. Random mutations allow individuals to cross barriers of lower fitness and move to other maxima, and initiate a population at or near this new maximum.

For simplicity, we define a species as a group of individuals in the vicinity of the same fitness peak. The basic evolutionary step in our theory is the transformation of one species to a similar, more fit, species. We call this step a "mutation" of the species, following Gould [6]. The detailed mechanisms making this step possible is not of our concern: we refer to the work of others for its motivation. For instance, the diffusion of a species from one state to another has been described by Lande [21], and by Newman et al. [22]. The mechanism is mutation and differential selection of the fitter variant, causing the whole population to evolve to this variant. Figure 1 shows how this step may take place in a laboratory experiment. The figure shows the mean scutellar bristle number in female Drosophila Melanogaster according to MacBean et al. [23], as simplified by Parsons [24]. The number jumps to a "fitter" value in response to selective pressure.

In principle, the "fitness" landscape for a given species can be expressed as a function of its genetic code [25,14], and also of the genes of certain other species that it depends on. However, we ascend another step in the level of abstraction: we assume that the fitness landscape is sufficiently rough that escape of a species from one local fitness maximum to another results in the replacement of one effective barrier towards further evolution with another effective barrier. We further assume that these subsequent barriers are uncorrelated. With this we have done away with many details of any specific model and obtained generality and mathematical simplicity in return. Our choice between what to leave out, and what to leave in, in the simplified model is an expression of what we consider quintessential mechanisms of evolution, mechanisms so fundamental that they should also characterize life on Mars, if it existed, and Ray's computer life in the Tierra environment [17].

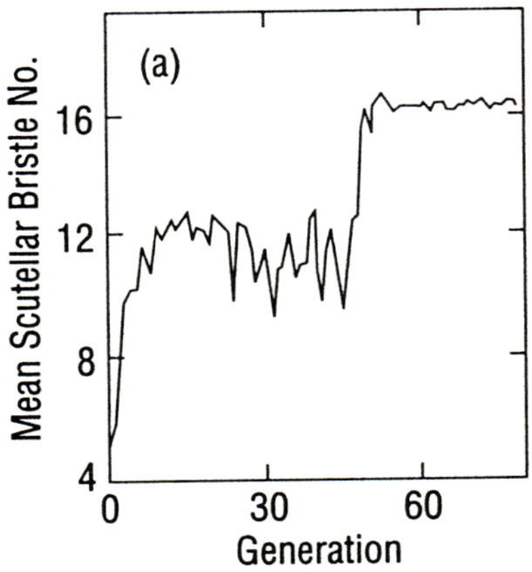

Fig. 1. "Punctuation" in mean scutellar bristle number in female Drosophila Melanogaster as observed in the laboratory by MacBean et al. [23].

The probability of jumping from one state to a better one is $p = e^{-B/T}$, where B is a random number expressing the barrier, i.e. number of single base pair mutations, separating the two states, and T is an effective mutation parameter defining the timescale of mutations. Notice that although the typical time between jumps from one state to another is large, i.e. $1/p$ small, the jump itself is very fast [21]. A species with a high fitness is unlikely to evolve to even higher fitnesses, so its barrier B is high, whereas a species with low fitness have an easy time doing this. Thus, the barrier of stability B can be thought of as a measure of fitness.

All of the above describes the evolution of a single species in a given landscape. However, as pointed out by Kauffman and Johnsen [14], the fitness landscape experienced by one species depends on other species in the ecology: it is a "rubber" landscape, changing with the physical properties of other species and therefore with their genes. The interacting species can for instance be consecutive links in a food chain. As the fitness of one species improves, the fitness of its neighbors is affected, typically making some of them likely to evolve. Thus species co-evolve.

We model this by assuming that the fitnesses of these neighbors assume new values, either directly as a consequence of the different environment, or because they quickly move to a new local fitness maximum, with a new, by assumption random, barrier value B towards further evolution. But in the course of doing so they may have induced yet other species to a fast evolution to new local fitness maxima with new random barrier values towards further evolution. We assume that the interactions between species are sufficiently weak or dilute to allow this primary chain reaction of co-evolution to die out fast. Thus, in such a brief chain reaction only a finite number of species, say K on the average, are affected and we choose a time-scale in our model in which the chain reaction is represented by a single step.

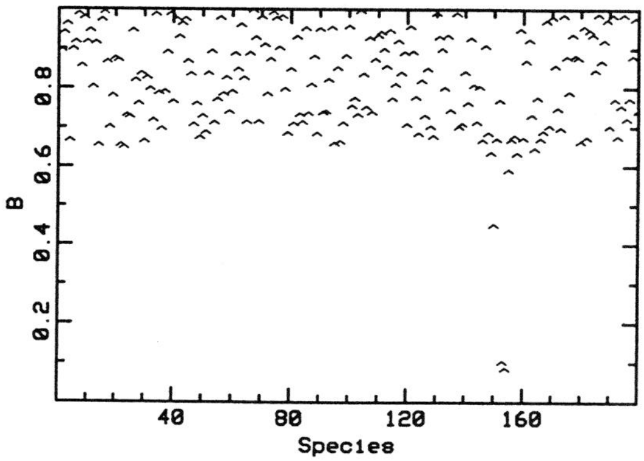

Fig. 2. Illustration of the "food-chain" version of our model. A snapshot of the barrier values for the different species are shown. Most barriers are above the critical value $B_c = 0.667$. In the next step, the species with the lowest barrier, here number $i = 790$, will mutate.

For the sole purpose of simplifying the bookkeeping of the network of interactions among species, let us assume first that the network forms a simple chain. We may then think of species as placed on a straight line, each species interacting with its two nearest neighbors on the line (Figure 2). Initially we choose random values B between 0 and 1 for the barriers towards further evolution experienced by the species. The choice of initial condition does not affect the outcome of evolution in this model. In the limit where the mutation parameter T is low, the first species to mutate is always the one with the lowest barrier. I.e. the first species to succumb is the least fit, as envisioned by Darwin.

Thus, the dynamics is as follows [26]: The species i with the lowest barrier is assigned a new random value of B between 0 and 1. Its two neighbors are also to be assigned new random numbers. Therefore, even if any of these neighbors had a high fitness, this property is likely to be lost. In the next step, the species which now has the lowest fitness mutates, affecting the fitness of *its* neighbors, and so on.

The actual rate of evolution varies enormously with the value of the minimum barrier representing the next species likely to mutate spontaneously. As the system evolves it turns out that no species with barrier values above a certain self-organized threshold, B_c (equal to 0.667 in the case considered here), will ever mutate spontaneously: They evolve only when their environment has changed sufficiently to lower their barriers. No high barriers towards evolution are ever transgressed. No highly improbable events are required in evolution. Instead, the improbable is made probable by changes in the environment; the "rubber" landscape of a species co-evolving with other species is modified by them to allow successive small, but rapid, evolutionary steps.

3 Random Neighbor and Mean Field Model

Here, for mathematical convenience, we select the $K-1$ interacting species at random among the N species in the ecology. This *Random Neighbor Model* is a first step towards a solvable mean field theory [27]. We also assume this randomness to be "annealed", i.e. the next time the same species triggers $K-1$ other species to evolve, they are chosen at random anew. A mean field theory can be constructed by neglecting correlations between barrier values. Then the ith smallest barrier value, call it x_i, is distributed as the ith smallest number out of N drawn from the distribution $p(x,t)$ of *all* barrier values in the ecology. If we let p_i denote the distribution for x_i then our *mean field approximation* is the assumption that

$$p_i(x) = \frac{N}{(i-1)[(N-i)]} P^{i-1}(x) p(x) Q^{N-i}(x), \tag{1}$$

where we have introduced

$$P(x) = \int_0^x dx' p(x') \tag{2}$$

$$Q(x) = \int_x^1 dx' p(x') \tag{3}$$

Normalization of p gives

$$\int_0^1 dx' p(x') = P(x) + Q(x) = 1 \ \ \forall x. \tag{4}$$

We can easily write down the evolution equation for $p(x,t)$,

$$\begin{aligned} p(x,t+1) = p(x,t) - \frac{1}{N} p_1(x,t) \\ - \frac{K-1}{N-1}\left(p(x,t) - \frac{1}{N}p_1(x,t)\right) + \frac{K}{N}, \end{aligned} \tag{5}$$

where Eq. (1) gives the distribution for the smallest barrier,

$$p_1(x) = N p(x) Q^{N-1}(x), \tag{6}$$

whose removal from the set of N barrier values is represented by the second term on the right-hand side of Eq. (5). The third term on the right-hand side of Eq. (5) represents the removal of $K-1$ of the $N-1$ barrier values remaining after the smallest has been removed from the set of N values. These $K-1$ values can be any of the $N-1$ values remaining, hence are distributed as these, i.e. as $(Np(x,t)-p_1(x,t))/(N-1)$. The last term on the right-hand side of Eq. (5) represents the addition of K new equi-distributed barrier values, replacing the K values that were removed with the preceeding terms. Notice that probability is conserved by Eq. (5).

Our mean field dynamics is an approximation to the master equation for the Markov process of the random neighbor model, both having one unique attractive fixed point. At this fixed point Eq. (5) is an integral equation fo $p(x)$, or, equivalently, an ordinary differential equation for $Q(x)$. It is solved by the positive root $Q(x)$ of the polynomial equation

$$\begin{aligned}(N-K)Q^N(x) + N(K-1)Q(x)\\ + (N-1)K(x-1) = 0.\end{aligned} \tag{7}$$

In the limit where $N \gg K > 1$, the first term in this equation is small relatively to the second term for such values of x where $Q(x)$ is less than 1 by more than $\mathcal{O}(1/N)$. Consequently we have

$$\begin{aligned}Q(x) &= \frac{(N-1)K}{N(K-1)}(1-x) - \frac{N-K}{N(K-1)}Q^N(x)\\ &\simeq \frac{K}{K-1}(1-x) \text{ for } x - 1/K \gg \mathcal{O}(1/N).\end{aligned} \tag{8}$$

Conversely, where $Q(x) \simeq 1$ we have

$$\begin{aligned}Q(x) &= \left(\frac{(N-1)K}{N-K}(1-x) - \frac{N(K-1)}{N-K}Q(x)\right)^{1/N}\\ &\simeq (1-Kx)^{1/N} \text{ for } 1/K - x \gg \mathcal{O}(1/N).\end{aligned} \tag{9}$$

Using $p(x) = -\frac{d}{dx}Q(x)$, we have

$$p(x) \simeq \frac{K}{N} \text{ for } 1/K - x \gg \mathcal{O}(1/N) \tag{10}$$

$$p(x) \simeq \frac{K}{K-1} \text{ for } x - 1/K \gg \mathcal{O}(1/N). \tag{11}$$

The exact solution of Eq. (7) is easily obtained numerically by iteration of Eqs. (8) and (9) for $x > 1/K$ and $x < 1/K$, respectively. It is shown in Figure 3a together with the resulting distribution of the smallest barrier, $p_1(x)$, both as dashed lines. The random neighbor model is easily simulated and its equilibrium distributions $p(x)$ and $p_1(x)$ are shown in Fig. 3a as full lines.

In the limit $N \to \infty$ we see that $p(x)$ has a discontinuity at $x = 1/K$; it vanishes below this threshold and is constant above it. It is easy to understand this result in approximate terms: Suppose $p(x) \simeq K/N$ for $0 \leq x \leq 1/K$ and $p(x) \simeq K/(K-1)$ for $0 \leq 1/K \leq x$. Then the smallest of N barrier values distributed according to p will be equi-distributed below the threshold value $1/K$, and the other $N-1$ will be larger than $1/K$, typically. Thus, when the smallest barrier value is removed none is left below threshold. Consequently, the $K-1$ additional barrier values which are randomly selected and removed, must be taken from above the threshold, hence are equi-distributed

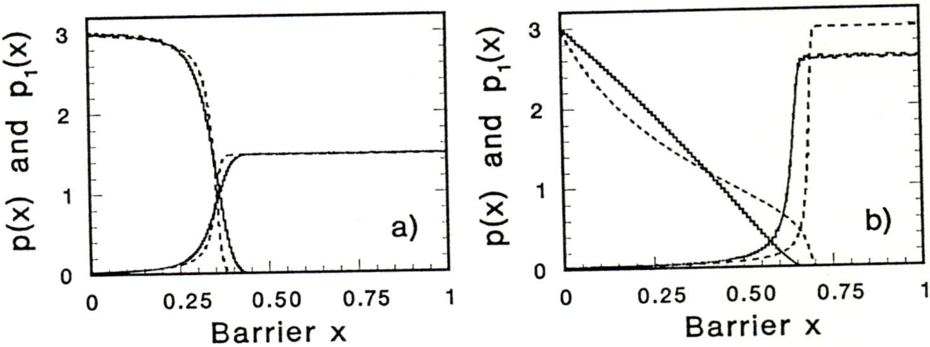

Fig. 3. (a) Equilibrium distribution of barrier values $p(x)$ and distribution of smallest barrier value $p_1(x)$ for simulated random neighbor model (full curves) and corresponding mean field theory (dashed curves). (b) Same distribution for 1d model (full curves) and mean field theory removing *two* lowest barrier values (dashed curves). All cases have $K = 3$ and $N = 100$.

by assumption about p. When we replace these K barrier values with K equi-distributed values, one of these typically falls below the threshold and the other $K - 1$ above. Since all of them are equi-distributed, p is left unchanged, as it should be.

This explanation points to another aspect of the asymptotic dynamics: If we trace in time which species trigger the bursts of evolutionary activity, then it is usually one of the species participating in recent activity. So at any given *late* time, the very species which acquired their current properties most recently are also the ones most apt to change them again. Thus, according to our model, the cockroach, which is much older than the human race, will resemble itself long after humans, as we know them, have disappeared.

4 Avalanches

In order to express the causal connections between bursts of evolutionary activity, we define an *avalanche* as a causally connected sequence of activity associated with barrier values below the self-organized treshold $1/K$. Suppose that at some time all barrier values are above the threshold value. Then the next burst will, on the average, result in one barrier value below threshold, which for its part will result in another barrier value below threshold, etc. The number of barriers below threshold remains constant equal to one, on the average. The actual number of barriers below threshold fluctuates and may become zero again, terminating the avalanche.

A more realistic value for the average number of barrier values below threshold can be obtained from our mean field approximation. It gives

$$NP(1/K) = \ln N - \ln\ln N - \ln(K-1) + \mathcal{O}(\ln\ln N/\ln N) + \mathcal{O}(1/\ln N) + \mathcal{O}(\ln N/N) \tag{12}$$

where $P(1/K) = 1 - Q(1/K)$, and $Q(1/K)$ is the solution to Eq. (7) with $x = 1/K$. With an average of $NP(1/K)$ barrier values below threshold, the fluctuation in this number needed to terminate an avalanche becomes increasingly rare with increasing N. Thus the *sizes* of avalanches, defined as the number of bursts they contain, grow with N, to diverge as $N \to \infty$.

In the limit $N \to \infty$, an avalanche defined as above can be identified with a critical branching processes with branching ratio K [29]. This is done by identifying each burst with a node, and each of K new barrier values resulting from a burst with either a branch rooted in that node (if the barrier value is less than the threshold value), or with a leaf rooted in the same node (if the barrier value is above threshold.) The limit $N \to \infty$ is necessary to obtain the tree structure. This identification tells us that avalanches come in all sizes s, and the larger ones are distributed according to a power-law with mean field exponent,

$$D(s) \propto s^{-3/2}, \tag{13}$$

showing that there is no average size to avalanches. The avalanches are critical, because the branching process is. Since the *medium* through which these avalanches propagate—the set of N barrier values—is transformed by the avalanches and driven by them to the unique asymptotic fixed point distribution that makes the avalanches critical, our model for biological evolution is a self-organized critical dynamical system.

By analytical means one can obtain a number of exact results for the Random Neighbor Model which was described here in the mean field approximation. The mean field description is sufficient for our purposes here, i.e. for demonstrating qualitative features shared by the Random Neighbor Model and finite dimensional models. Readers interested in exact results for the Random Neighbor Model are referred to [30].

5 One-dimensional Model

So far, we have seen criticality only in the mean field approximation. Now let us study a finite dimensional case. We have simulated the dynamics of the one-dimensional ecology and measured a number of its properties in the equilibrium state. Figure 3b shows the distribution of barriers, $p(x)$, and the distribution for the lowes barrier value, $p_1(x)$, as full curves. Both are for $K = 3$, N=100. They do not resemble the random neighbor and mean field results for $K = 3$, N=100, shown in Fig. 1a. The dashed lines in Fig. 1b show results from a different mean field model, obtained also with $K = 3$ and $N = 100$, but by replacing the *two* smallest barrier values plus *one* randomly selected value with random numbers in each time step. It is easy to understand why this latter algorithm gives results much closer to the 1d results: low barrier values are clustered in one dimension, so the replacement of the smallest barrier value together with the values on its nearest neighbor sites amounts to replacing the lowest value plus 0–2 other low values. Actually, some of the difference between the mean field and 1d results shown in Fig.1b is due to finite-N effects being more pronounced in the 1d results. For example the value of $p(x)$ for $x > 0.7$. It will approach 3 as $N \to \infty$, while the mean field value for $p(x)$ is already very close to 3 for $x > 0.7$.

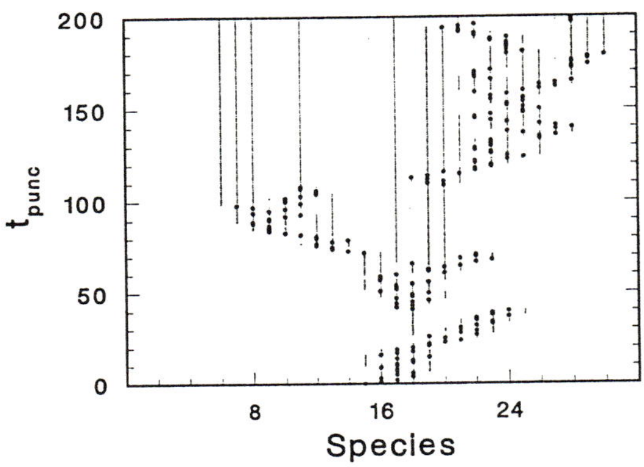

Fig. 4. Space-time map of an avalanche in the self-organized critical state. At any time the site with minimum barrier value is shown as a large dot. Sites with barrier values below the threshold value 0.67 are shown with a small dot. The activity is seen to always return to sites below threshold.

Figure 4 shows a space-time map of those sites on which species change barrier values in the time interval covered. Whenever the lowest barrier value is found among those $K = 3$ last renewed, the site of lowest barrier value performs a random walk, because those 3 sites have equal probability for being the one with smallest barrier value. The figure shows that this is what happens most frequently. When the site of lowest barrier value moves by more than one lattice spacing (jumps), it most frequently backtracks by two lattice spacings to a site that was updated in the next-to-last time step. But longer jumps occur, too; actually jumps of any length occur, as indeed they must in order to be consistent with results below. These jumps always take the walker back to a site that was updated recently, the longer jumps typically to a less recently updated site. In popular terms, the site of lowest barrier value is a "jumpy random walker with a repetitious compulsion".

Figure 5 shows three aspects of this repetitious random walk which differ from a truely random walk: The root mean square of the distance traveled vs time and the number of different lattice sites visited as a function of time both

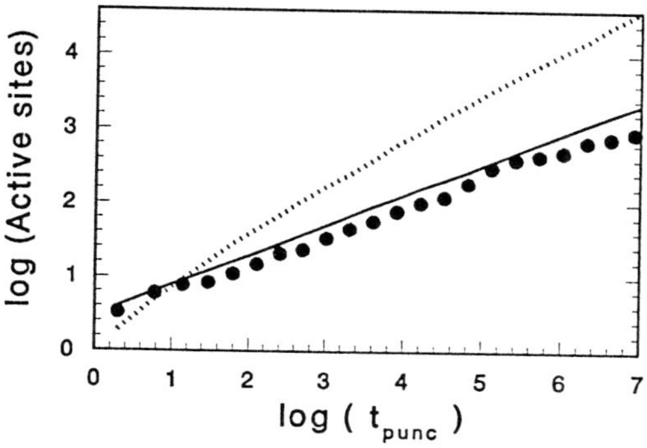

Fig. 5. Large dots: Displacement of the activity Large dots: Displacement of the activity as a function of time, starting at an arbitrary time in the critical state. This quantity grows as $t^{0.43}$ [28].

grow with exponent 0.40, in contrast to the exponent 1/2 obtained for the random walker. The maximum number of updates of any state as a function of time appears to grow as $t^{0.6}$. This exponent may have a simple explanation in terms of the others: the total number of updates is proportional to time, and the number of different sites visited to $t^{0.4}$. So the number of visits to a given site (in particular to the one most often visited) should grow with exponent $1 - 0.4 = 0.6$. For the random walker this relationship reads $1 - 1/2 = 1/2$ for the number of visits to any site, for instance the origin of the walk.

The biological implication of this correlated spreading of evolutionary activity is that species that evolved recently are also most likely to change again. cf. humans vs cockroaches above. The actual values of the exponents, here 0.4 and 0.6, depend on the dimension, here chosen equal to one. If our model has an upper critical dimension above which mean field theory is exact, and this dimension is a small integer, the mean field version of our model is probably the most relevant one to use in an analysis of historical biological data.

6 The Theory vs. the Fossil Record

Figure 6a shows a time series for the relative number of species becoming extinct in consecutive intervals of approximately 5 million years, as presented by J. Sepkoski, Jr. [4]. Note the intermittent behavior with a few large peaks representing mass extinction events and many small peaks and valleys representing periods with smaller relative numbers of extinctions. Figure 6b shows similar results from a numerical simulation of our model.

Figure 7a shows a histogram of essentially the same data as those presented by Raup [2]. The histogram shows the number of genera becoming extinct in each of 106 time intervals covering approximately 5 million years

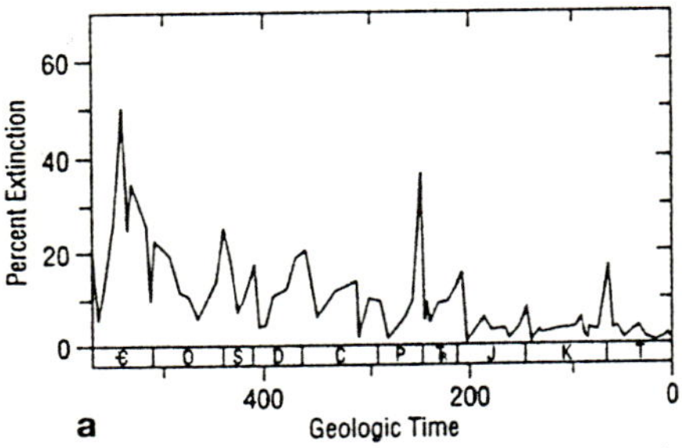

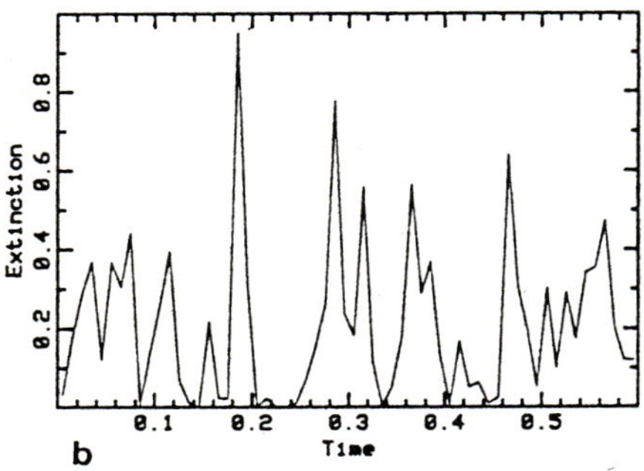

Fig. 6. Temporal evolution of extinctions recorded over the last 600 million years, as given by J. Sepkoski (4). The ordinate shows estimates of the percentage of species that went extinct within intervals of 5 million years. (b) Temporal evolution of the "mutation" activity of species recorded in a 1-dimensional model ecology with 200 "species" and a mutation rate parameter $T = 0.01$.

each. The distribution is highly skewed, with 52 of the intervals having less than 10 percent extinction, and a few large extinction events with up to 60 percent extinction. The histogram varies fairly smoothly between the periods with small activity and those with large events. This smoothness suggests that a common mechanism is responsible for all events.

Figure 7b shows a similar histogram for a subset of families, including 2316 marine animal families. Again, a smooth variation is observed. Of course, because of the small number of periods there are very large statistical variations, in particular for the few large events. Actually, these large events are so few that they each have names.

Figures 7c shows further results from numerical simulations of our model. The quantity measured in these simulations is the number of "mutations" of species occurring, rather than extinction events. The time series shows the characteristic behavior with long periods of little activity, interrupted by peaks of large evolutionary activity. The histogram shows the smooth transition from the many periods with little activity to the few large catastrophic events. Mathematically, the histogram of events of size s follows a power law, $N(s) = s^{-\tau}$, with the exponent τ ranging from 1.1 to 1.5 depending on a certain feature of the simulated, theoretical ecology (its so called "imbedding dimension"). The histogram shown is for randomly connected species (infinite imbedding dimension). Our theory confines the value of the activity exponent

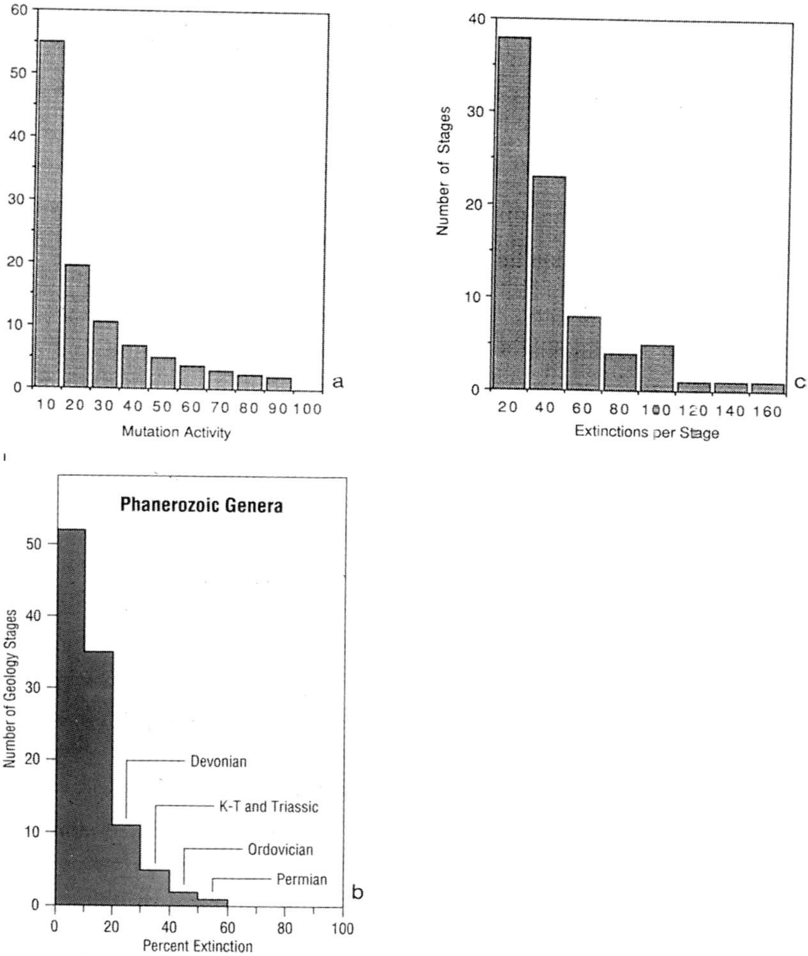

Fig. 7. (a) Histogram of extinction events from Fig. 6a, as shown by Raup. The extinctions are binned in 106 intervals of approximately 5 million years each. (b) Distribution of extinction events based on recorded time of extinction of 2316 marine animal families. (c) Histogram of mutation activity as predicted by the random neighbour version of our model.

τ to this very narrow range, whereas actual data for extinction events appear slightly steeper than an ecology of randomly connected species allows for. But the data is too scanty to allow for a real quantitative test of the theory.

Figure 8a shows the time series for a different type of data, namely the variation of the morphology of a single species. The figure shows the increase in thoracic width of the Antarctic radiolarian *Pseudocubus vema* during 2.5 million years according to Kellogg [19]. The figure shows punctuated equilib-

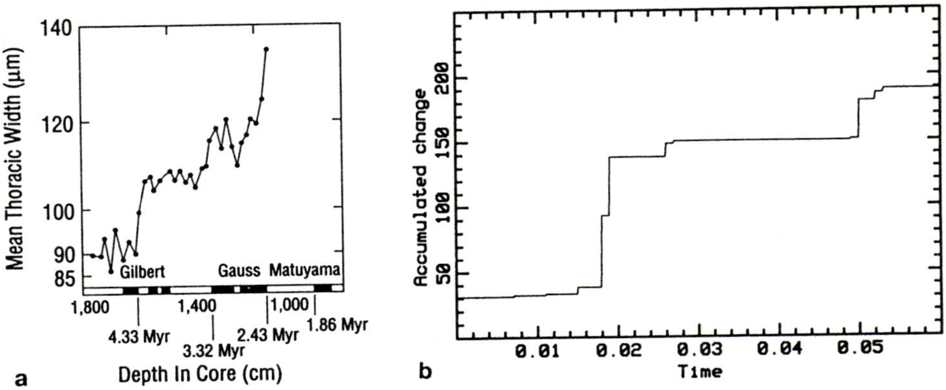

Fig. 8. (a) Time series for the variation of the morphology of a single species. The figure shows the increase in thoractic width of the Antarctic radiolarian Pseudocubus vema during 2.5 Myr according to Kellogg [16]. (b) Model prediction for time series for change of single species morphology, estimated as its accumulated mutational activity.

ria with a series of three plateaux, i.e. periods of stasis, separated by periods of rapid change. Figure 8b shows similar behavior from our simulations. Again, the distribution of jumps can be shown as a histogram, with power law distribution of bursts.

In our theory, the evolution of single species is coordinated. Figure 9 shows the time series of global extinctions together with the evolution of a single species. The bursts of rapid variation take place during periods of large biological activity.

The most interesting feature of evolution is, perhaps, the existence of periods of stasis. During such periods species in an ecology seem to be in balance. Figure 10a shows a histogram of the lifetimes of species based on data on 17505 genera tabulated by Sepkoski, as presented by Raup [1]. The distribution varies smoothly from very many species with short life span, to few species with long life span, up to several hundred million years. The number of species N, with a life-time t can be fitted quite well to a power law, $N(t) \propto 1/t^{\alpha}$ with $\alpha \approx 2$. Figures 10bc shows the results of our simulation from the one-dimensional version of the model. Our measure of lifetime is the interval between successful mutations of a given species. A mutation event can be thought of as an extinction event followed by the replacement by another species. The theoretical distribution of lifetimes from our model is also a power law, however with an exponent $\alpha \approx 1$ for small times, and much steeper when measured over times that are comparable to the times between the largest events.

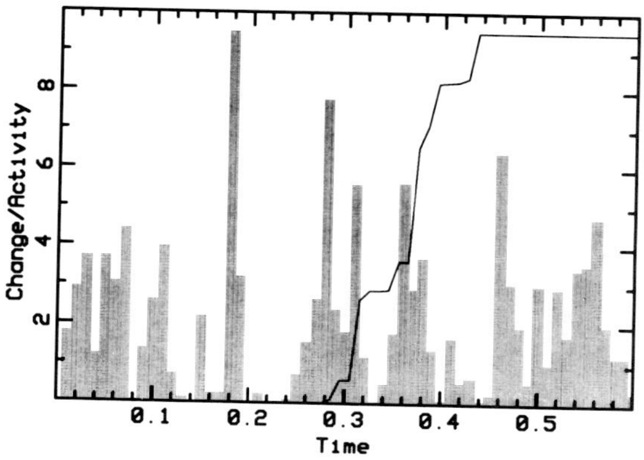

Fig. 9. Model predicted time series of global extinctions together with the evolution of a single, randomly chosen, species. The bursts of rapid variation takes place during periods of large biological activity.

7 Life in the Self-Organized Critical State

As evolution proceeds, the ecology organizes itself into a state where further evolution takes place as avalanches of hectic co-evolutionary activity, or "punctuations" in the language of Gould and Eldredge. An avalanches may be defined rigorously as the total activity during a period where at least one species has a barrier below the critical threshold B_c. During an avalanche, there are several species with relatively low fitness. Between avalanches, in periods of stasis, the fitnesses of all species are above a threshold, so nature appears to be in balance.

The magnitude, or size, of an avalanche, or punctuation, is defined as the total number of successful evolutionary moves constituting the avalanche. If one plots a histogram of the number of avalanches of a given size s, one finds that the size distribution is a power law, $N(s) = s^{-1.1}$ [28]. This power law distribution indicates that the system is in a *critical state,* with a terminology borrowed from theoretical physics. Avalanches of all sizes occur, including large catastrophic one. This is in contrast to normal, or Gaussian distributions which have exponentially small tails that effectively prohibit large events.

So the catastrophic events occur due to a collective behavior of the interacting species, each of which is trying to gain maximum fitness in its landscapethe replacement by another species. The model says nothing about the actual nature of the interaction leading to extinction of a specific specie. It could be the result of falling prey to other species, starvation from change in habitat or competition from other species, epidemics, in short, the usual fates of the weak.

The important point is that in this model there is no need for external causes, such as climatic changes, to explain the observed pattern of extinction

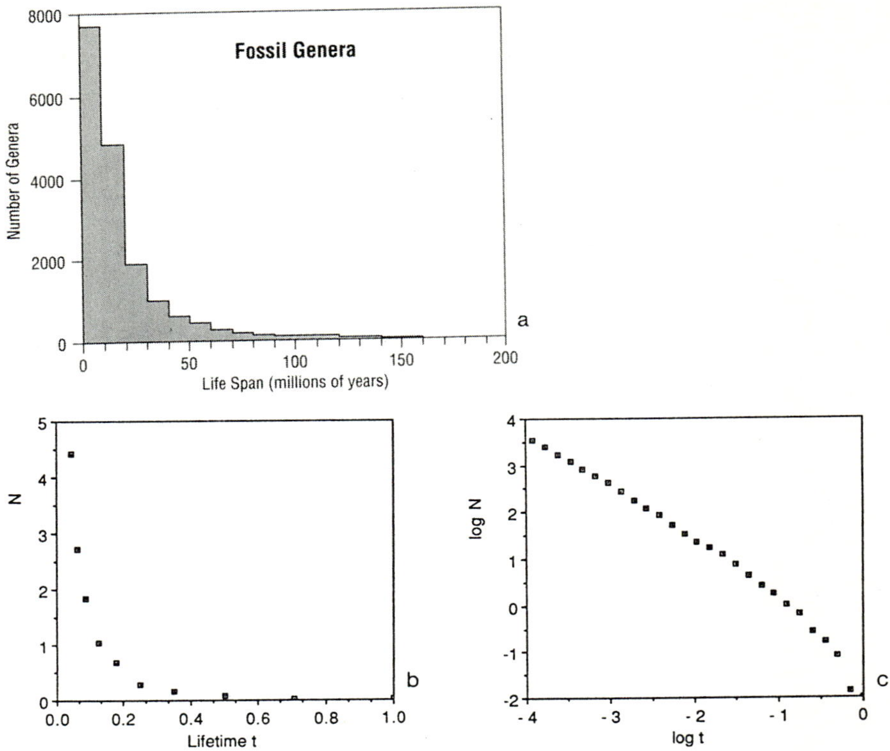

Fig. 10. (a) Life time distribution for species as recorded by Sepkoski et al. [1–4]. The distribution can be well fitted by a power law $N(t) \propto 1/t^2$. (b) Distribution of life times for the 1-d model for a mutation parameter $T = 0.001$. Time is measured in units of $\exp(B_c/T)$. (c) Log-log plot of the same distribution. For small times the distribution is $\propto 1/t$.

events, even when these cut across functional, physiological and ecological lines. In particular, no external cataclysmic impact is necessary in order to generate large events in evolution. The late Permian extinction event with an estimate of up to 96% species going extinct [1], or the late Cretaceous event opening for the early Tertiary evolutionary radiation of mammals, may thus have been endogenous events in evolution.

Returning to our artificial lattice arrangement of the network of interactions between species, the exponent of the power law depends on the dimension of the lattice [26–28]. On a two dimensional lattice the exponent is 1.27..., in higher dimensions it is yet higher, but the exponent does not exceed 3/2, actually has that value for all lattice dimensions larger than 4. If the network of interactions is chosen to be random—a choice that presumably resembles reality more than any lattice network— the exponent can be calculated analytically to be exactly 3/2 [27], and a number of other properties can be derived analytically as well [29]. Figure 7c shows a histogram of the size of avalanches as calculated here.

Since the minimum barrier value fluctuates much, the actual time scale represented by a single mutation in our computer simulation varies enormously. In order to represent the evolution on a real time scale, simulations have been performed at a low value of the mutation parameter, $T = 0.01$. At each time step, a given species with barrier B_i mutates with the probability $p = \exp(-B_i/T)$. When a species mutates, we assign both it and its two neighbours new random barrier values B between 0 and 1. Figure 11 shows a space-time plot of the activity. The horizontal axes is the "species axis," and the time where a given species mutates is shown as a black dot. The avalanches are shown as connected black areas. On the time resolution of the plot, the avalanches appear as almost horizontal lines.

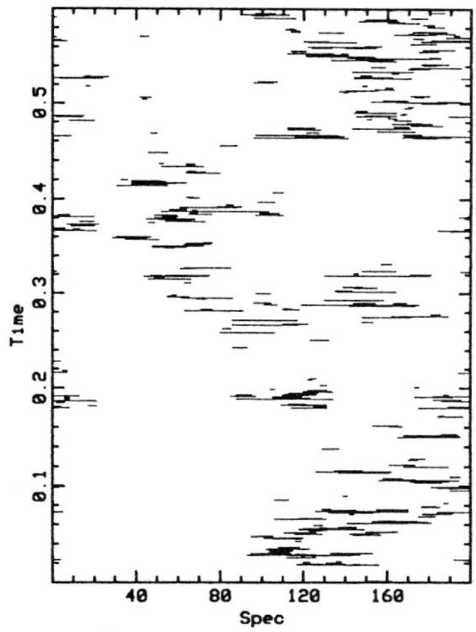

Fig. 11. Space-time plot of the activity. The horizontal axes is the "species axis." The time at which a given species mutates is shown as a black dot. The avalanches appear as connected black areas. On the time resolution of the plot, the avalanches appear as almost horizontal lines. Calculation was done for a value of the mutation parameter $T = 0.01$.

During an avalanche, individual species may undergo many mutations, so their properties, or "morphologies," can change significantly even if the effect of individual mutations is small. Thus, on a timescale that is large compared with that of the avalanche, evolution sometimes appears to take place in terms of saltations. There is a close connection between the punctuations of evolution (and the periods of stasis) of the individual species, and the global intensity of evolutionary activity. They are two sides of the same coin: the self-organized critical state. The punctuated equilibrium behavior of single species is a collective effect. The macroscopic jumps between "useful" or highly fit states are effectuated by the cumulative small jumps through intermediate states, which only could exist in the temporary environment of a burst.

For example, a grounded species may develop wings rapidly trough a sequence of smaller steps in a turbulent environment, where such development would be impossible in periods of stasis due to competition from other highly adapted species. Maybe this mechanism can soften Mivart's [30] critique of Darwin's theory for the "incompetency of natural selection to account for the incipient stages of useful structures."

Figure 8b shows the accumulated number of mutations for a single species versus time in the model. The number of mutations might be seen as a measure of the change in morphology for that particular species. The evolution shows punctuated equilibrium behavior. For more data see also Ref. [6] and references therein, e.g. the well documented Elephant lineages of Ref. [31].

If the evolutionary activity, as measured for instance in terms of the number of extinctions, is measured for an extended period, there will in general be several smaller and larger avalanches taking place during that period. One can prove mathematically that the distribution of activity in such intervals will converge towards a Pareto-Levy distribution function in the limit where there are many avalanches. This Pareto-Levy distribution has a power-law tail with an exponent equal to that of the individual avalanches. In Figure 6b, we have coarse-grained the time scale of an ecology of 128 species evolving over a total of 2^{14} steps, in 60 equal time intervals. The large fluctuations seen in the distribution of individual avalanches remain. The Pareto-Levy distribution is not a power law for small events. In order to extend the power-law downwards it is important to coarse grain over small intervals. Thus it would be nice to see the histogram of fossil extinctions measured on a finer time scale, say a million years.

If the value of the mutation parameter T is increased, some regions of the ecology decorrelate, and the predicted power laws will be limited to a scaling regime which diminishes with increasing values of T. For this reason a detailed study of scaling behaviour for fossil records of subsets of the ecology may help to determine better the avalanche exponent. At higher values of the mutation parameter T, i.e. for higher mutation rates, well defined punctuations gradually cease to exist, and the evolutionary activity changes nature from intermittent to continuous. Well-defined periods of stasis disappear, and evolution as we know it with ecologies in apparent balance most of the time, ceases to exist.

8 Discussion and Conclusions

For the present model to have any chance of representing evolution in Nature, it is important that those of its properties that we have focussed on here are robust and fundamentally unchanged by essentially any modification of the model that leave its defining elements unchanged. We have simulated many versions of the model with many different representations of the interactions between species. In all cases we found punctuated equilibrium with exponents depending only on the dimension of the lattice.

Our results demonstrate the advantage of simple "toy" models over more complicated, and supposedly more realistic, models. Not only are the toy models numerically tractable, but they are also exactly mathematically solvable in some respects. For the *Random Neighbour Model* described in [27] one can explicitly prove that it self-organizes to the critical state and find a number of properties of that state [27,29]. The mechanism for self-organized criticality in the general case of models with assigned neighbors has been identified by Paczuski et al. [28]. Having completed the present study, one could in principle return to the more elaborate representations of the landscape for the single species, as for instance the spin-glass model [25], or the NKC model [14]. Based on our observation of robustness, we conjecture that our conclusions, including the specific values of the exponents, will remain unaltered.

One important observation from this study is that Darwinian evolution acting on the level of the individual does not converge towards a state where every species is maximally fit, i.e. a state in which all barrier values B are maximal. In a non-interactive ecology, this would eventually happen, but the time-scale would be enormous since one would have to wait for the occurrence of states with very high B-values, and therefore very low transition rates. The ecology discussed here evolves relatively fast to the globally correlated critical state, and once it has arrived there, it keeps evolving forever, alternating between periods of stasis and intermittent spikes of co-evolutionary activity of all sizes. According to this scenario, Life in its normal state is synonymous with volatility, not with stability and fitness. Darwins principle does not translate directly to the whole ecosystem, which in fact does not evolve towards higher fitness or stability. The critical state is not "a nice place to be," contrary to what Kauffman suggested. As the least fit species mutates to improve its fitness, other species find their fitnesses reduced and soon mutate, too, possibly triggering changes throughout the ecology, as we have seen. As the fitness of any species is no more durable than the state of the species with which it interacts, all species experience a "Red Queen" effect: they are forced to keep evolving towards higher fitness just to maintain their fitness.

Can any of this be studied in the laboratory? One possibility is to study the dynamics of a limited ecology of very simple species on the molecular level [32] in order to identify the interplay between local punctuations at the level of single species and the evolution of the ecology as a whole.

Acknowledgments

PB is supported by Brookhaven National Laboratory, the Division of Materials Science, U.S. Department of Energy under Contract No. DE-AC02-76CH00016. HF is supported by the Danish Natural Science Research Council, Grant No. 11-0244-1. KS is supported by the Danish Natural Science Research Council, Grant No. 11-0608-1. PB and HF thank the Isaac Newton Institute for its hospitality.

References

1 Raup, D. M. & Sepkoski, J. J. Jr. (1982) Science **215**, 1501-1503.
2 Raup, D. M. (1986) Science **231**, 1528.
3 Raup, D. M. & Boyanjian, G. E. (1988) Paleobiology **14**, 109-125.
4 Sepkoski, J. J., Jr. (1993) Paleobiology **19**, 43-51.
5 Gould, S. J. & Eldredge, N. (1977) Paleobiology **3**, 114-151;
6 Gould, S. J. & Eldredge, N. (1993) Nature **366**, 223-227.
7 Darwin, C. (1859) The Origin of Species by Means of Natural Selection. 6th ed. John Murray, London, D Appleton, London, 1910.
8 Hallam, A. (1986) Nature **319**, 765-768.
9 Newell, N. D. (1952) J. of Paleontology **26**, 371-385.
10 Vrba, E. S. (1985) Suid-Afrikaanse Tydskrif Wetens **81**, 229-236.
11 Alvarez, L. W., Alvarez, F. A., & Michel, H. V. (1980) Science **208**, 1095-1108.
12 Bak, P., Tang, C., & Wiesenfeld, K. (1987) Phys. Rev. Lett. **59**, 381- 384; (1988) Phys. Rev. A **38**, 364-374. Bak, P. & Chen. K. (1991) Scientific American **264**(1), 46-53. Bak. P. & Paczuski, M. (1993) Physics World **6** (12), 39-43.
13 Bak, P., Chen, K., & Creutz, M. (1989) Nature **342**, 780–781.
14 Kauffman, S. A. & Johnsen, S. J. (1991) J. Theo. Biology **149**, 467-506.
15 Bak, P. (1993) "Self-Organized Criticality and Gaia" in Thinking about Biology, Santa Fe Institute Studies in the Science of Complexity, Lec.N. Vol. III (ed. W. Stein and F.J. Varela) Addison-Wesley, 255-268.
16 Flyvbjerg, H. & Lautrup, B. (1992) Phys. Rev. A **46**, 6714-23; Bak, P., Flyvbjerg, H., & Lautrup, B. (1992) Phys. Rev. A **46**, 6724-6730.
17 Ray, T. S. (1992) in Artificial Life II, Santa Fe Institute Studies in the Sciences of Complexity, Proc. Vol. X. (ed. C. G. Langton) Addison Wesley, Redwood City, CA, 371-408.
18 Adami C. (1994) Self-Organized Criticality in Living Systems, Caltec Preprint.
19 Kellogg, D. E. (1975) Paleobiology **1**, 359-370.
20 Wright, S. (1982) Evolution **36**, 427-443.
21 Lande, R. (1985) Proc. Natl. Acad. Sci. USA **82**, 7641-7645.
22 Newman, C. M., Cohen, J. E., & Kipnis, C. (1985) Nature (London) **315**, 400-401.
23 MacBean, I. T., McKenzie, J. A., & Parsons, P. A. (1971) Theor. Appl. Genet. **41**, 227-235.
24 Parsons, P. A. (1983) The Evolutionary Biology of Colonizing Species (Cambridge University Press, New York).
25 Anderson, P. W. (1985) in Emerging Synthesis in Science. Proc. of the Founding Workshop of the Santa Fe Institute. (ed. D. Pines).
26 Bak. P. & Sneppen, K. (1993) Phys. Rev. Lett. **71**, 4083-4086.
27 Flyvbjerg, H., Sneppen, K. & Bak, P. (1993) Phys. Rev. Lett. **71**, 4087-4090.
28 Paczuski, M., Maslov, S., & Bak, P. (1994), Europhys. Lett. in press.
29 Harris, T. E. (1963) *The Theory of Branching Processes,* (Springer, Berlin).
30 de Boer, J., Derrida, B., Flyvbjerg, H., Jackson, A. D., & Wettig, T. (1994), submitted to Phys. Rev. Lett.
31 Mivart, S. G. J. (1871), On the Genesis of Species (D. Appleton & Co., New York).
32 Maglio, V.J. (1973) Amer. Philos. Soc. Trans. **63**, 1-149; For an excellent survey of evolution diagrams and macroevolution in general, see Stanley, S.M. (1979) Macroevolution, Patterns and Process (W.H. Freeman and Co., San Francisco).
33 Eigen, M. & Schuster, P. (1993) *The Hypercycle.* (Springer-Verlag, Berlin, Heidelberg, New York, 1979.)
34 Eigen, M. & Schuster, P. (1977) Naturwissenschaften **65**, 541; (1978) idem 65, 341.

Index

Springer Series in Synergetics

Editor: Hermann Haken

Synergetics, an interdisciplinary field of research, is concerned with the cooperation of individual parts of a system that produces macroscopic spatial, temporal or functional structures. It deals with deterministic as well as stochastic processes.

1 **Synergetics** An Introduction 3rd Edition By H. Haken
2 **Synergetics** A Workshop Editor: H. Haken
3 **Synergetics** Far from Equilibrium Editors: A. Pacault, C. Vidal
4 **Structural Stability in Physics** Editors: W. Güttinger, H. Eikemeier
5 **Pattern Formation by Dynamic Systems and Pattern Recognition** Editor: H. Haken
6 **Dynamics of Synergetic Systems** Editor: H. Haken
7 **Problems of Biological Physics** By L. A. Blumenfeld
8 **Stochastic Nonlinear Systems** in Physics, Chemistry, and Biology Editors: L. Arnold, R. Lefever
9 **Numerical Methods in the Study of Critical Phenomena** Editors: J. Della Dora, J. Demongeot, B. Lacolle
10 **The Kinetic Theory of Electromagnetic Processes** By Yu. L. Klimontovich
11 **Chaos and Order in Nature** Editor: H. Haken
12 **Nonlinear Phenomena in Chemical Dynamics** Editors: C. Vidal, A. Pacault
13 **Handbook of Stochastic Methods** for Physics, Chemistry, and the Natural Sciences 2nd Edition By C. W. Gardiner
14 **Concepts and Models of a Quantitative Sociology** The Dynamics of Interacting Populations By W. Weidlich, G. Haag
15 **Noise-Induced Transitions** Theory and Applications in Physics, Chemistry, and Biology By W. Horsthemke, R. Lefever
16 **Physics of Bioenergetic Processes** By L. A. Blumenfeld
17 **Evolution of Order and Chaos** in Physics, Chemistry, and Biology Editor: H. Haken
18 **The Fokker-Planck Equation** 2nd Edition By H. Risken
19 **Chemical Oscillations, Waves, and Turbulence** By Y. Kuramoto
20 **Advanced Synergetics** 2nd Edition By H. Haken
21 **Stochastic Phenomena and Chaotic Behaviour in Complex Systems** Editor: P. Schuster
22 **Synergetics – From Microscopic to Macroscopic Order** Editor: E. Frehland
23 **Synergetics of the Brain** Editors: E. Başar, H. Flohr, H. Haken, A. J. Mandell
24 **Chaos and Statistical Methods** Editor: Y. Kuramoto
25 **Dynamics of Hierarchical Systems** An Evolutionary Approach By J. S. Nicolis
26 **Self-Organization and Management of Social Systems** Editors: H. Ulrich, G. J. B. Probst
27 **Non-Equilibrium Dynamics in Chemical Systems** Editors: C. Vidal, A. Pacault
28 **Self-Organization** Autowaves and Structures Far from Equilibrium Editor: V. I. Krinsky
29 **Temporal Order** Editors: L. Rensing, N. I. Jaeger
30 **Dynamical Problems in Soliton Systems** Editor: S. Takeno
31 **Complex Systems – Operational Approaches** in Neurobiology, Physics, and Computers Editor: H. Haken
32 **Dimensions and Entropies in Chaotic Systems** Quantification of Complex Behavior 2nd Corr. Printing Editor: G. Mayer-Kress
33 **Selforganization by Nonlinear Irreversible Processes** Editors: W. Ebeling, H. Ulbricht
34 **Instabilities and Chaos in Quantum Optics** Editors: F.T. Arecchi, R.G. Harrison
35 **Nonequilibrium Phase Transitions in Semiconductors** Self-Organization Induced by Generation and Recombination Processes By E. Schöll

Springer-Verlag and the Environment

We at Springer-Verlag firmly believe that an international science publisher has a special obligation to the environment, and our corporate policies consistently reflect this conviction.

We also expect our business partners – paper mills, printers, packaging manufacturers, etc. – to commit themselves to using environmentally friendly materials and production processes.

The paper in this book is made from low- or no-chlorine pulp and is acid free, in conformance with international standards for paper permanency.

Printing: Saladruck, Berlin
Binding: Buchbinderei Lüderitz & Bauer, Berlin